animals without backbones

without backbones

an introduction to
the invertebrates

BY RALPH BUCHSBAUM

Second Edition

New Impression, with Additions and Revisions

THE UNIVERSITY OF CHICAGO PRESS
CHICAGO AND LONDON

THE UNIVERSITY OF CHICAGO PRESS, CHICAGO 60637
The University of Chicago Press, Ltd., London

Copyright © 1938, 1948, 1976 *by The University of Chicago*
All rights reserved. Published 1938. *Second Edition* 1948
Eighteenth impression 1976, *with additions and revisions.*
Printed in the United States of America

International Standard Book Number: 0–226–07870–1
Library of Congress Catalog Card Number 48-9508

TO W. C. ALLEE

WHO INTRODUCED ME TO
THE INVERTEBRATES

PREFACE TO THE SECOND EDITION

THE BASIC features of this book as outlined in the Preface to the original edition have been reconsidered and, on the basis of almost ten years of experience, have been retained. The temptation to expand each chapter with technical material has been resisted; instead, the chapters have been changed only to bring them up to date in fact and theory.

A NEW chapter has been added to give the student an orientation in the literature on the invertebrates. To encourage the student to read scientific articles and to help him to realize the significance of the scientific paper as the basic unit of zoölogical literature, three short scientific articles, of three types, are reproduced in the book. A selected bibliography has been included to give the student some general leads into the various types of literature on the invertebrates.

THE PHOTOGRAPHIC section has been increased by thirty-two completely new pages, and many of the original photographs have been replaced by better ones. In all, there are about a hundred and fifty new photographs. As in the first edition, the photographs are intended to support the text, to stimulate interest in the invertebrates, to impress the student with the variety of forms, and to give the student a vicarious field experience. Several text figures have been added and a number have been replaced; all were drawn by Miss Catherine O'Brien.

I WISH to express my thanks to the many individuals who have written to me about various features of the book, and I welcome comments on this edition. I wish to acknowledge my special thanks to Douglas P. Wilson, of the Plymouth Laboratory of the Marine Biological Association, England, for a number of superb photographs from his book, *They Live in the Sea*. His photographs of the developmental stages of *Aurelia* deserve special attention. Several new photographs by the late Percival S. Tice are added to the fine collection he contributed to the original edition. I am grateful to Dr. Ralph Wichterman for a series of photographs made from his motion picture of the mating reaction in *Paramecium bursaria*. Miss E. J. Batham kindly allowed me to reproduce her picture of the

Plymouth Laboratory. Three new photographs were kindly supplied by the General Biological Supply House. The Marine Biological Laboratory, Woods Hole, Massachusetts, furnished a picture of the laboratory. I also wish to thank my fellow-investigators at the Mount Desert Island Laboratory, Salsbury Cove, Maine, and at the Barro Colorado Island Laboratory, Panama, who helped to collect many of the specimens which I photographed. As in the first edition, I am indebted most of all to my wife, Mildred, who helped in every stage of the preparation of this revision.

RALPH BUCHSBAUM

UNIVERSITY OF CHICAGO

A NOTE ON THE 1976 PRINTING

Various obscure groups were purposely omitted in previous printings because it was thought that they were not needed to give the student for whom this book was intended an overall view of the main groups of invertebrates. However, there has been some demand for mention of these groups. Hence, in this printing there is added to the text, or to the appendix, some material on the mesozoans, placozoans, gnathostomulids, kinorhynchs, priapulids, pauropods, symphylids, pogonophorans, monoplacophorans, pentastomids, and tardigrades.

The scientific papers reproduced in previous printings have been replaced by an appendix giving a version of classification of the invertebrates that is probably generally acceptable to most zoologists.

A number of figures have been revised to incorporate newer knowledge. These were executed by Miss Mildred K. Waltrip.

And an updated bibliography has been substituted for the old one.

RALPH BUCHSBAUM

PACIFIC GROVE, CALIFORNIA

ELEMENTARY and general accounts of the invertebrates, suitable for the beginning college student or layman, have been limited to two sorts of books: natural histories, which describe the habits of a great many animals but are lacking in descriptions of basic structure and in theory, and formal textbooks, which are packed with morphological detail and technical terminology. This book is an attempt to present the main groups of invertebrate animals in simple nontechnical language. Each group is used to illustrate some principle of biology or some level in the evolution of animals from simple to complex forms. The material is divided into two kinds of chapters. The basic or *indispensable chapters* (1, 2, 3, 4, 6, 7, 10, 11, 14, 17, 19, 22, 25, and 28) present a continuous story of the more general and elementary aspects of the main invertebrate groups and can be read consecutively without reference to the others. They are recommended for the general reader or for students in any introductory college course in biology. They are essentially the ones read by our students for the invertebrate section of the "Introductory General Course in Biology" at the University of Chicago. The remaining chapters are advanced or *optional chapters,* and these treat each group more fully and present additional principles. They are intended to serve as optional reading for the invertebrate section of a general course. The book as a whole is designed as a textbook for a college course in invertebrate zoölogy.

IN THE preparation of the manuscript I was extremely fortunate in having the advice and criticism of Dr. Libbie Hyman, whose extensive knowledge of the literature on the lower invertebrates has helped to bring the material up to date. She has contributed many valuable suggestions to the organization of the book and has corrected many errors. To Dr. Merle Coulter, of the Department of Botany and director of the Introductory General Course in Biology at the University of Chicago, I am indebted for the idea of dividing the book into two sets of chapters. Dr. Coulter has read the manuscript; and his suggestions, free from the natural bias with which a zoölogist approaches a book on animals, have been especially helpful in improving the clarity of many sections. Dr. W. C. Allee, of the Department of Zoölogy, has read the manuscript critically, much to its advantage, and has supplied encouragement throughout the preparation of the book. From Dr. Alfred Emerson, of the Department of Zoölogy, I have adopted many ideas for the organization of the material. Dr. Emerson has read the manuscript and has made many suggestions, particularly in the chapters on the arthropods. Dr. Thomas Park, of the Department of Zoölogy, contributed a valuable criticism of the chapters on the arthropods. Dr. Harry Andrews, of the Chicago Junior Colleges, has read some of the early chapters. To all of these colleagues and friends, who have given so generously of their time and effort, I am deeply indebted and sincerely grateful.

THE drawings are mostly diagrammatic and have been designed to convey ideas about structure, function, or habit, rather than to show the details of any particular species of animal (except where specifically labeled). As far as possible, the same symbols have been used for corresponding structures in different drawings; and it is hoped that this will aid in their ready interpretation. The large number of new drawings, and the adaptation of the borrowed ones to the style followed in this book, required a close collaboration between author and artist made possible only by the fortunate circumstance that the artist was my sister, Miss Elizabeth Buchsbaum. To her skillful and artistic execution of the drawings the book owes much of its attractiveness.

THE numerous photographs, unusual in a textbook, supply the elements of specific form and texture which are missing from the diagrammatic, stylized drawings. They are intended as a sort of laboratory exhibit and vicarious field experience.

The source of every photograph is acknowledged in the legend which accompanies it, but I am especially indebted to Dr. Douglas Wilson, of the Plymouth Laboratory of the Marine Biological Association, England, for several excellent pictures from his book, *Life of the Shore and Shallow Sea;* Dr. W. K. Fisher, of the Hopkins Marine Station, Pacific Grove, California; Mr. Richard Westwood, managing editor of *Nature Magazine;* Mr. A. S. Windsor, of the General Biological Supply House, Inc., Chicago; Dr. C. M. Yonge for photographs from his book, *A Year on the Great Barrier Reef;* Dr. James A. Miller, of the Department of Anatomy, University of Michigan; Mr. Albert Galigher, of Berkeley, California; Mr. Louis Diamond, Chicago; and Mr. Leon Keinigsberg, Chicago.

Mr. Percival S. Tice, of Chicago, deserves special mention for his outstanding photographs, many of which he made especially for this book.

I value them particularly because some of them, like the hydra series, are of subjects extremely difficult to photograph and not usually attempted.

All photographs not otherwise acknowledged are by myself, and I take this opportunity to express my thanks for facilities and assistance offered me at the various marine biological stations where I have made photographs. I am especially indebted to Dr. W. K. Fisher, director of the Hopkins Marine Station, Pacific Grove, California, and Dr. Bolin of the same station; Dr. J. F. G. Wheeler, director of the Bermuda Biological Station, St. George's, Bermuda; Dr. A. Tyler of the Kerckhoff Marine Laboratory, Corona del Mar, California; and officials of the Marine Biological Laboratory, Woods Hole, Mass.

Finally, I am indebted most of all to my wife, who has assisted in every phase of the preparation of the book, from writing the text and reading proof to designing drawings and making photographs.

RALPH BUCHSBAUM

UNIVERSITY OF CHICAGO

CONTENTS

CHAPTER 1

BY WAY OF INTRODUCTION

ANYONE can tell the difference between a tree and a cow. The tree stands still and shows no signs of perceiving your presence or your hand upon its trunk. The cow moves about and appears to notice your approach. This striking difference in the behavior of plants and animals is related to the fundamental difference in plant and animal nutrition.

Plants make their own food from simple constituents of the air and soil. With the aid of a green pigment, chlorophyll, the tree utilizes energy from the sun to combine carbon dioxide and water into food—a process known as photosynthesis. The cow cannot stand in the sun and soak up energy with which to make nutritive substances, but must get its food by eating the bodies of the food-manufacturers, the independent plants. To find a constant supply of plants the cow must move from place to place and must react rapidly to changing conditions in its environment.

Not all animals move about. The sponges, for example, grow attached to the substratum. They have internal moving parts which create currents in the water, thus drawing food toward the sponge. Since this was not apparent to the early naturalists, they classified sponges and many other stationary animals as plants. While there is no longer any doubt that the sponge in the bathroom or the piece of coral decorating the mantelpiece are the skeletons of animals, the question "What is an animal?" is not always easy of exact answer.

Flagellates.

As we examine simpler and simpler forms of life, distinctions of behavior and of nutrition grow less and less obvious. Eventually we find microscopic organisms that exhibit some characteristics possessed by both plants and animals. These "plant-animals," the **flagellates,** swim about by lashing long threadlike extensions called **flagella** (singular, flagellum—the Latin word for "whip"). Some flagellates carry on photosynthesis, but they move about and show the same sensitivity and rapidity of response as do typical animals. Some flagellates not only photosynthesize but also feed like animals, thereby seeming to make doubly sure of a source of nourishment and aligning themselves with neither plants nor animals. Other flagellates have lost their chlorophyll and feed only in an animal manner. The existence of such organisms indicates that, in the beginning, there were no differences between plants and animals and that life was restricted to very simple forms. What these forms were and how they originated are questions on which we have no direct evidence.

One of the most plausible of the hypotheses advanced to account for the origin of life states that at some time in the earth's history, in suitable places, as in ponds or on the seacoasts, there were, as there are now, simple compounds of the elements which compose the living substance, protoplasm. With the energy of the sun or the heat of warm springs, various chemical combinations were formed. Some of these possessed the power of self-propagation, that is, the ability to manufacture additional combinations like themselves. An analogy to such a state of living matter may be found in a group of substances which are so small that they are just under the limits of visibility of the ordinary microscope and pass through the pores of the finest porcelain filters. Because of this filter-passing character, and because they are responsible for diseases such as smallpox, yellow fever, mumps, infantile paralysis, and the mosaic diseases of plants, these substances are called *filterable viruses* ("poisons"). The viruses contain nucleic acids and proteins, and several different viruses already have been prepared in pure crystalline form. Even after repeated crystallizations, a treatment no obviously living substance has ever been able to survive, viruses resume their activities and multiply when returned to favorable conditions. While no one has yet succeeded in growing them in the absence of living matter, it is clear that viruses help to bridge the gap that was formerly thought to exist between nonliving and living things. No longer can it be said that there is some sharp and mysterious distinction between the nonliving and the living, but rather there seems to be a gradual transition in complexity.

If we imagine that the earliest self-propagating substances were something like viruses, it is not difficult to suppose that an aggregation of virus-like proteins could lead to the development of larger bacteria-like organisms, independent, creating their own food from simple substances, and using energy from the sun.

Such a level of organization may be compared to present-day forms like the *independent bacteria*, some of which conduct photosynthesis without chlorophyll, using, instead, various green or purple pigments. Others utilize the energy derived from the oxidation of simple salts of nitrogen, sulphur, or iron. These, for instance, can oxidize ammonia to nitrites, or hydrogen sulphide to sulphates, with the release of energy which is utilized in forming carbohydrates.

From primitive bacteria-like forms to the simple chlorophyll-bearing organisms is a relatively short jump in complexity, however long it may have taken in time.

There is evidence that both plant and animal kingdoms originated from primitive flagellates. By losing locomotor flagella and assuming a rounded form, some flagellates become indistinguishable from the simplest plants, the algae; and, in fact, many of the green flagellates regularly pass into such an immotile state when they reproduce. By loss of chlorophyll, other flagellates become purely animal types which capture and ingest food.

Arising from primitive flagellates, animals have evolved into a bewildering variety of forms of ever increasing complexity of structure. When these ani-

It is thought that both plant and animal kingdoms arose from primitive flagellates.

mals are carefully studied and compared, it is found that many of them resemble man in various ways, notably in the presence of a row of bones (vertebras) along the middle of the back, as well as in the presence of bones inside the limbs and head. The animals having internal bones, including a backbone or vertebral column, are known as **vertebrates** and comprise all the fish, the frogs, toads, and salamanders, the lizards, snakes, turtles, and crocodiles, every kind of bird, and all the hairy animals known as mammals, such as elephants, lions, dogs, bats, and mice. These more or less familiar animals have a highly exaggerated importance in our minds because they are closely related to man, because they are mostly of large size, and because, like man, they usually manage to make themselves conspicuous. Actually, in terms of number of living species they comprise only about 3 per cent of the animal kingdom.

The remaining 97 per cent consists of animals without backbones. We are all aware of the difference between these two groups of animals when we indulge in fish and lobster dinners. In the fish the exterior is relatively soft and inviting, but the interior presents numerous hard bones. In the lobster, on the contrary, the exterior consists of a formidable hard covering, but within this initial handicap is a soft edible interior. A similar situation exists in the oyster, lying soft and defenseless within its hard outer shell. The lobster and the oyster are but samples of a tremendous array of animals which lack internal bones and which are, from their lack of the vertebral column in particular, called **invertebrates.**

A distinction between vertebrates and invertebrates was first recognized by Aristotle, although he did not use these terms but divided animals into those with blood (vertebrates) and those without blood (invertebrates). Unfortunately, Aristotle's neat distinction had little to do with the facts, since many invertebrates possess red blood and a great many other invertebrates have colorless blood, which he did not recognize as blood at all. Although Aristotle did about as well as one might expect from the limited knowledge of animal structure available in his time, it was partly because of the weight of his authority that his error was not corrected for over two thousand years. With the development of the fruitful independent scientific spirit of the early nineteenth century, Lamarck and Cuvier finally made the correct distinction, based upon the fundamental plan of organization of the animal body.

There is a popular but vague recognition of the difference between vertebrate and invertebrate animals in the expression "spineless as a jellyfish." In this book we shall be concerned not only with the jellyfish, which is seldom seen by inland dwellers, but also with many animals without backbones, like clams, crayfish, earthworms, and fleas, which are supposed to be already familiar to most people. In addition, many forms will be presented which generally pass unnoticed because they are too small to be seen without a microscope, because they live under water or in the ground, because they inhabit remote parts of the world, or simply because they escape the unobservant eye.

CHAPTER 2

LIFE-ACTIVITIES

TO KEEP alive and healthy, all animals from the lowest to the highest must carry on certain **life-activities.** Because these activities center about the utilization of energy, they have often, and very appropriately, been compared to the functioning of a combustion engine. But there is a point at which this analogy breaks down. For not only is the living machine a self-feeding, self-tending, and self-perpetuating one, but by its very nature is a machine which must operate at all times. A stalled motor may easily be repaired. Not so a "stalled" organism, in which the failure of certain functions automatically brings on the disintegration of the machinery itself. A machine can be oiled, covered, and put away on a shelf until ready for use, while an organism must be kept running—sometimes rapidly, sometimes very slowly—but continuously from the start until the natural or accidental finish.

This continuity of living processes appears, on first thought, to be contradicted by our ordinary observations of animals. For example, we know that crayfish may lie dormant in the mud of a dried-up pond and then resume their activities when spring rains fill the pond. What we see here is only a temporary cessation in the easily observable activities of the crayfish. Going on all of the time at a greatly slowed rate is a much more fundamental activity, the liberation of energy. The energy liberated comes from the burning of food stored by the crayfish during the months preceding dormancy. This food was obtained directly from plants or "second hand" from other animals that had eaten plants. To catch, bite off, and chew up parts of other animals or of plants, the crayfish has special structures and behavior which make possible an activity—lacking in plants and so characteristic of animals—the capture and ingestion of food.

Locomotion, or motility, of some type enables most animals to rove about in search of food or to escape from their enemies, who are also on the move hunting their prey. *Sessile,* or fixed, animals, which cannot move about but live firmly attached to the substratum, have moving parts that propel food their way. These include sponges, corals, barnacles, certain bivalves, and other animals. In addition, there are *sedentary* animals, like clams or web-building spiders, which feed while remaining for some time in one spot but which can and do move about to escape danger or to take up a new and more profitable feeding station.

Ingestion is the taking-in of food. Animals differ strikingly in their mode of ingestion. The differences are related partly to the size and complexity of the animal and partly to the diversity of the food itself. Mouth parts that tear flesh will not do for chewing wood; sucking sap is not the same as sucking blood. In its essentials, however, the feeding machinery involves: a set of *sensory receptors* to get information about the external environment, *mechanisms for locomotion and ingestion,* and a *means of co-ordinating* the locomotory and feeding mechanisms with the information received from the surroundings so that the net result will be the getting of something to eat and the avoidance of being eaten.

Digestion is the chemical alteration of raw food into a form in which it is usable as a source of *energy* for the life-activities and as a source of *materials* for growth and replacement of worn-out or damaged parts. The raw food consists of water, carbohydrates, fats, proteins, inorganic salts, and some other substances. The water and some of the salts need not be digested; they are immediately available for incorporation into the living animal body. The other substances must be broken down into simpler units because they are too large to pass through the living membranes and because they are too complex to be used directly in growth or in other living processes. Digestion, then, is the breakdown of raw food into smaller units. To facilitate this breakdown, the living organism has a digestive apparatus into which are poured a number of kinds of chemical substances, or secretions.

Secretion is the manufacture of special chemical substances out of materials obtained from the surrounding environment. These substances, or secretions, may be used where produced or may be carried to other parts of the organism. Silk, sponge fibers, calcareous shells, and mucus are well-known animal secretions. Less easily seen, but more essential, are the secretions which enter into the chemistry of the basic life-activities. Of these the most important are the **enzymes.**

Enzymes are complex substances, manufactured only by living organisms, which speed up chemical reactions. They make possible all kinds of chemical activities which in the nonliving world take place so slowly that they could not possibly serve the needs of a constantly changing living organism. Life as we know it could not exist without enzymes.

Enzymes are proteins, or protein-like, as we know from experiments that show enzymes to be inactivated by the same physical and chemical agents (heat, alcohols, heavy metals, etc.) that coagulate proteins. Enzymes are notably specific. Most enzymes accelerate only one particular reaction. For example, some digestive enzymes act only on carbohydrates, others only on fats, and still others only on proteins.

Elimination is the ejection from the body of indigestible food or other accumulated solid wastes. Most plant-eaters do not have the enzymes needed to digest completely the woody tissues of the plants they feed upon. Most insect-feeders cannot break down the complex substance that forms the hard outer skeleton of insects. These indigestible portions of the food constitute the solid wastes, or *feces*, and must be removed lest they clutter up the digestive machinery.

Assimilation is a constructive chemical process by which materials derived from digestion are incorporated into the living substance, protoplasm. After the food is converted from its condition as the structural part of one kind of animal or plant into smaller, simpler units, it is suitable for building the kinds of carbohydrates, fats, and proteins peculiar to the structure of the animal concerned. Just as innumerable useful objects can be fashioned from combinations in various proportions of only a few dozen building-materials, so a few dozen kinds of food units can be built into an almost infinite variety of organisms—each kind of organism with its own specific composition.

Respiration is a destructive chemical process by which food is burned in the release of energy. The energy stored in the food through the photosynthetic action of green plants is released in somewhat the same way that man releases, by burning, the energy stored in coal. The high temperatures involved in the burning of coal are not necessary in respiration because the chemical reactions are accelerated not by heat but by special respiratory enzymes. The burning of coal, or of almost anything else, requires air, or, more exactly, the oxygen of the air. The release of energy in the living organism may be described in three steps.

The first is the bringing of oxygen to the fuel. In man this is accomplished by *breathing* and by the circulation of the blood from the lungs to the tissues. The second step is the actual burning, or the chemical union of oxygen with the fuel, resulting in the liberation of energy; this is known as *oxidation.* The third step is the removal of the by-products of respiration —water and carbon dioxide. These wastes usually pass in the reverse direction along the same route by which the oxygen entered.

Excretion is the separation from the living protoplasm of the nongaseous waste products of assimilation and other chemical activities. The by-products of oxidation of carbohydrates and fats are carbon dioxide and water. The burning of proteins yields other wastes in addition to carbon dioxide and water, namely, compounds of nitrogen, which require special methods of disposal. These nitrogenous compounds are poisonous, and their prompt removal is indispensable to life. In man and in other vertebrates the kidneys filter out these wastes in the formation of urine, which is eliminated to the outside; various devices do this same work for other animals.

Reproduction is the production of new individuals to take the places of the old ones which die because their machinery wears out or because they are eaten or destroyed by their enemies. Unlike the other life-activities, reproduction is not necessary to maintain life in any single individual; it is essential only for the continued existence of the group.

Metabolism is the total of the chemical changes that go on in the animal body. There are two phases of metabolism: building up, or *constructive metabolism,* and breaking down, or *destructive metabolism.* Assimilation is a constructive metabolic process; respiration is a destructive one. Secretions may result from either of the two types of metabolism. In young and vigorous organisms the constructive phase overbalances the destructive phase, and there results an increase in the total amount of living substance, which we call **growth.** In most mature animals the two metabolic phases are just about balanced. Though there is constant tissue repair and often a regeneration of lost or injured parts, active growth seldom or never occurs. In old age or disease the balance is tipped the other way, and the destructive metabolic processes predominate. One kind of history of an animal might be told in terms of the metabolic changes or events that befall it from beginning to end.

Although all the life-activities are essentially chemical transformations, living substance has a special capacity, not found in nonliving systems, for altering its complex chemical activities either in character or in the rate at which they proceed so as constantly to adjust to changes in the environment. This responsiveness of all living substance is called *irritability* or *modifiability*, or **reactivity.** This last term perhaps best expresses the full nature of the relationship, in which the animal not only modifies its own internal activities but may react on the environment itself, changing the conditions in the surroundings, with results that, on the whole, lead to the continued existence of the organism. The favorable adjustment of an organism to its environment is known as **adaptation.**

WHILE the mechanisms employed by animals for carrying on their life-activities differ considerably, they are all made from variations of a basic living substance called **protoplasm** ("primitive form"). The exact chemical composition of protoplasm, or, rather, of any of the various protoplasms, is not yet known. But it is at least 75 per cent water and can be loosely described as a watery mixture of proteins, carbohydrates, fats, salts, and various other substances. Most of the salts are in true solution, while most of the complex organic constituents are thought to be suspended in the watery medium in particles of a size larger than the particles of a true solution. Such a suspension is called a *colloidal suspension*. And protoplasm acts most like those colloidal suspensions in which the particles are covered with layers of water molecules that can be set free or taken up again, so changing the colloid from a more fluid, or *sol*, state to a more solid, or *gel*, state. Such a reversible change is familiar to all of us in the setting and liquefaction of gelatin. The *reversible* solation and gelation of protoplasm is one of its most fundamental and important properties, making possible much of the unique chemistry of living things. How it plays a role in protoplasmic movements we shall see later.

The protoplasm of larger animals does not exist as a continuous mass but is divided up by partitions into minute units called **cells.** The cell is the unit of living structure and activity. This is the same as saying that the cell is the smallest part of an animal which can carry on all the life-activities. When viewed through the microscope, each cell appears as a mass of protoplasm bounded by a surface or **cell membrane** which separates it from adjacent cells. This membrane is a thin layer composed mostly of fat and protein. It possesses the unique property of *selective permeability*, admitting certain substances and excluding others and so enabling the cell to maintain a constant chemical composition which may be different from that of its environment and to some extent

independent of changes in the environment. The protoplasm within the
membrane is visibly differentiated into a rounded body, the **nucleus,**
and the relatively unspecialized **cytoplasm,** in which the nucleus lies
imbedded. The nucleus is bounded by a nuclear membrane, also thought
to act selectively, regulating the exchange of materials between nucleus
and cytoplasm. The nucleus contains a special material, the chromatin,
found mainly in nuclei. Microchemical tests show the chromatin to be
composed of nucleic acids (DNA) and nucleoproteins. There are many
kinds of cell organelles besides the nucleus, but some occur only during cell
division and others are specializations found only in particular kinds of
cells or in particular animals. Several kinds of such organelles, which
are rather consistently seen in cells carrying on their ordinary metabolic
activities, are shown in the photograph of a living cell.

There are free-living organisms whose bodies are not divided up into
cells. Instead, they consist of a continuous mass of protoplasm inclosed
in a single membrane and containing one or more nuclei. Whatever differentiation is present has occurred within the protoplasm of a single microscopic cell.

BEGINNING with one of the least complex of these unicellular animals, we shall see, in the chapters that follow, the ever increasing
complexity and efficiency of the living machinery with which the various
kinds of invertebrates carry on their life-activities. The details of animal
structure which will be presented may be interesting in themselves but are
meaningless unless we view them in relation to their function in the life of
the animal and as a stage in the evolution from simple to complex forms.

In the chapter that follows, most of the classes that comprise the
phylum Protozoa are briefly described. A few green flagellates are included to stress certain similarities between primitive plants and primitive
animals. The difficulty of separating such plants and animals at the
most primitive levels has led many biologists to remove the protozoans
from the kingdom Animalia and to place them in the kingdom **Protista,**
which also includes the green flagellates and the unicellular green algae.
Protistans *all have nuclei but they are not divided up into cells.* In such a
scheme, shown in the appendix at the end of this book, the classes of
protozoans (as described in the next chapter) are elevated to the rank of
phyla.

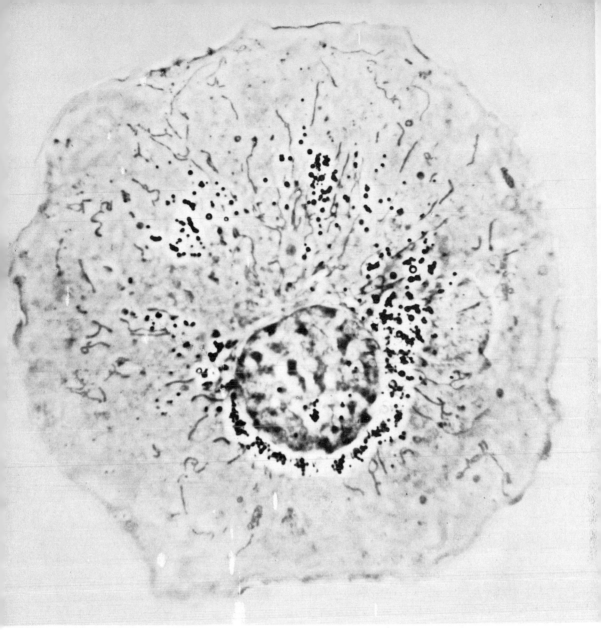

The living cell shown above is a macrophage, a cell of a general type which can be found wandering about in the tissues of many kinds of animals. It was prepared for observation by cutting out a fragment of the spleen of a salamander and mounting it in a drop of nutritive salt solution on a glass cover slip inverted over a hollowed glass slide. Such a method for keeping living cells in glass containers so that they can be easily observed is called "tissue culture." Cells so maintained go on living for a long time if the nutritive medium is changed regularly. The cell shown here is fairly typical and relatively unspecialized and illustrates the essential structure of an animal cell. The *protoplasm* of the cell is bounded by a *cell membrane* and is further differentiated into a number of cell organelles. The most important of these is a centrally located body, the *nucleus*, which is bounded by a *nuclear membrane* and contains solid dark-appearing material, the chromatin. The nucleus lies imbedded in the relatively unspecialized protoplasm or *cytoplasm*, which is strewn with numerous bodies, the *mitochondria* (rods and filaments distributed in the cytoplasm). In the indentation on one side of the nucleus can be seen a clear gray area, emphasized by radiating lines of mitochondria. It contains at its center a dark granule, the *centriole*. The centriole functions during cell division. The prominent black globules are fat. Actual size of cell, 85μm (1/300 inch) in diameter. (Photomicrograph, phase-contrast microscope)

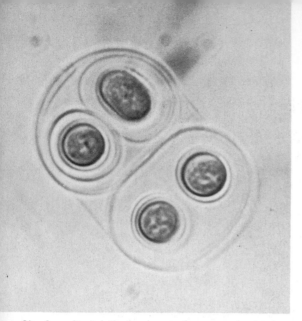

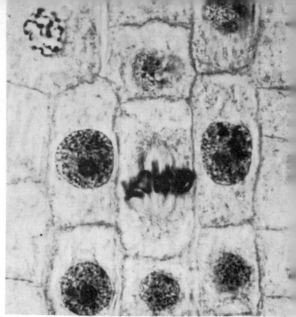

Single-celled plant, *Anacystis rupestris*, a blue-green alga, has no definite nucleus, but the material in the center of the cell is similar to the chromatin of differentiated nuclei. (Photomicrograph of living algae)

Cells of multicellular plants such as these in the tip of an onion root show the heavy cellulose walls that separate the cells from one another. The cell in the center is in the midst of division. (Photomicrograph of stained preparation, courtesy Gen. Biol. Supply House)

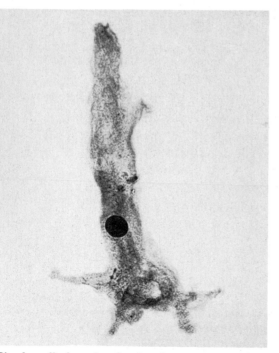

Single-celled animal, *Amoeba proteus*, stained with a dye that was taken up more readily by the nucleus than by the cytoplasm, shows the same essential structure as do the cells of multicellular animals. (Photomicrograph of stained preparation, courtesy General Biological Supply House)

Cells of multicellular animals often are crowded together into tissues, like these from the intestine of an earthworm, and they show a marked deviation from the nearly spherical shape of freely moving cells. (Photomicrograph of stained preparation, courtesy General Biological Supply House)

CHAPTER 3

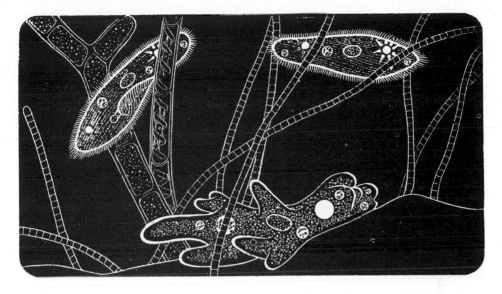

THE FIRST TRUE ANIMALS

THE **PROTOZOA,** whose name means "first animals," are nucleated, non-cellular, mostly microscopic organisms that feed on live prey or organic matter. They comprise the first of the large groupings of invertebrate animals. Of the thousands of species of protozoa that occur everywhere in fresh and salt waters, damp soils, and dry sand, and that live upon or inside the bodies of other animals as parasites, only two will be discussed in this chapter—one simple and one more complex form.

The relation between a simple animal and a complex animal may be compared to the relation between the behavior of prehistoric man and the behavior of modern man. The first man, we may imagine, lived without mechanical aids of any kind, captured his food with his bare hands, sought shelter in a tree or a cave, and existed at the mercy of natural forces. Modern man, on the other hand, has invented mechanical devices to assist him in every way and is gradually learning to protect himself from the caprices of nature. In an analogous way, a simple animal represents a level of organization in which the protoplasm has not evolved many special devices making for greater efficiency in the business of life. Such an animal lives on what we may call the **protoplasmic level of construction;** the protoplasm performs all the life-activities and does not have any very complex structure correlated with any particular activity.

The appearance of specific structures which add to the efficiency of performance of some special activity is called *differentiation* or *specialization*. The ameba is an animal that displays a minimum of differentiation.

A SIMPLE PROTOZOAN—AMEBA

THE common ameba of fresh-water ponds is a microscopic animal; but some of the largest known amebas may reach a diameter of half a millimeter, being visible to the naked eye as white specks. Each ameba is a little mass of clear gelatinous protoplasm containing many granules

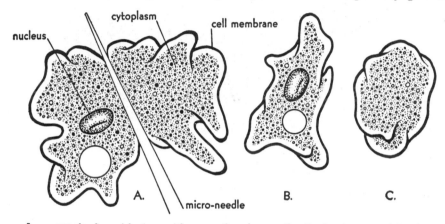

A, an **ameba is cut in two** with a very fine glass needle. **B,** the piece containing the nucleus. **C,** the piece without the nucleus.

and droplets. The surface of the ameba's protoplasm forms a delicate **cell membrane** through which materials pass in and out of the animal. Water passes freely through the cell membrane; but the proteins, carbohydrates, fats, and salts of the protoplasm are prevented from escaping into the surrounding water by this same membrane. If an ameba is cut in two, each piece rounds up and immediately produces a complete membrane, thereby preventing the loss of the interior protoplasm. The formation of a surface membrane appears to be a general property of protoplasm.

The protoplasm of the ameba, as in almost all cells, is differentiated into nucleus and cytoplasm. The **nucleus** occupies no fixed position. The ameba furnishes excellent material for the study of the function of the nucleus, since the animal may be cut into two pieces, one with and the

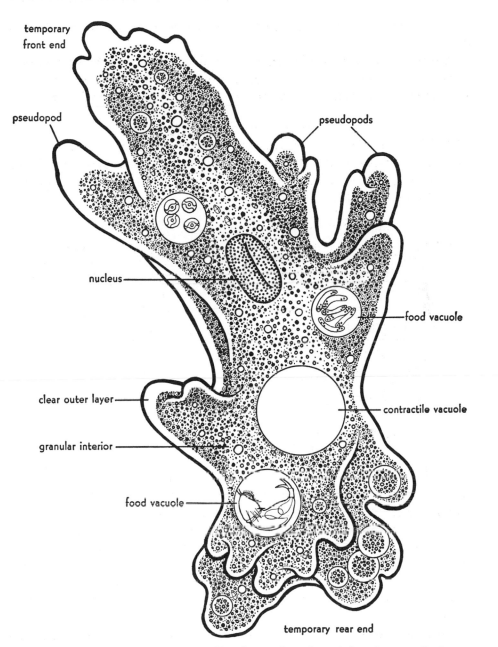

temporary
front end

pseudopod

pseudopods

nucleus

food vacuole

clear outer layer

contractile vacuole

granular interior

food vacuole

temporary rear end

An **ameba,** showing principal structures. Vacuoles are shown in optical section; actually they are imbedded in the cytoplasm (compare with the figure on page 18).

other without the nucleus. The piece with the nucleus behaves like an entire animal, soon grows to its previous size, and finally reproduces. The piece without the nucleus moves about in more or less normal fashion for a time and may sometimes feed but is unable to digest food, to grow or to reproduce; and it dies after the food stored in the protoplasm is used up. From such experiments it is concluded that the nucleus is largely concerned with constructive phases of metabolism and with reproduction. The **cytoplasm** of the ameba is distinguishable into a *clear, outer layer* and a more or less *granular interior*. This interior contains various sorts of crystalline granules, fat droplets, and food bodies in process of digestion, besides droplets containing a watery fluid.

Motility is one of the striking characteristics of animals as contrasted with plants. The type of movement exhibited by the ameba is called, natu-

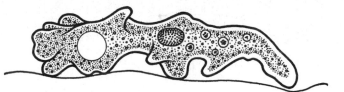

An **ameba in profile.** When the microscope is arranged so that the animal is viewed from the side, it can be seen that only the tips of the pseudopods are in contact with objects, the general mass being free in the water. The pseudopods appear to act like little legs put out one after another, but the "legs" are temporary and soon flow back into the general cytoplasm. (Based on Dellinger)

rally enough, "ameboid movement," and has always excited great interest because it is presumed to be one of the most primitive types of animal locomotion. It appears to be totally different from the muscular movements of complex animals; but what goes on in a muscle when it contracts to move a limb may prove to be similar to the chemical and physical changes that go on in a moving ameba. Furthermore, some of the cells in the tissues of all higher animals, including man, are ameboid.

For these reasons **ameboid movement** has been the object of intensive study. Amebas have no distinct head or tail ends but have a surface which is everywhere the same, and any one point on this surface may flow out as a blunt projection or **pseudopod** ("false foot"). This pseudopod continues to advance for some time through the passage into it of some of the mass of the ameba, but sooner or later another similar projection forms at an adjacent point, and then the cytoplasm flows into the new pseudopod. In this manner the animal progresses in an irregular fashion—flowing first to one side, then to the other. It often alters its course by putting out pseu-

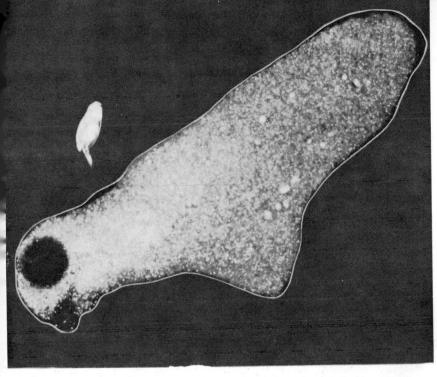

An ameba. Under dark field illumination granular protoplasm and outer membrane appear white on a dark background. The clear outer protoplasm, lacking visible granules, appears black, as does the contractile vacuole. *Pelomyxa carolinensis* is unusual in having numerous nuclei (large white dots among smaller granules). Although one-celled, amebas are not the smallest animals. This species, a giant among amebas, dwarfs the many-celled rotifer just above it. (Photomicrograph of living animal by P. S. Tice)

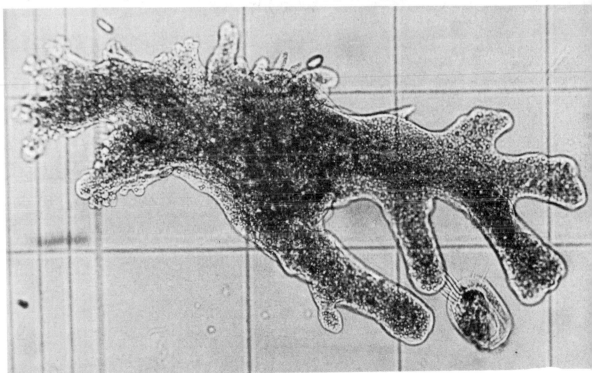

Actively moving ameba, same species as above, seen under bright field illumination. The cell membrane and the granular protoplasm appear dark on a white background. As the picture was taken, the ameba was extending long pseudopods in the direction of its small prey, a ciliated protozoan. The size of this large ameba can be estimated from the lines ruled onto the glass slide upon which the ameba is moving. Each of the largest ruled squares is 250 μm (1/100 inch) on a side. (Photomicrograph of living animal).

14-1

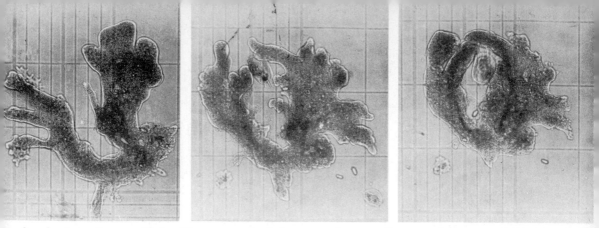

Ameba ingesting a ciliate. The same animal shown on the preceding page here successfully surrounds a ciliate by extending pseudopods in a wide circle around the sides of the animal, which above and below is trapped between slide and cover slip. (Photomicrographs taken at 1-minute intervals)

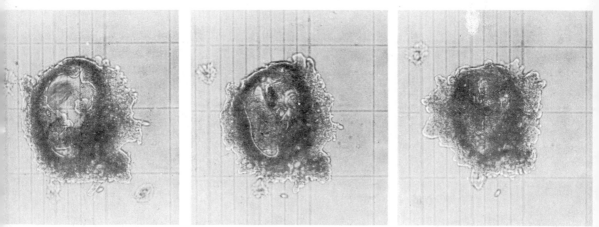

Left: The ciliate appears blurred as it moves back and forth very rapidly without finding an opening through which to escape. *Center:* The ameba finally extends a pseudopod over the top of the prey and (*right*) completely incloses it in a food vacuole. While incorporating the food vacuole, the ameba is rounded up.

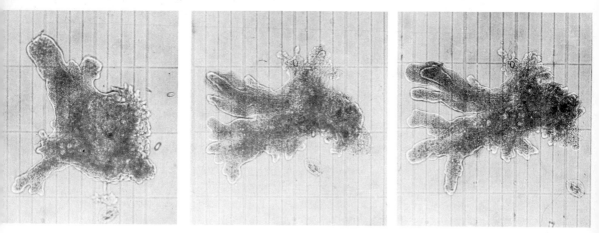

With the ingested ciliate undergoing digestion within a food vacuole, the ameba stretches out and again moves actively along. Anyone who has watched a moving ameba, with its shape never the same from moment to moment, will admire the appropriateness of the name "ameba," derived from a Greek word meaning "change."

dopods on the side opposite the previous advance. As new pseudopods form, the old ones flow back into the general mass. Ameboid movement is very slow, and the animal does not proceed for long in any one direction.

Of the various explanations of ameboid movement that have been advanced, the one which seems most acceptable at present is based on changes in the consistency of the cytoplasm. The cytoplasm, as already noted, is visibly differentiated into a clear, outer layer and a granular, inner one. In the inner layer we further distinguish an outer jellylike region, the **plasmagel,** and an inner fluid region, the **plasmasol.** In a moving ameba, plasmasol flows in the direction of movement; as it reaches the tip of the pseudopod and is deflected to the sides, it changes to plasmagel, while more plasmasol flows forward into the moving tip. This process can be compared to a stream of liquid cement in which the flowing cement hardens on the outside, building ahead of itself a tube of hard cement through which comes more liquid cement from behind. A constant flow of plasmasol is maintained because the plasmagel liquefies at the rear and again flows forward

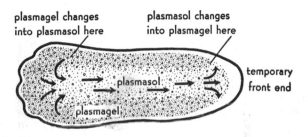

The **movement of an ameba** depends upon reversible changes in the consistency of the cytoplasm. The arrows indicate the path of flowing cytoplasm. (Modified after Mast)

as plasmasol. The flow occurs probably because the tube of plasmagel contracts, exerting pressure on the fluid interior and forcing it forward at the point where the plasmagel is thinnest. Any influence which causes the plasmagel to thicken at the forward end will change the direction of movement. Thus, if pressure is exerted against the end of the leading pseudopod, a thick layer of plasmagel forms at this end and the flow of the animal is reversed. An ameba that is irritated simultaneously on several sides rounds up into a motionless ball, presumably because a heavy layer of plasmagel forms on all sides.

Pseudopod formation occurs not only in locomotion but also in the capture and **ingestion** of food. Although they have no special cell organs of taste or smell, amebas are able to distinguish inert particles from the minute plants and animals upon which they feed. Pseudopods are thrown out around the sides and over the top of the object. In this way the food is held against the substratum, then completely surrounded by cytoplasm, and finally incorporated into the main mass of the ameba's body. The behavior of the ameba varies somewhat with the kind of food. If the food organism is active, the pseudopods are thrown out widely and do not touch or irritate the prey before it has been surrounded. When the ameba is ingesting a quiescent object, such as a single alga cell, the pseudopods surround the cell very closely.

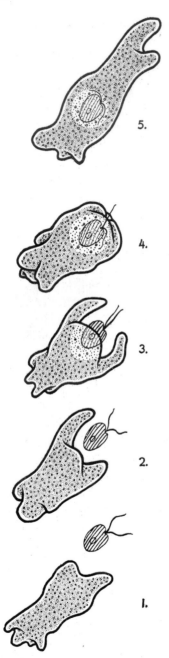

The food body usually lies in the ameba's cytoplasm in a drop of water which was taken in when the food was inclosed by the pseudopods. This drop of water, inclosed by a membrane, contains the food and is called a **food vacuole.** The food very soon begins to undergo **digestion** by enzymes that enter the food vacuole from the surrounding cytoplasm.

In the newly formed vacuole the ingested prey can be seen to struggle for some time, its metabolism probably being responsible for the increasing acidity of the vacuole fluid as shown by the change in color of an indicator dye added to the culture medium. When the acidity of the vacuole is at its peak, the vacuole begins to increase in size through the entrance into it of fluid from the cytoplasm. That this fluid is alkaline is shown by another change in the color of the indicator dye. That it contains digestive enzymes is presumed from the fact that the food body now begins to undergo a perceptible dissolution. This reminds us, in a general way, of the situation in the human intestine, where the digestive enzymes can act only in an alkaline medium.

As the food body gradually dissolves, the dissolved substances pass into the general cytoplasm, where they are **assimilated.** The indigestible fragments are **eliminated** in the simplest fashion possible. They are gradually shifted to the temporary rear end of the animal and are left behind as the ameba flows away.

Respiration requires no special breathing mechanism in a minute creature like an ameba. There is a free exchange of gases with the surrounding water, and the ameba does not "breathe" in the same sense as this expression is used with regard to man, that is, its sides do not heave in and out. Yet it carries on all

An ameba ingests a flagellate. **1,** the ameba moves toward the prey. **2,** pseudopods begin to extend. **3,** pseudopods are thrown out around the sides, and a thin sheet of cytoplasm extends over the top. **4,** a sheet of cytoplasm extends below the prey, which is now completely inclosed except for one flagellum. **5,** the food organism lies in the ameba's cytoplasm in a food vacuole. (Based on Schaeffer)

the essentials of respiration in that energy is liberated from food and made available for other life-processes.

The oxygen dissolved in the surrounding water passes into the cytoplasm of the ameba by diffusion. (*Diffusion* is the tendency of the particles, of which matter is composed, to disperse equally throughout any space which is available; that is, the particles tend to move from regions of higher to regions of lower concentration.) Since the concentration of oxygen in the water is higher than that in the ameba's cytoplasm, oxygen constantly enters and is immediately used up in the burning of foods. Thus the concentration of oxygen within the animal always remains lower than that in the outside water, and oxygen continuously enters the animal and is available for energy requirements.

The water that results from the burning of carbohydrates and fats is a normal constituent of the animal body, and its rapid disposal is not necessary. The carbon dioxide is harmful if allowed to accumulate, and must be removed more promptly. It diffuses to the outside because it is always at a higher concentration within the ameba than in the surrounding water. This method of respiratory exchange (diffusion in of oxygen and diffusion out of carbon dioxide) will work successfully only if the animal is very minute, so that its exposed surface is large in proportion to its bulk or mass, and if it is not covered with a thick protective layer which would interfere with the free diffusion of gases.

Burning protein, as we already know, yields not only carbon dioxide and water, but also nitrogenous substances which are poisonous and which must be rapidly **excreted.** In the ameba no special parts have been shown to excrete harmful wastes and these are thought to diffuse through the cell membrane into the surrounding medium.

Near the rear of a moving ameba is a large spherical water vacuole, called the **contractile vacuole,** which contracts at regular intervals, discharging its contents to the exterior. It then forms again from one or more minute droplets and gradually swells to a maximum size, whereupon it again collapses, ejecting its contents through a temporary pore to the outside.

The role generally assigned to the contractile vacuole is like that of a pump in a leaking ship, in which the pump must be kept going all of the time to keep pace with the incoming water. The water pumped out by the vacuole may come from several sources: it may be produced as a result of respiration, it may be included when food particles are engulfed, or it may enter (by osmosis) because the salts and certain other substances **in** the protoplasm of the ameba are more concentrated than those in its fresh-water environment. (In a living system *osmosis* is the diffusion of water through a membrane which is permeable to water but not to most dissolved substances. When two solutions of different concentrations are separated by such a semi-permeable membrane, the water diffuses in both directions, but more rapidly from the less concentrated to the more concentrated solution than in the reverse direction, because in the more concentrated solution the greater number of dissolved particles interferes with, and in other ways decreases, the outward diffusion of the water. As a result, water tends to accumulate on the side with a greater concentration of dissolved substances.) Experimentally increasing the concentration of salts outside of the ameba causes the vacuole to contract less and less frequently and finally to

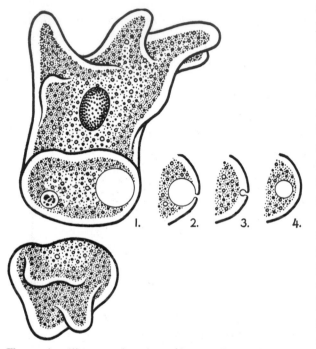

The **contractile vacuole** as it would appear if we could make a crosswise cut through the ameba at the level of the vacuole and move the rear piece back without the contents of the ameba spilling out. **1,** the vacuole at maximum size. **2,** ejecting its contents to the outside through a minute pore. **3,** almost completely emptied. **4,** the vacuole forms again and increases in size. The pore is a temporary structure, formed anew for each ejection.

vanish altogether. Conversely, some marine amebas develop contractile vacuoles when placed in fresh water. Thus, it is probable that the chief function of the contractile vacuole is to *regulate the water content* of the ameba.

An apparatus of this kind seems admirably suited for expelling nitrogenous wastes, but no one has yet been able to produce convincing experimental evidence that the vacuole serves an important excretory function.

On the other hand, there is evidence against the excretory role of the contractile vacuole. When the fluid in the vacuole is withdrawn by means of a micropipet, chemical analysis fails to show a concentration of urea (nitrogenous waste) high enough to indicate that the vacuole has a special excretory function. Analysis of vacuole fluids of other protozoa shows that, even though small amounts of nitrogenous materials may be dissolved in the water expelled from the vacuole, the amount is not great enough to account for the total amount of waste excreted by these organisms, and most of the materials must diffuse out through the cell membrane. Perhaps improved chemical methods will settle this problem in the future.

Less direct evidence for the water-regulatory role of the vacuole is the fact that many marine and parasitic protozoa, which live in environments having a salt concentration higher than that in their own protoplasm, do not possess vacuoles. Since water moves through a membrane from a region of low salt concentration to one of higher salt concentration it does not accumulate within these animals. In the parasitic protozoa that have vacuoles, most of the water expelled probably enters in the feeding process.

The ameba **reproduces** by the simple process of dividing into two amebas. When an ameba has fed well for some time, it rounds up into a spheri-

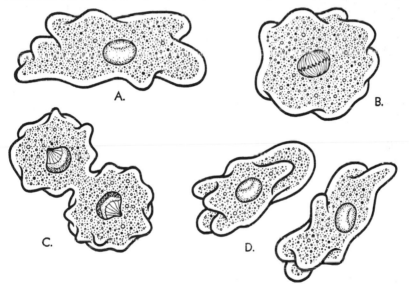

An **ameba divides.** A, old, well-fed ameba. B, the ameba rounds up, and the nucleus undergoes a change preparatory to division. C, the nucleus divides (by a process known as "mitosis"), and the cytoplasm constricts. D, the two young "daughter" amebas, each with a nucleus and half the parent's cytoplasm. (Based on Dawson, Kessler, and Silberstein)

cal mass, the nucleus divides, and the cytoplasm constricts until the slender strand that connects the two halves ruptures. The entire process requires less than an hour. Each half, called for no particular reason, a "daughter" ameba, behaves just like the parent and soon increases to maximum size. Since an ameba thus continues to exist in its offspring, it may be said to be "immortal"; and every ameba which now exists is directly continuous through the ages with the first ameba. However, the protoplasm of every cell is continually being destroyed and renewed, and we may be almost certain that no part of the original ameba is present in any modern ameba.

The ameba can carry on its routine activities only when immersed in water. If conditions of life become unfavorable, as when the pond dries up or when the food supply runs low, the ameba rounds up and secretes on its outer surface a hard and impervious protective shell called a **cyst.** Within the cyst the animal's rate of metabolism falls to a level just above that necessary to maintain the organization of the protoplasm. Replaced in a suitable environment, the cyst breaks open and the inclosed animal emerges and resumes its usual activities.

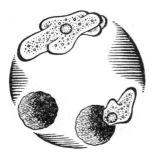

An **encysted ameba** survives unfavorable conditions. With the return of proper conditions, the animal emerges from its cyst.

Although the ameba has nothing comparable to our organs of special sense, it can distinguish food from particles of no food value. As mentioned before, it uses different tactics in approaching plant and animal food, probably because the movements of animal prey create disturbances in the water that stimulate the ameba. The ameba flows away from a bright light, injurious chemicals, or mechanical injury. When poked with a glass rod, it contracts, reverses the direction of flow, and moves away. Extreme disturbance or injury causes it to take on a spherical shape and remain motionless for some time.

The **behavior** of an ameba, particularly its ability to select food, has been used by some as evidence that an ameba exhibits "conscious" behavior or possesses some trace of those powers which in man are vested in the brain and which have been called the "psychic property" of protoplasm. Others maintain that the simple activities of an ameba imply no psychic attributes, and point to the fact that it is possible to duplicate practically all of the activities of an ameba with (nonliving) **mechanical models**—not only the chemical changes, such as those involved in respiration and digestion, but also the more characteristically living activities. Ameboid movements are produced by injecting a little alcohol into a droplet of clove oil in water. The alcohol changes the surface film of the oil droplet and causes it to send out "pseudopods" and to flow about like an ameba. A drop of chloroform in water appears to be quite as "finicky" in its "eating habits" as an ameba. When offered small pieces of various substances, such a drop will "refuse" sand, wood, and glass and will even eject them if they are forcibly introduced. On the other hand, bits of shellac or paraffin are "eagerly" enveloped. If we play a trick on the chloroform drop by feeding it a piece of glass coated with shellac, it will engulf this "delicacy," dissolve the shellac, and then "eliminate" the glass. There are many other mechanical models which simulate growth and even reproduction. The resemblances are usually quite superficial, and most of them throw little light on the real mechanisms involved in the living systems which they apparently imitate. But these experiments do suggest that much of the "mystery," which some writers attribute to the behavior of the living ameba, might be explained if we knew enough about the purely physical phenomena involved.

The ameba differs from the mechanical models in that several models are required to demonstrate the activities that are displayed by a single ameba, a fact which emphasizes the complexity of this "simple animal." A more important difference is that the behavior of the ameba is *adaptive*, that is, it is of a type likely to result in survival of the animal.

In the ameba we have emphasized simplicity and the ability of protoplasm to perform all the necessary life-activities without the aid of highly specialized structures. We shall see now that in another member of the large and varied phylum Protozoa, many specializations are possible even within the limits of the protoplasm of a single cell. These specializations have the same use as the inventions and machines of human construction: they enable the animal to carry on its activities with greater efficiency.

A COMPLEX PROTOZOAN—PARAMECIUM

PARAMECIA are found everywhere in fresh waters and can be obtained in enormous numbers by letting a bit of food decay in pond water. Like an ameba, a paramecium consists of a microscopic mass of protoplasm which is differentiated into a semifluid granular interior and a more dense, clear, outer layer. But many differences between the two animals are at once apparent. Instead of a delicate outer membrane, a paramecium is covered by a stiff but flexible **outer covering.** This covering gives the animal a definite **permanent shape,** somewhat like that of the sole of a slipper. Also, a paramecium has **distinct front and rear ends,** the front rounded, the rear pointed—a good example of streamline form. And most striking of all is the rapid rate at which a paramecium swims about, as compared to the slow creeping of an ameba.

Beneath the outer covering, and imbedded in the clear outer cytoplasm, are small oval bodies called **trichocysts.** These bodies reach the surface through pores and can be discharged to the exterior. During the process of discharge they become greatly elongated into fine threads. It is not clear what function the trichocysts serve. It is thought that they afford a means of protection, since a paramecium discharges them when touched by injurious chemicals or when attacked by an enemy. It has also been suggested that a paramecium uses the trichocysts to anchor itself while feeding on bacteria.

The paramecium has put on speed by developing accessory structures for **locomotion** which are not unlike the oars of a racing shell. This small animal is covered with about twenty-five hundred short "hairs," which are

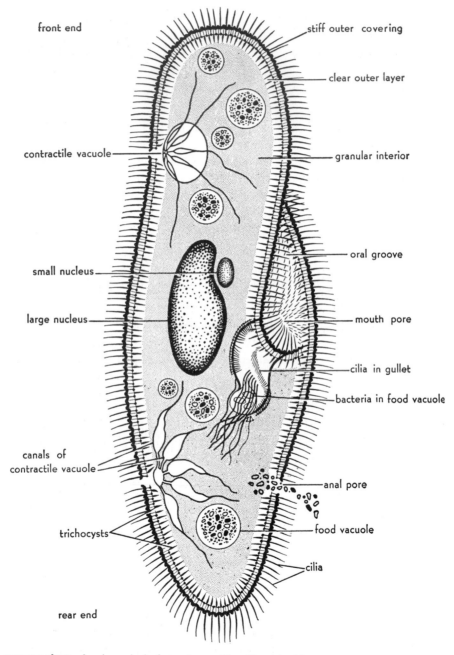

front end

stiff outer covering

clear outer layer

contractile vacuole

granular interior

oral groove

small nucleus

mouth pore

large nucleus

cilia in gullet

bacteria in food vacuole

canals of contractile vacuole

anal pore

food vacuole

trichocysts

cilia

rear end

A **paramecium,** showing principal structures. The cilia and trichocysts, shown only at the edge, really occur all over the surface of the body. (Certain details based on Lund and on King)

The **cilia in a single row** do not beat all at once but one after another, so that they appear to beat in waves. Ordinarily, they beat so fast that all we see is a flickering at the edge of the paramecium. (After Gelei)

really protoplasmic extensions through minute holes in the stiff surface covering. These protoplasmic extensions, called **cilia,** beat in somewhat the same manner as the arms are moved in the crawl stroke in swimming; they reach forward in the relaxed part of the stroke and then give a strong backward lash. The combined effect of all the cilia, rhythmically stroking backward, is to drive the animal forward. The cilia do not beat simultaneously but in a wave beginning at the front end of the animal and progressing backward. Further, they beat obliquely rather than straight backward. The oblique stroke causes the animal to revolve on its long axis so that, as it swims through the water, it revolves continually and describes a spiral path. A paramecium can swim backward by a reversal of the ciliary stroke and can turn in any way.

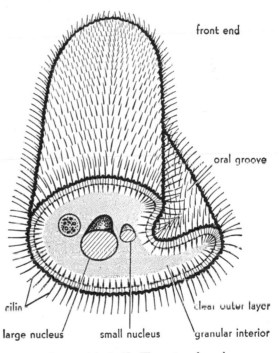

Paramecium cut in half. The cut surface shows some of the relations that are not clear from the large diagram.

The **food-catching apparatus** of the paramecium is much more specialized than in the ameba. Food is taken in only at a definite place on the surface. One side of the paramecium is strongly depressed, forming a concavity, the **oral groove,** as if a piece had been cut out of the animal. This concavity leads backward to an opening, the **mouth pore,** from which a funnel-like tube, the **gullet,** extends down into the cytoplasm.

When a paramecium stations itself near a bit of decaying material, the beat of the cilia in the oral groove drives bacteria and other minute organisms toward the gullet. The bacteria are whirled around by special ciliary tracts and are concentrated into a ball at the bottom of the gullet. The finished ball then passes as a **food vacuole** into the cytoplasm. A paramecium that has found a suitable bit of debris and is feeding actively will soon become filled with food vacuoles. These vacuoles are moved about in the interior cytoplasm in a more or less definite course by a slow circulation of the semifluid cytoplasm, and in the meantime their contents undergo **digestion** essentially as described for the ameba.

The few indigestible remnants in the food vacuoles are finally **eliminated** from the body at a definite **anal pore** in the outer covering.

Respiration and **excretion** take place, as in the ameba, by diffusion through the surface and are essentially the same as in all other animals— that is, oxygen is taken in and used for the burning of foods; and carbon dioxide, water, and nitrogenous wastes (said to be ammonia and urea) are given off.

Two **contractile vacuoles** occupy fixed positions near the surface on the side opposite the oral groove, one near the front end, the other near the rear. The apparatus is more complicated than in the ameba, for each vacuole is surrounded by a circle of canals which radiate from the vacuole for some distance into the cytoplasm. At short intervals these canals fill with fluid, then discharge their contents to form the vacuole, which in turn ejects the fluid to the exterior. While the contractile vacuole of an ameba appears to be a temporary structure which re-forms before each contraction, in a paramecium the vacuolar canals and the pore through which the vacuole discharges are probably permanent structures, even though the vacuole is probably not. Concerning the function of this vacuole mechanism, there is nothing to add to what was said about the ameba. Apparently in the paramecium also the system serves primarily for regulating the water content of the animal.

In a paramecium the two contractile vacuoles can eliminate a volume of water equivalent to its body volume in about half an hour, as compared with four to thirty hours required by an ameba. An average man eliminates a volume of urine equal to his body volume in about three weeks, but he also disposes of excess water through the lungs and sweat glands.

The high degree of co-operation displayed by the cilia in the swimming movements or in food-taking suggests that some **co-ordinating mechanism** resembling nervous control in higher animals is present, and such has, indeed, been discovered in the paramecium. Near the surface of the animal is a **system of protoplasmic fibers** (fine threads) which run longitu-

dinally and connect the rows of small granules at the bases of the cilia. There seems little doubt that this system of fibers constitutes the mechanism which regulates the activity of the cilia; for, when the mechanism is injured, as has been done experimentally in certain ciliates, the cilia no longer beat in co-operation.

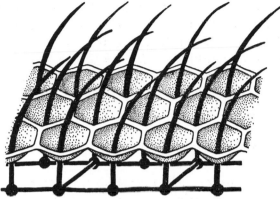

A small portion of the surface of a paramecium showing the **co-ordinating fibers** connecting the bases of the cilia. (Modified after Lund)

The large number of minute cilia in a paramecium makes for a co-ordinating system of fibers, or neuromotor apparatus, that is not easy to experiment upon. However, a ciliated protozoan like *Euplotes* has only a few large cilia, and it is possible to cut the fibers which

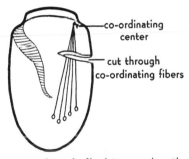

co-ordinating center

cut through co-ordinating fibers

A cut through *Euplotes* severing the co-ordinating fibers, inco-ordinates the large rear cilia to which these fibers run. The cilia are shown in the figure of Euplotes on the next page. (After C. V. Taylor)

run to the large rear cilia from the center at which the co-ordinating fibers meet. After the operation these large rear cilia do not beat in co-ordination with the others, and certain of the swimming movements are interfered with. A cut of equal size, which does not injure the fibers, does not influence the co-ordination of the cilia.

The innermost granular cytoplasm, as in an ameba, is more fluid than the surface layer, and contains food vacuoles, fat droplets, and other food bodies, as well as **two nuclei,** one large and one small nucleus (there are several of the small nuclei in some species of *Paramecium*). The **large nucleus** appears to be concerned with the ordinary business of the cell, while the **small nucleus** is especially active during reproductive processes.

The experimental **removal of the small nucleus** to determine its function is difficult in a paramecium but can be done conveniently with *Euplotes*, which stands microsurgery well and has the added advantage of a small nucleus that is easily visible and can be removed with a micropipet. The operated animal appears to be uninjured otherwise and may even divide once or twice; but after two or three days it dies, without ever reforming the small nucleus. That death was not due to injury other than the absence of the small nucleus is proved by the complete recovery of animals undergoing "control" opera-

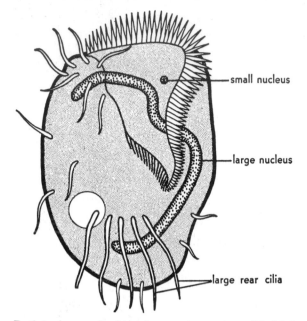

tions in which a portion of the cytoplasm, near the small nucleus, is removed, and another operation in which the small nucleus is first removed and then immediately replaced. Also, animals in which a portion of the large nucleus is removed, the small nucleus being left undisturbed, recover from the operation and produce a large number of normal descendants. It is apparent that the small nucleus, probably more so than the large nucleus, is necessary for continued life in this animal.

small nucleus

large nucleus

large rear cilia

Euplotes is a good subject for experimentation. (Modified from a photomicrograph by C. V. Taylor)

It is not at all clear why a paramecium (and other protozoans related to the paramecium) should have two sorts of nuclei, a condition unknown in other animals. We can only say that among these protozoans the functions of a nucleus have been subdivided between the two different bodies.

A paramecium **reproduces** by dividing in two in a manner similar to that described for the ameba. Both kinds of nuclei elongate and pull apart into two halves, one of which remains in each daughter-cell. A constriction forms around the middle of the animal; and as the constriction deepens, the cytoplasm divides into two daughters. The front and rear halves of a paramecium are not exactly alike; but even before separation occurs, each half forms the parts necessary for a complete paramecium. Thus the gullet, which is behind the middle, falls to the rear daughter; the front daughter early in the process of division forms a new gullet.

When well fed, paramecia may divide two or three times daily, so that enormous numbers of them can be obtained in a short time. For this reason, paramecia, as well as other protozoa, have been used in many population studies. The results of such studies contribute to our understanding of the laws of growth of human populations.

The beginnings of **sexual processes** occur in the paramecium, although the animals themselves do not show visible differentiation into males and females. However, sex differences can be distinguished physiologically. Only when individuals of certain strains are placed with individuals of certain other strains do they adhere in pairs and unite by their oral grooves.

A.

Two animals unite by their oral grooves.

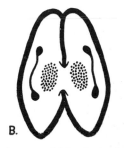

B.

The large nuclei begin to degenerate; the small nuclei divide twice, three degenerate, the remaining one divides again.

C.

One part of each small nucleus migrates to the opposite paramecium.

D.

The small nuclei fuse. The animals separate.

E.

The fusion nucleus divides several times.

F.

The animal divides twice, resulting in four small paramecia.

Conjugation in one species of the paramecium. Not all stages are shown. For clarity, the large nuclei are omitted from **C**; actually they do not degenerate completely until after the conjugants have separated.

While two individuals are so united, their nuclei undergo complicated changes, the result of which is the passage of a portion of the small nucleus from each animal into the other. Each migrating nucleus fuses with the opposite remaining nucleus. The two paramecia separate and undergo a series of divisions; the animals resulting from these divisions then continue their usual activities. This sexual process is called **conjugation.** Although more typical sexual reproduction, involving differentiated sperms

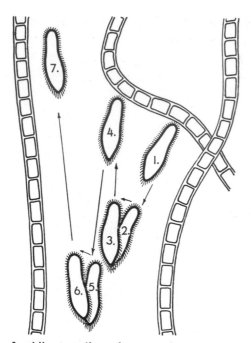

Avoiding reaction. 1, paramecium encounters an obstacle. **2,** the animal backs up. **3,** shifts its position. **4,** again meets resistance. **5, 6,** backs up and turns. **7,** finds a free path. (Based on Jennings)

and eggs, occurs in other protozoa, conjugation in paramecia has the essential features of the sexual process in all animals, that is, the *transfer of nuclear material having new hereditary possibilities from one animal to another.*

The **behavior** of a paramecium is exactly what one would expect of an animal that has no specialized sense organs to direct its movements. When not quietly feeding on bacteria, it roams about ceaselessly, bumping "head on" into obstacles in its path. After such a collision, the paramecium backs up by reversing the beat of the cilia, turns to one side, and goes off in a new direction. If this second path results in another collision with the same obstacle, the set of movements is repeated. Finally the animal encounters a free path and

continues on its course. The set of movements with which a paramecium backs up, turns, and swims off in a new direction is called the **avoiding reaction.** Mechanical obstacles, excessive heat, excessive cold, irritating chemicals, unsuitable food, a predaceous enemy—all elicit the avoiding reaction, which may be said to constitute most of the behavior of a paramecium.

In its constant explorations the paramecium may swim by

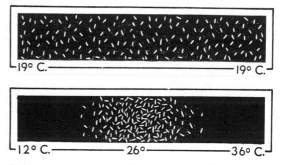

Behavior in relation to temperature. When the temperature is uniform throughout the tank, the paramecia are uniformly distributed. If one end of the tank is cooled to 12° C. and the other end is heated to 36° C., the paramecia avoid the extremes of temperature and accumulate in the region of optimum (most favorable) temperature. (After Mendelssohn)

chance into a region rich in bacteria. Each time that it crosses the boundary of this region into a less favorable area, it gives the avoiding reaction; thus it remains in the more favorable region. This general method of finding the best conditions of existence is called **trial-and-error behavior** and is employed to some extent by all animals, including man.

A paramecium need not actually enter an unfavorable region before it can react negatively. The beating of the cilia in the oral groove draws a constant

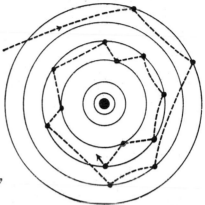

Trial- and-error behavior in relation to chemicals. A small amount of a chemical is placed in the center of a drop of water, and it slowly diffuses outward. The concentric circles indicate zones of diminishing concentration of the chemical from the center. The broken line shows the path of a paramecium placed in the drop. As the animal swims about, it gives an avoiding reaction whenever it enters a zone less favorable than the one it is in; it thus appears to be "trapped" in the most favorable region. (After Kühn)

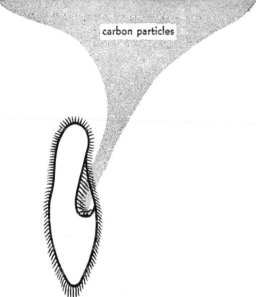

carbon particles

Paramecium "samples" the water ahead, as can be seen by placing in the path of the animal a drop of India ink containing visible particles. (After Jennings)

stream of water, in the form of a cone, toward the oral groove. If there is an irritating chemical in the water ahead or if the water is hotter or colder, a portion of the water from the new region will be drawn backward into the oral groove. Thus the paramecium constantly receives "advance information" of the environment ahead and responds with the avoiding reaction without actually entering an unfavorable region.

Paramecia have only a poorly developed ability to discriminate between foods, since they very readily take in and form food balls of almost any minute particles, such as carbon grains, dye suspensions, and the like. However, after a time they will reject these inert particles while still accepting bacteria. Paramecia avoid strong acids; but they give the avoiding reaction when passing from dilute acids to ordinary water, and therefore tend to aggregate in regions of low acidity. This behavior aids the animal in feeding, because bacteria are most likely to be present near decaying organic matter, which renders the surrounding water slightly acid. On the whole, it may be said that the behavior of a paramecium is remarkably adaptive for an animal that has to find its way about simply by keeping out of trouble.

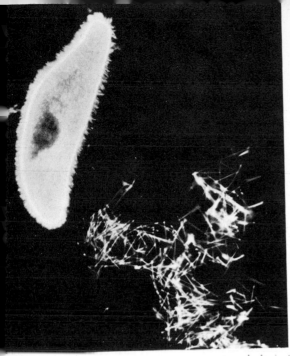

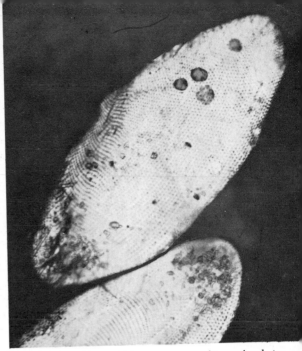

Trichocysts discharged from a paramecium irritated and later killed by a drop of ink added to water in which the animal was swimming. Said to be protective, trichocysts appear more useful in anchoring the animal while feeding. (Photo by P.S. Tice)

Outer coverings of two paramecia stained to show the surface markings which correspond to the positions of the cilia in the living animal. The lower covering shows the pattern of cilia around the oral groove. (Photo of stained preparation by P. S. Tice)

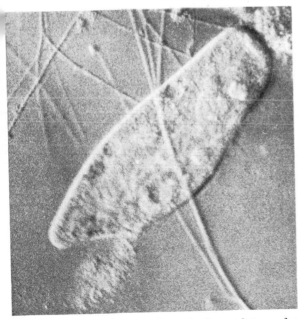

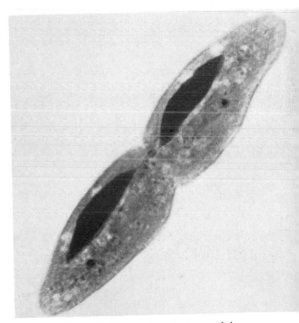

Ejection of solid wastes through the anal pore of a paramecium. The pore has a fixed position posterior to the oral groove and is thought to be a weak spot in the outer layer of the animal. (Photo of living animal from motion picture by P. S. Tice)

Paramecium dividing. The large nuclei appear as two large, dark, elongated bodies joined by a delicate strand. The small nucleus is already completely divided into two small, dark bodies. The cytoplasm has begun to pinch in two. (Photo of stained preparation)

30-1

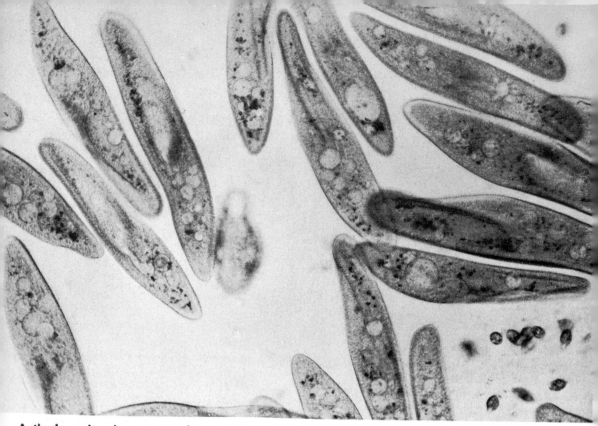

Actively swimming paramecia, "stopped" by a photomicrograph taken at 1/30,000 second, are seen in various positions. On the left are two individuals which present the full oral surface. The oral groove occupies the center of the upper portion of the animals, and at its lower end can be seen the outline of the gullet. All the individuals show food vacuoles in various stages of digestion. At least three individuals show the canals of the contractile vacuole in the process of filling.

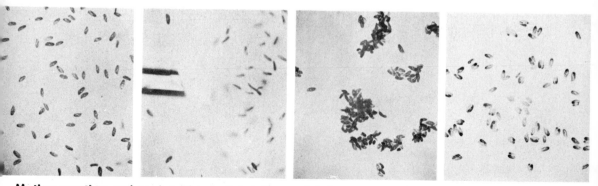

Mating reaction and conjugation in living *Paramecium bursaria. From left to right:* **1** Individuals of one mating type swimming about. **2** A pipette introduces into the same drop of water individuals of another mating type. **3** Within 5 minutes after mixing, the paramecia are stuck together in large clumps. These clumps then become smaller, and after about 6 hours only small groups and conjugating pairs remain. **4** 23 hours after mixing there are only conjugating pairs and a few single individuals that were unable to obtain mates. Pairs usually remain joined for about 36 hours. Conjugation occurs only under certain conditions. The individuals must be mature, in a certain state of nutrition, and be mixed at a certain time of day. In the culture shown here the mating reaction occurred from about 10:00 A.M. to 3:00 P.M. In *Paramecium bursaria* several separate groups of mating types are known. The types of any one group do not conjugate with any types from groups other than their own. (Photomicrographs from a motion picture by R. Wichterman)

CHAPTER 4

CLASSIFIED KNOWLEDGE

TO THE casual observer porpoises and sharks are kinds of fish. They are streamlined, good swimmers, and live in the sea. To the zoölogist who examines these animals more closely, the shark has gills, cold blood, and scales; the porpoise has lungs, warm blood, and hair. The porpoise is fundamentally more like man than like the shark and belongs, with man, to the mammals—a group that nurses its young with milk. Having decided that the porpoise is a mammal, zoölogists can, without further examination, predict that the animal will have a four-chambered heart, bones of a particular type, and a certain general pattern of nerves and blood vessels. Without using a microscope they can say with reasonable confidence that the red blood cells in the blood of the porpoise will lack nuclei. This ability to generalize about animal structure depends upon a system for organizing the vast amount of knowledge about animals.

This knowledge was not always so well classified. Only a few hundred years ago the study of animals was in a crude descriptive stage. The accumulations of facts about animal structure and animal behavior were almost entirely useless because little attempt had been made to relate one fact to another and because the facts were arranged in categories based on superficial distinctions. Biological science made little progress until it was realized that animals must be grouped according to their fundamental similarities in structure and then arranged into a workable **system of classification.**

The heterogeneous assortment of organisms that compose the animal kingdom are first divided into large groups, called **phyla,** which are based on *radically different plans of organization.* The members of a phylum may live in every kind of habitat, may vary in size and body form, and in their methods of locomotion and feeding—but they have a common basic structure. In the Protozoa this underlying structure is the differentiation of protoplasm within a single cell—other phyla have other body plans. The chief phyla are well recognized and agreed upon by zoölogists, though there is debate over the classification of some of the rarer and more highly specialized or degenerate animals. Unfortunately, there is always a certain amount of arbitrariness in any system of classification, and the exact number of phyla that we name depends upon our criteria of a radically different plan of organization.

Within each phylum the members are further divided into groups, called **classes,** on the basis of a *significant variation* in the fundamental plan, usually in adaptation to a *special way of life.* To take a crude analogy from everyday experience: if we regard all vehicles driven by gasoline engines as belonging to the same category or "phylum," significant variations are automobiles, airplanes, and motor boats. Each vehicle has the same fundamental plan, a gasoline engine—yet each is constructed for a significantly different kind of travel. Similarly, protozoa that move by pseudopods, like the ameba, are grouped in a separate class from protozoa that move by means of cilia, like the paramecium.

Each class is further divided into smaller categories called **orders.** In terms of the analogy given above, we can subdivide the class automobiles into "orders": passenger cars, trucks, racing-cars. Order differences are still of such magnitude that they can be recognized easily. For instance, the class Insecta has orders such as the beetles, the flies, the butterflies and moths, the fleas, and others.

Each order consists of a number of **families.** The anatomical distinctions between families are still important enough to be of survival value to the groups concerned. That is, the structures which serve as a basis of classification are likely to be the ones that enable a member of one family to live in a place that is uninhabitable for a member of another family. A diving beetle that has its legs modified for swimming and its jaws for seizing prey would have difficulty getting along in a forest, the home of a leaf-eating beetle.

The anatomical criteria used to divide a family into groups called **genera** (singular, genus) are usually so small that they would not be noticed by most people and in general have less adaptive value for the animals concerned than have family distinctions. Crayfish of the genus *Cambarus* have seventeen pairs of gills, while those of the genus *Astacus* all have eighteen pairs of gills. The possession of the extra pair of gills probably makes no difference in the success with which *Astacus* meets its environment as compared to *Cambarus*. *Astacus*, which is found only west of the continental divide, would probably do well east of the Rockies in the territory of *Cambarus*. But the two groups of crayfish have long been separated by an impassable barrier, the Rocky Mountains.

When we divide a genus into **species,** we are at last dealing with a category which is somewhat less arbitrary and which represents what the scientist means by a *kind of animal.* There are borderline cases that do not clearly fit this definition; but for the vast majority of animals a species may be defined as a natural population of organisms which has a heredity distinct from that of any other group, and the members of which breed only with one another to produce fertile offspring. To return to the example of the crayfishes, the genus *Cambarus* can be divided into distinct species on the basis of minor details, particularly the shape of the first pair

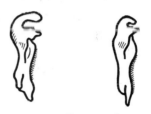

Species differences in two crayfish—the first abdominal appendage of the male. *Cambarus limosus,* left; *C. diogenes,* right. (After Ortmann)

of appendages on the abdomen of the male. One of these species lives in rivers and is called *Cambarus limosus.* Another, called *Cambarus diogenes,* lives in swamps, where it burrows in the mud. The minor anatomical difference in the appendages of the male, by which we can distinguish the two, has no practical bearing on the more deep-seated physiological differences which really make these crayfishes two species or populations that do not breed together. Physiological differences are not easy to measure or define, and they are difficult or impossible to determine from dead specimens. Since animals must often be classified long after they are collected, it is desirable, whenever possible, to base the criteria for identification of the species on some easily observable anatomical character which is not changed by the death of the animal. Thus, a few distinctive characters are selected for identification of species. However, the individuals of two species differ in not one but in a great many minor characteristics of form and behavior, and these differences are due to the accumulations of minor changes in the course of evolution.

Species cannot be described in all cases by clearly visible anatomical differences; sometimes physiological characteristics must be studied. For example, two species of a single-celled green alga (*Chlorella*) can be distinguished only by measuring their average rates of respiration. There are two species of termites that can be distinguished most easily by identifying the other animals that live associated with them. Except for this difference in the "guests" which they harbor, the termites can be distinguished only on the basis of average differences in certain body measurements made on a large number of individuals. A single termite, found away from the nest, cannot be assigned definitely to either species. Such cases emphasize the need for dealing with populations, rather than with single individuals, in describing a species.

In a widespread species there may be a gradual change in the characteristics of the population from one end of its realm to the other, so that in some instances widely separated individuals from the same population would be regarded as different species if there were no intergrades. When there are intergrades, such widely separated forms are regarded as belonging to different **subspecies.**

If we refine our observations and criteria, we can often distinguish, among the members of a species, individuals that can, but usually do not,

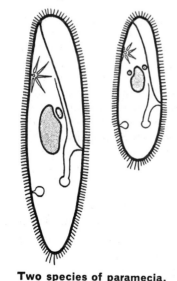

Two species of paramecia. *Paramecium caudatum*, left; *Paramecium aurelia*, right. (After Wenrich)

breed together or live together in the same region. They exhibit minor differences, such as the color variations in the skin of man. The differences are usually more superficial than those which distinguish a species, and the groups classified on this basis are termed **varieties, races,** or **strains.** Still less significant are differences between the individuals of a variety or race.

The double name by which we referred above to the river and swamp crayfishes is called the **scientific name** and consists of, first, the genus name (written with a capitalized initial letter) and, second, the species name (not capitalized). Man's scientific name is *Homo sapiens*. A common paramecium is *Paramecium caudatum*. The name *caudatum* means tail and refers to the long cilia at the rear. *Paramecium caudatum* is distinguished from *Paramecium aurelia* by the possession of one small nucleus instead of two, the more pointed rear end, and larger body size.

The scientific name is frequently written with the name of the scientist who first adequately described and named the species. For example, "*Balanus balanoïdes* Linnaeus" is the common barnacle, which was first described by Linnaeus (Carl von Linné), a Swedish naturalist who established the system of giving two names to animals. In his time (1707–78) Latin and Greek were the accepted languages for scientific writings, and Linnaeus gave all the animals Latin names. This system has become universally adopted, and it is now obligatory for scientists to produce a Latinized name for a newly described animal. Almost a million animal species have already been described and named by this system, and it is probable that several times this number are still to be discovered. Every year several thousand new species are added to the list of catalogued animals.

The assigning of scientific names is governed by a definite set of regulations, internationally recognized. According to these rules, the valid name of a species is the name which was given to it by the first person who published for the species a binomial Latinized name, together with a description of the species. This seems clear enough. But many decades ago the descriptions of species often appeared in obscure publications; and communications between scientists all over the world were not very good. It often happened that two or more people independently named and described the same species. In addition, many of the old descriptions are so incomplete or unclear that it is impossible to determine exactly to which animal the description applies. Or, an old genus may be shown to be a collection of several diverse groups of species, each of these groups distinctive enough to be a separate genus.

When such difficulties arise, either the generic name or the specific name, or both, may have to be changed, even though they have been in long and familiar use. A scientific name cannot be changed simply because the name is less appropriate than another that might be suggested. The change can be made only for reasons clearly defined in the rules of nomenclature. Cases in which a proposed change is controversial, or for which a motion is made to suspend the rules, are decided by an international commission of scientists.

The rules of nomenclature do not apply to taxonomic names above the family level. That is why the larger groupings of animals have been renamed many times, and the classification into orders, classes, and phyla may differ considerably in the various publications. As our information about the structure and the embryology of the groups increases, there is less need for substituting speculation for facts, and the major groupings of animals are becoming more stabilized.

When zoölogists mention animals in scientific papers, they do not use common names, because these vary from place to place and from time to time. What is called a "crayfish" in one locality may be an entirely different species or a different genus from what is known by the same common name in another locality. The scientific name is international, recognized the world over as referring only to one clearly defined species of animal. Sometimes a species is so distinctive that it can be recognized at sight or by reference to a catalogue. But, as a rule, the number of similar species is so great and the differences between them so small that they can be accurately identified only by a specialist in classification, a taxonomist, who knows the detailed characteristics of the group. There is no virtue in giving the full name of an animal if the name is not correct. Rather than commit this "scientific crime," one refers to a common animal by using its genus (or generic) name. Thus we have said, in chapter 3, that one could operate on a *Euplotes* as if it were a particular animal. Yet the name actually refers to a genus consisting of several species of animals. Sometimes we do not even capitalize the generic name, if the animal to which it refers is so well known that its generic name has become a common name. We may refer to an individual of any species of the genus *Paramecium* simply as a "paramecium."

THE system of classification represents a **scheme of animal relationships.** When we see two men who are strikingly similar, we are likely to say: "They must be brothers." If they resemble each other somewhat less, we may say: "No doubt they are cousins." If they bear the same name but resemble each other hardly at all, we guess that they are only distant relatives. In other words, we judge the closeness of their relationship by their degree of similarity. Two brothers have a pair of very recent common ancestors, their parents. Two cousins have only a pair of grandparents in common, a less recent common origin. In the same way, two species of the same genus have had a fairly recent common ancestor. Species of two different genera have had a more remote common ancestor. And species of two different families are still more distantly related. Thus, the position of an animal in the scheme of classification indicates our idea of its relationship to other animals.

CHAPTER 5

A VARIETY OF PROTOZOA

PROTOZOA are cosmopolitan—the same species may be found on every continent. In sharp contrast with the provincial habits of most larger animals, which are limited in their spread by ocean or land barriers, protozoa are readily swept along in ocean and river currents, or in the encysted state may be blown by the wind or transferred from one pond to another in the mud that clings to the feet of birds. Encystment not only furthers the distribution of protozoa but also enables them to live in habitats they otherwise could not invade. Within the heavy walls of their cysts, some protozoa can resist the heat and drought of summer in the desert. After the first rain the cysts break open and the protozoa move about, frequently encysting again after only an hour of activity. Animals of such versatility are well adapted to exploit the rich possibilities offered by the moist and nutritious interiors of other animals. Humans are common hosts, but protozoan parasites are rather impartially distributed among all animals.

About thirty thousand species of protozoa have already been described, though it is probable that many more remain to be discovered. Several other phyla have many more species; but when we consider that all larger animals, and even some protozoa, harbor one or more species of protozoa, we are led to the interesting conclusion that there are many more individual protozoa than individuals of all other animals combined.

The protozoans display such an enormous range of differences in body symmetries, in nuclear states, and in other structural and physiological characteristics, that their diversity is thought to be comparable with that seen among the many phyla of multicellular invertebrates. In recognition of this, the phylum Protozoa is often raised to the rank of a subkingdom, and its traditional classes to rank of phyla. The various many-celled phyla are then grouped under the subkingdom **Metazoa.** However, the classification of protozoans is at present undergoing so much rethinking that it seems prudent here to retain, tentatively, the traditional phylum Protozoa, divided into classes based, in most cases, on the chief means of locomotion.

THE FLAGELLATES

THE flagellates (class **Flagellata**) are protozoa that have one or more long filamentous protoplasmic extensions, **flagella,** by which they swim. They have a more or less definite shape, usually oval, and a definite front end from which the flagella arise. The flagella beat in whiplike fashion, and the locomotion of flagellates is usually slower and more irregular than that of a paramecium and its ciliated relatives.

The flagellates are divided into two groups, the more primitive **plant-like** types which make their own food by means of the chlorophyll they possess, and the **animal-like** types which capture and ingest other organisms.

One of the simplest of the plantlike flagellates is *Chrysamoeba,* which has a yellow photosynthetic pigment. This form sometimes temporarily loses its single flagellum. It then moves about and ingests solid food by means of pseudopods, like an ameba. Since such flagellates are both inde-

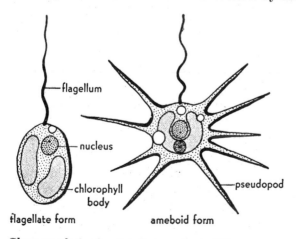

flagellum

nucleus

chlorophyll body

pseudopod

flagellate form   amoeboid form

Chrysamoeba is a flagellate that can also feed like an ameba. (Modified from Lang)

pendent and capable of feeding like animals, they are regarded as being more primitive than the amebas, which never show flagellate stages and feed only like animals.

Euglena is one of the most common of the green flagellates of fresh waters and is often so numerous that it produces a green scum on the surface of ponds. The name applies to a number of species, all of which have a very elastic outer covering and specialized protoplasmic contractile fibers which permit contraction and elongation of the body in a characteristic squirming called **euglenoid movement**. At the front end there is a flask-shaped depression, the gullet, from which springs one very long delicate flagellum. By lashing this flagellum, the animal swims slowly forward, at the same time revolving on its long axis, as does a paramecium. Since Euglena has never been seen to feed, it is probable that the gullet is not used to take in food but serves mainly for the attachment of the flagellum. A large contractile vacuole discharges its contents into the gullet at frequent intervals.

Beside the gullet is a bright-red granule, called the **eye-spot**, which is thought to be associated with a light-sensitive apparatus. The specialized eyespot of Euglena is more sensitive and therefore more efficient as a light-detecting apparatus than the undifferentiated protoplasm of the ameba, which also reacts to light. This specialization is important since euglenas depend upon photosynthesis for

Euglenoid movement. 1–5, a succession of characteristic shapes.

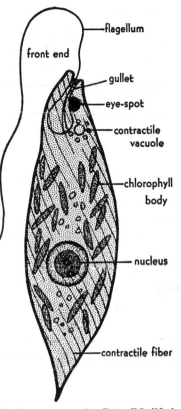

Euglena, a green flagellate. (Modified after Doflein)

their food, and it is advantageous that they should expose themselves to the light as much as possible. When placed in a dish, euglenas quickly aggregate on the side of the dish nearest the light. As long as they are exposed to light, they maintain themselves by photosynthesis and store up carbohydrates in storage bodies which are conspicuous in the cytoplasm. However, euglenas can live in the dark if they are placed in a nutrient solution. Under these conditions the chlorophyll degenerates, the animals become colorless, and the nutrient materials are absorbed through the surface membrane. It is doubtful if this mode of nutrition is ever used in nature.

Volvox is seen in fresh-water ponds as a small, green sphere, which may be $\frac{1}{10}$ inch in diameter. The sphere is composed of thousands of flagellates

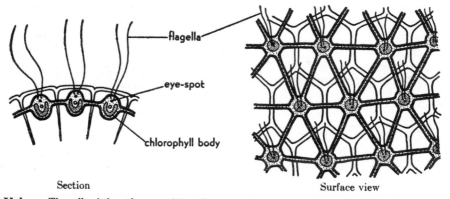

Section   Surface view

Volvox. The cells of the colony are joined by protoplasmic connections. (Modified after Janet)

imbedded in the surface of a jelly ball. Each flagellate resembles Euglena and has two flagella, a red eyespot, two contractile vacuoles, chlorophyll bodies, but no gullet. Volvox is a **colony** of single cells rather than a many-celled organism, because even the simplest many-celled organisms have considerably more differentiation between cells than appears among the cells of Volvox. The colony swims about, rolling over and over from the action of the flagella; but remarkably enough, the same end of the sphere is always directed forward, and thus we can distinguish front and rear ends. Its behavior can be explained only by supposing that the activities of the numerous flagellates are subordinated to the activity of the colony as a whole. If the flagella of each member of the colony were to beat without reference to the other members, the sphere would never get anywhere. In such subordination of the individual cells of a colony to the good of the colony as a whole, we see the **beginnings of individuality** as it exists in the higher plants and animals, where each behaves as a single individual although composed of millions of cells.

The **co-ordination** of the Volvox members in swimming implies some means of transmission between them. They are, in fact, connected by protoplasmic strands extending through the jelly. The co-ordination of numerous components into an individual is usually followed by the specialization of different individuals for different duties. Only the slightest degree of specialization is seen in the Volvox colony; the flagellates of the back part of the colony are capable of reproduction, while the front members never reproduce but have larger eyespots and serve primarily in directing the course of the colony.

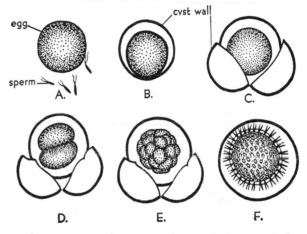

In **asexual reproduction** (that is, not involving any sexual processes), a cell enlarges, loses its flagella, and divides a number of times until a small daughter-ball is produced. A Volvox sphere usually contains in its hollow interior sev-

Sexual reproduction and development of *Volvox*. **A**, fertilization. **B**, the zygote forms a cyst wall. **C**, an inner cyst wall forms, and the outer wall is discarded. **D**, the zygote divides into two cells. **E**, a small ball of cells. **F**, a young colony within the cyst wall. (Modified after Kircher)

eral such daughter-colonies in process of formation. They are liberated when the mother-colony breaks open.

In the **sexual reproduction** of Volvox one of the stages in the evolution of sex is illustrated. In the early stages of sexual differentiation the two fusing cells are alike, and such a condition is seen in some of the relatives of Volvox. In Volvox this evolution has reached completion; and fully differentiated male gametes, or **sperms,** and female gametes, or **eggs,** are formed. In forming an egg, a cell of the colony increases greatly in size, takes on a rounded form, and becomes loaded with food, especially fatty substances. This food is contributed in part from adjacent cells and serves to give the young Volvox a good start in life. Another cell of the same or another colony, by repeated divisions, gives rise to numerous small flagellated sperms. These sperms swim about until they find an egg or die. Only one sperm penetrates the egg, and this union of egg and sperm is called **fertilization.** When an egg is fertilized, it undergoes a

A **dinoflagellate,** *Gonyaulax,* sometimes becomes so abundant along the coast of southern California that it colors the ocean red for miles. These outbreaks of "red water" cause the death of fish and other animals, which are cast up on the beach and decay. The stench has been compared to that of the Nile when, according to ancient writings, that river "turned to blood." (After Kofoid)

change which prevents the entrance into the egg of additional sperms. The fertilized egg, or **zygote,** of Volvox secretes upon its outer surface a hard spiny shell which protects the cell during unfavorable conditions, such as drying or freezing. Inside the shell, the zygote divides into two cells, these into four, and so on until a small ball has been produced. With the return of favorable conditions of heat and moisture, the shell breaks; and the young colony, indistinguishable from a daughter-colony produced asexually, emerges.

The **dinoflagellates** occur in enormous numbers in the surface waters of the ocean; there are also a number of fresh-water species. A typical dinoflagellate, *Gonyaulax,* is inclosed in a tight-fitting armor of cellulose plates and has two flagella, one lying in a groove encircling the animal and the other trailing downward. Many dinoflagellates have oil drops which help them to float. Some of them possess a brown pigment in addition to chlorophyll; others are colorless and feed on minute organisms. Some dinoflagellates produce light, as do many other marine and some land animals. The mechanism of light production in certain of the higher forms has been partially worked out (and will be given in chap. 18) but has not been proved to apply to these protozoa. One of the most abundant of the luminescent dinoflagellates is *Noctiluca,* which does not resemble the typical dinoflagellates but is colorless and ingests solid food. It is spherical, about a millimeter in diameter, with one short thick and one very delicate flagellum. It floats on the surface near shores, often in inconceivable numbers, and, being faintly

A **luminescent dinoflagellate,** *Noctiluca.* (Based on several authors)

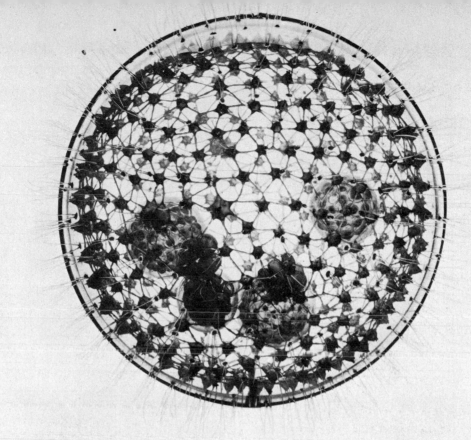

A protozoan colony, *Volvox*, is seen in ponds as a green sphere, about 1/25 inch in diameter. It swims by the action of flagella, co-ordinated by means of protoplasmic strands between neighboring cells. Within the hollow colony are several daughter colonies. (Glass model. Photo, courtesy American Museum of Natural History).

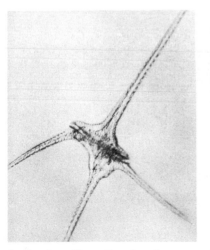

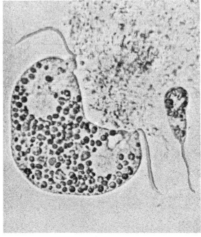

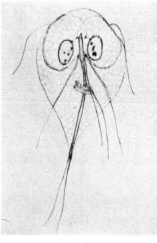

A dinoflagellate, *Ceratium*, has projections which aid in floating. Found in ponds and lakes, sometimes in tap water. (Photo of preserved specimen by P. S. Tice)

Euglena dividing. The flagellum and anterior part of the animal have already split. A small flagellate is seen at the right. (Photo of living animals)

A parasitic flagellate, *Giardia*, that lives in the small intestine of man, causing diarrhea. (Drawing, courtesy Army Med. Mus.)

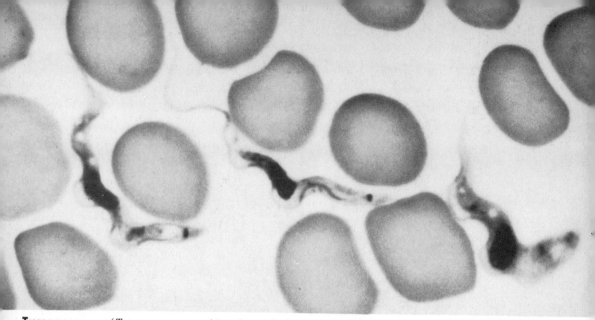

Trypanosomes (*Trypanosoma gambiense*) which cause African sleeping sickness are shown here in a blood smear, among the red blood cells. The parasites are about 1/1,000 inch long. Trypanosomes probably lived originally in the intestine of blood-sucking insects, where they were constantly exposed to the vertebrate blood ingested by their insect hosts. Accidentally introduced into the blood of vertebrates by the bites of insects, the flagellates became adapted to their new environment, and finally dependent for part of their life-cycle on development in the blood stream of a vertebrate. (Photo of stained preparation, courtesy Gen. Biol. Supply House)

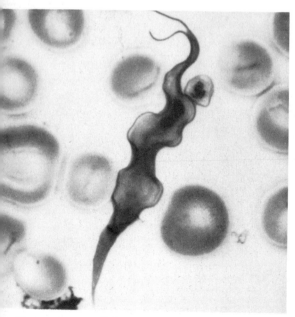

Trypanosome dividing in blood of rat. The flagellum is already divided but the nucleus is still single. This flagellate (*Trypanosoma lewisi*) does not harm the rat. The parasites multiply rapidly at first, but gradually the rat's blood develops the property of inhibiting their reproduction. (Photo of stained preparation, courtesy General Biological Supply House)

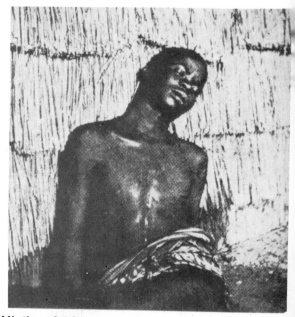

Victim of African sleeping sickness. Much of the laziness attributed to the African natives by the early European explorers was no doubt due to this disease, and slavetraders early learned not to accept slaves having swollen glands in the neck, a symptom of trypanosome infection. (Photo made in Zaire, courtesy Army Medical Museum)

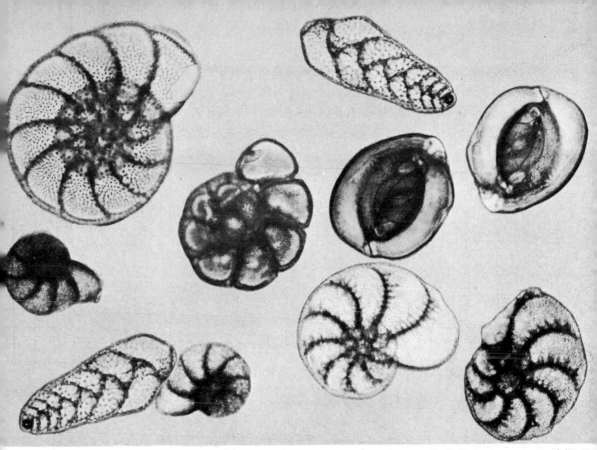

Foraminifer shells, when highly magnified under the microscope, look like snail shells but are secreted by ameboid protozoa. The minute pores which perforate the surface of most of the shells shown here are the openings through which the live animal extended pseudopods. The shell on the upper right has no pores, and its former occupant protruded all the pseudopods through a single opening at one end.

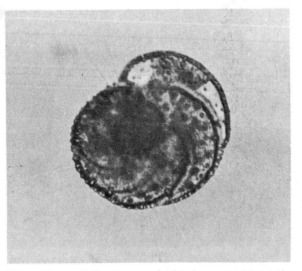

Living foraminifer in organic debris scraped from the surface of a bit of seaweed. Long delicate pseudopods, by which the animal feeds, can be seen extending through the shell. (Pacific Grove, California)

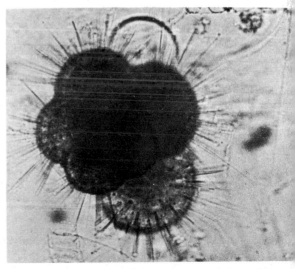

Globigerina is one of the most common foraminifers. The shorter radiating processes are spines on the shell. The pseudopods are mostly withdrawn. (Photo of living animal. Pacific Grove, California)

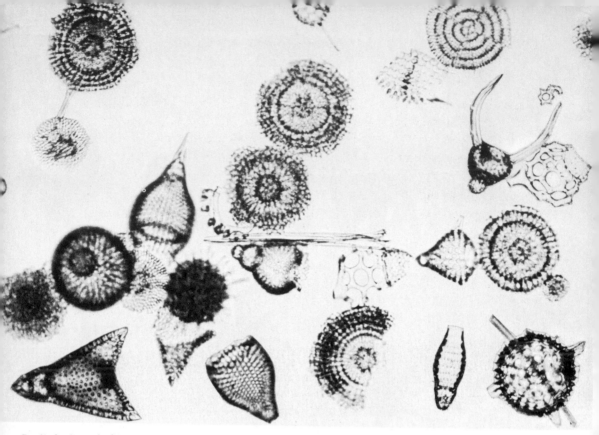

Radiolarian skeletons are made of silica secreted by the protoplasm and usually take the form of an intricate latticework, through the openings, of which are extended the pseudopods. These minute but hard skeletons constitute a part of "Tripoli stone" which is used in abrasive powders for polishing metals.

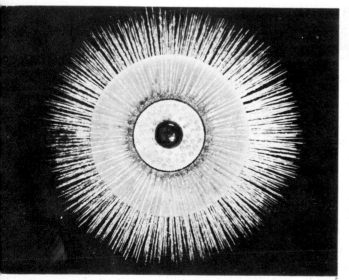

Glass model of living radiolarian. The black central body represents the nucleus. This is surrounded by a cytoplasmic capsule which, in turn, is imbedded in a frothy protoplasm. The pseudopods differ from those of foraminifers In being stiff. (Photo, courtesy Amer. Mus. Nat. Hist.)

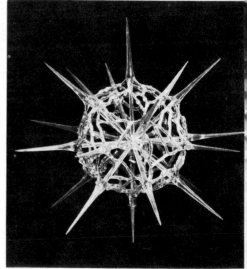

Glass model of radiolarian skeleton shows spherical symmetry characteristic of free-floating protozoa that cannot swim and must meet their environment on all sides. (Photo, courtesy Amer. Mus. Nat. Hist.)

pink, may in the daytime cause whole areas of the ocean to look like weak tomato soup. The minute flashes emitted by these dinoflagellates are seen at night when the animals are agitated as the waves strike rocks or other objects. Night swimming or boating in an area filled with Noctiluca causes these dinoflagellates to emit light, furnishing spectacular displays of what appears like fireworks being shot off under water.

Among the animal-like flagellates are the **collar-flagellates,** solitary cells that live attached by their stalks to the substratum. They become

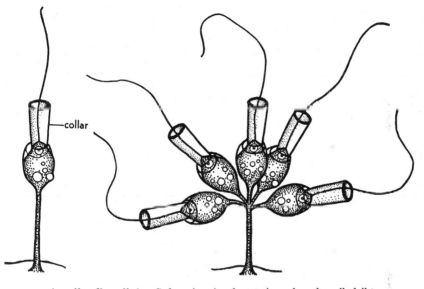

A **collar-flagellate,** *Codonosiga,* A colony arises when the cells fail to separate after division. (After Lapage)

colonial when the cells fail to separate completely after dividing. The cell has a delicate, transparent protoplasmic collar from the center of which emerges the single flagellum. The beating of the flagellum draws a current of water toward the cell. The food organisms in the current do not enter the collar but impinge upon the sides of the cell and are taken up into food vacuoles. The collar-flagellates are of special interest because a similar type of cell occurs nowhere else in the animal kingdom except in the sponges.

Many parasitic forms are included among the animal-like flagellates. Of these, the most notorious are the **trypanosomes** that cause African sleeping-sickness in man (not the same as the sleeping-sickness known as "encephalitis" and caused by a virus). The trypanosomes and their relatives live as parasites in insects, certain plants, and most vertebrates in Africa without causing much inconvenience to their hosts. But apparently they have not yet become adapted to living in man or in his domestic animals without producing an incapacitating and usually fatal disease.

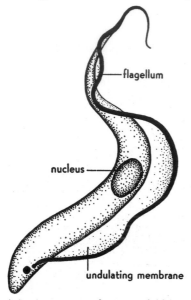

A **trypanosome,** the cause of African sleeping-sickness. The edge of the undulating membrane bears the flagellum, which extends free at one end of the animal.

The trypanosomes that cause African sleeping-sickness are transmitted by the blood-sucking **tsetse flies.** Practically all wild game in Africa harbor trypanosomes in their blood; and when a tsetse fly sucks the blood of an antelope, for example, or of an infected man, the blood drawn into the intestine of the fly will contain these trypanosomes. In the fly's intestine the trypanosomes undergo changes. From the intestine they invade the salivary glands of the insect, where they continue to change and to multiply. If such an infected fly bites a man, the trypanosomes are injected into the blood of the victim with the saliva of the fly. In the blood they multiply rapidly and wriggle about among the blood corpuscles, propelled by the undulations of a delicate ruffled membrane which extends along one side of the body.

The bite of an infected fly sometimes does not cause fever for weeks or even months, while the flagellates increase in number. When the attacks of fever begin, the victim becomes weak and anemic, probably because of poisonous by-products of metabolism given off by the flagellates. Finally, the parasites invade the fluid surrounding the brain and spinal cord, the person loses consciousness, and the "sleeping-sickness" goes to a fatal end.

Medical treatment of African sleeping-sickness consists, in the main, of injection of various drugs and is very effective in the early stages of the disease before the flagellates have invaded the nervous system. However, if the dose injected is not adequate to kill the parasites, they may become drug-resistant and then are unaffected by doses large enough to kill the patient.

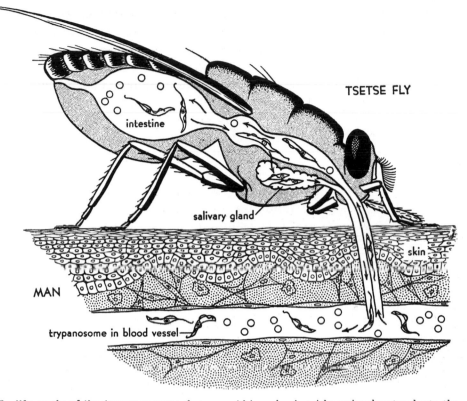

The **life-cycle of the trypanosomes** that cause African sleeping-sickness involves two hosts, the tsetse fly and a vertebrate, such as an antelope or a man. Here an infected fly is biting an infected man. Principal stages in the life-cycle are shown: trypanosomes in the salivary glands of the fly being injected into the blood of the man, forms living in the blood and entering the sucking tube of the fly, and those living in the intestine of the fly. (Based on several sources)

Control measures for this disease are complicated by the fact that even if it were possible to cure or isolate all human cases, the wild game would still act as a reservoir from which flies could be reinfected. An adequate solution for this problem seems to lie in wholesale destruction of the tsetse flies, which is an exceedingly difficult task, although not impossible.

Large regions in Africa are uninhabitable for men and their stock because of the tsetse fly, and it is safe to say that the past and present accomplishments of man in the larger part of Africa have been controlled by the fly. The future of that continent depends not upon military leaders or diplomats of state but upon medical and ecological investigators who are working to check the ravages of that murderous pair, the fly and the flagellate,

Shaded portion shows the extent of sleeping-sickness in Africa.

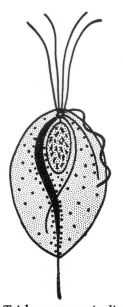

Trichomonas vaginalis is found in the vagina (passage leading from the uterus to the exterior) and is under suspicion as a contributing factor in death from childbirth. (After Powell)

The **trichomonads** are flagellates common in the digestive tracts of vertebrates. They are pear-shaped protozoans with several flagella springing from the anterior end. One of these flagella extends backward along the edge of an undulating membrane. The body is supported by an internal stiff rod which projects from the rear end and is frequently used to anchor the animal while it feeds. *Trichomonas buccalis* inhabits the mouth and may be involved in pyorrheal conditions; *Trichomonas hominis* and some other kinds of trichomonads live in the intestine and seem to be related to a type of diarrhea.

The flagellates are considered to be the simplest class of protozoa because certain of their members are very primitive. It must be realized, however, that no class of animals is made up only of simple forms. Practically all large groups have a few members which are more complex than the less developed members of the next higher group. The older an animal group, the longer has been its evolution and the time available for development of complexity. It is not surprising, then, to find certain flagellates that show an amazing degree of specialization. One of

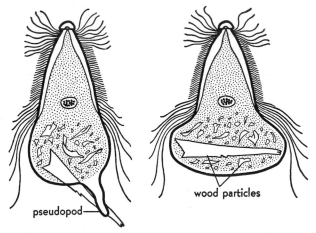

pseudopod

wood particles

Trichonympha engulfs minute bits of wood by pseudopods extended from the lower part of the animal. This method of ingestion is surprising, in view of the fact that the surface cytoplasm of the upper part of the animal is covered with hundreds of long flagella and is structurally more complex than in almost any other protozoan. (Modified after Swezy)

these, *Trichonympha,* is among the most complex of all protozoa. It is also interesting because it inhabits the intestine of wood-eating termites. It was for a long time difficult to understand how the termites were able to subsist on a wood diet, since wood contains only a very small amount of digestible protein and sugar, and most animals lack the necessary enzymes for digesting its chief constituent, cellulose. In recent years evidence has been presented to show that flagellates, such as Trichonympha, which lives sheltered in the termite's intestine, ingest the minute bits of wood occurring there and transform them to soluble carbohydrates, part of which can be utilized by the insect host. This relationship, in which the members of two species live together in an association of mutual benefit, is an outstanding example of **mutualism** (or symbiosis).

THE AMEBOID PROTOZOA

THE ameboid protozoa (class **Sarcodina**) include all those protozoa which move about and capture food by means of pseudopods. Certain of these resemble closely the typical ameba, described in chapter 3, and live free in fresh and salt waters and in damp soil.

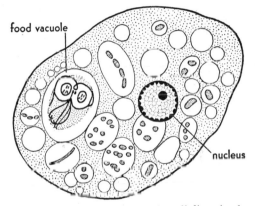

food vacuole

nucleus

In addition to the free-living amebas, there are a number of species which live in the interior of animals, particularly the digestive tract. Most of them are harmless to the animal they inhabit, living in the intestine and feeding on bacteria and food fragments, at no expense to their host —a relationship known as **commensalism.** A few species are definitely parasitic, that is, they attack the host itself and fre-

A **harmless ameba,** *Entamoeba coli,* lives in the intestine of man and feeds on particles in the intestinal contents. This particular specimen has just rendered a useful service to its host by engulfing a harmful parasitic flagellate called *Giardia.* The smaller food vacuoles contain bacteria. (After Doflein)

quently cause disease. These parasitic amebas resemble in appearance and activities the free-living species but lack contractile vacuoles.

About half a dozen species of amebas may live in man, of which only one, the **dysentery ameba,** *Entamoeba histolytica,* is definitely harmful. The dysentery ameba occurs in 5–10 per cent of the population of civilized countries and in as much as 60 per cent of the people in backward com-

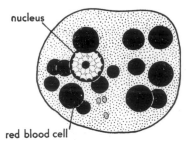

The **dysentery ameba,** *Entamoeba histolytica*, causes occasional epidemics in temperate regions, as when faulty plumbing allows sewage to get into the water supply. In tropical countries amebic dysentery is a constant menace and a serious drain on the energy of the natives. Travelers in the Orient, where human feces are used as fertilizer, should never eat uncooked vegetables or drink unboiled water. (After Doflein)

munities. It is passed on from one person to another by eating food or drinking water contaminated with the excreta of individuals already infected with the ameba. Since many individuals harbor these amebas without showing obvious ill-effects, carriers of amebic dysentery are more common in the population than is generally realized. The dysentery ameba inhabits the large intestine, where it feeds upon the living cells and tissues, leading to the formation of abscesses and bleeding ulcers. The active amebas cannot live outside the body and therefore do not serve to transmit the disease, but they pass readily into the encysted stage, in which the nucleus divides twice to form four nuclei. These four-nucleated cysts pass out

in the stools and, when swallowed by other persons, infect them with the disease. Several drugs are successfully used in the treatment of amebic dysentery. Sanitary disposal of fecal matter and cleanliness in preparing food are essential in the prevention of the disease. Parasitic amebas, like most disease-producing organisms, are better avoided than killed.

The **mouth ameba** (*Entamoeba gingivalis*) is found in a large part of the human population (perhaps 75 per cent or more) in the mouth, where it feeds on bacteria and loose cells. These amebas occur chiefly in the pockets formed between the teeth and the gums in the disease pyorrhea. It is not certain that the amebas play any role in the initiation of pyorrhea, but it is probable that they aggravate the disease, once it is started. This ameba is not known to form cysts and is spread from mouth to mouth in eating and in kissing. The best one can do to reduce

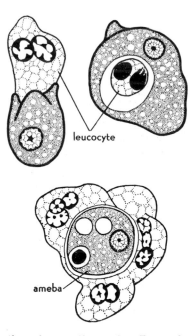

Above, the **mouth ameba**, *Entamoeba gingivalis*, engulfs a leucocyte. *Below*, the tables are turned; three leucocytes have joined forces and are engulfing an ameba. (Modified after H. Child)

the probability of infection is to take ordinary care of the mouth.

Difflugia is a free-living ameba which gathers sand grains, cements them together with a sticky secretion, and thus builds a kind of house which it carries about with itself and into which it withdraws when disturbed. The various species of Difflugia can be recognized by the specific shapes of the protective coverings which they construct. Thus, while we appear to be classifying these amebas on the basis of differences in structure, we are also classifying them by their differences in behavior. The common *Arcella* of fresh-water ponds secretes a hard shell about itself. When the animal divides, one daughter retains the shell; the other has to construct a new one.

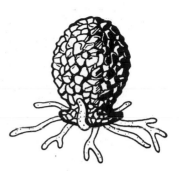

Difflugia.

Arcella is inclosed by its hemispherical shell. The pseudopods extend from the hole on the under side.

The **foraminifers** ("hole-bearers") are ameboid protozoans that secrete many-chambered calcareous (chalky) shells. Nearly all occur in marine waters, some species floating near the surface, but most of them living on the mud of the ocean bottom. A young foraminifer resembles an ameba and secretes a shell about itself. As growth continues, the protoplasm flows out of the shell opening, spreads over the surface of the shell, and secretes another shell—thus making a second chamber. This process is repeated as the foraminifer grows, until sometimes there are more than a hundred communicating chambers. In many common species the chambers are added on in a spiral pattern, resulting in a shell that resembles a

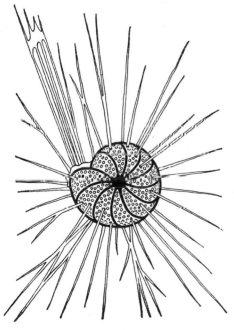

In a **living foraminifer** the pseudopods extend through the pores in the walls of the shell as well as out of the main opening.

miniature snail shell. The animal occupies all of the chambers and sends out long, delicate pseudopods through the pores in the walls of the shell, as well as out of the main opening. The pseudopods exhibit constant streaming movements and unite into pseudopodal networks in which food is captured and digested.

Aside from their architectural accomplishments the foraminifers are remarkable among protozoa for the large size that some species attain. Living species may be $\frac{1}{2}$ inch in diameter, and certain extinct fossil foraminifers were as large as 4 inches across. However, most forms are just visible to the naked eye.

It is astounding to contemplate the probable number of individual shells composing the chalk cliffs of Dover, England, or the large chalk beds (1,000 feet thick in places) of Mississippi and Georgia. The presence of chalk beds indicates that these regions were once covered by the sea, and such information is useful to geologists. The deposition of the shells, if we are to judge by the present rate of accumulation at the bottom of the Pacific, was about 2 feet in a hundred years. Most of the shells now being deposited are those of the foraminifer *Globigerina*, which floats in the surface waters. These animals are constantly dying and their skeletons sinking in a slow but steady rain to the ocean floor, where they form a gray mud called "globigerina ooze." About 30 per cent of the floor of the ocean (40,000,000 square miles) is covered with globigerina ooze. Some deposits form chalk, with as much as 90 per cent calcium carbonate. Nearer the shore the deposits contain sediments which have been washed from the land, and the resulting rock is a fossiliferous limestone such as the famous Indiana building-stone.

Extinct foraminifers preserved as fossils in rocks are of great value in developing oil fields. Borings at different depths are examined for their fossil foraminifers, and from the species present a great deal can be learned about the underlying rock structure.

The **radiolarians** are ameboid protozoans that secrete elaborate skeletons composed mostly of *silica*. They extract the silica from the sea water just as foraminifers extract calcium carbonate. Food organisms are caught on the stiff pseudopods that radiate out through holes in the skeleton. Like many other marine protozoa, they have no contractile vacuoles. The cytoplasm is filled with (noncontractile) vacuoles that give it a frothy appearance and enable the animals to float.

The skeleton of a **radiolarian** may be relatively simple, with only a few large spines like this one, or may be an almost unbelievably intricate latticework of silica. (After Haeckel)

Radiolarian skeletons can be found in marine deposits in shallow water; but it is only in very deep regions that they occur in a concentration of at least 20 per cent and the deposit can be classed as "radiolarian ooze." This ooze occurs in the Pacific and Indian oceans, where it covers an area of almost 3,000,000 square miles. Hardened radiolarian deposits are found in other rocks as siliceous inclusions, flint or chert.

The **heliozoans** are the only fresh-water ameboid protozoa which are comparable to the exclusively marine radiolarians. Some forms have perforated siliceous skeletons which resemble those of radiolarians, but others have only a gelatinous covering or a skeleton of loosely matted needles of silica. Fresh-water species have a contractile vacuole. The radiating

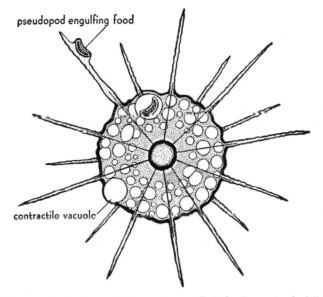

A common **heliozoan**, *Actinophrys sol*, is sometimes called the "sun animalcule" because its stiff radiating pseudopodia, whose firm protoplasmic axes run through the cytoplasm and converge around the centrally located nucleus, suggest the rays of light from the sun.

pseudopods are even stiffer than those of the radiolarians and show little movement besides a streaming of the granules in the protoplasm. When a large organism is being engulfed, several pseudopods work together. The firm protoplasmic axes of the pseudopods are absorbed, and the pseudopods wrap themselves around the prey and inclose it in a food vacuole.

THE SPORE-FORMERS

THE spore-formers (class **Sporozoa**) have no special mode of locomotion and are all parasites. Some of them are responsible for human disease. They are called sporozoans because they reproduce by "spores," in the formation of which the nucleus of a protozoan divides many times until a number of nuclei have been produced. A little cytoplasm then gathers around each nucleus, and the protozoan falls apart into a number of offspring, corresponding to the number of nuclei. The products of such sporulation may be naked or they may be inclosed in a resistant wall. Such multiple fission increases enormously the reproductive potential, a decided asset for parasites, which run great risks in transferring from one host to another.

There are several groups of spore-formers, and they are probably not closely related, having arisen independently from primitive protistan stock. They are usually grouped as sporozoans for the convenience of zoologists. The **cnidosporidians** cause diseases of salmon, halibut, and other fishes, and epidemic death in cultivated honeybees and silk-worms. The **gregarines** (photo, page 56-1) infest the digestive tracts and body cavities of many invertebrates, especially annelid worms and insects. The **coccidians** parasitize the liver and blood; only one host is involved. Coccidians occur in humans and perhaps cause diarrhea, but their bad reputation is based on the heavy toll they take of domestic animals, especially chickens, rabbits, and cattle. They may also infect dogs, cats, and canaries.

The **hemosporidians** are often included with the coccidians, but the life cycle involves two hosts, a vertebrate and an arthropod. The most studied of the hemosporidians is *Plasmodium*, the **malarial parasite,** which causes malarial fever in millions of humans and in other warm-blooded vertebrates. The several species that affect humans are transmitted by anopheline mosquitoes. The sexual stages occur in the body of the mosquito, and the asexual stage in the human host. Both phases involve multiple fissions and these may be repeated before going on to the next stage in the life cycle.

Malarial disease in humans begins with the bite of an anopheline mosquito, but the sporozoites that invade the blood stream leave so promptly to enter tissue cells, that blood smears do not reveal them at this "hidden" stage, and there are no symptoms. The shivering chills and burning fevers (up to 106° F.) of malaria do not occur until a sufficient number of blood cell merozoites are liberated into the circulating blood, together with pigment granules and other toxic wastes from the red cells. In *Plasmodium vivax* infections, the most common form of human malaria, the parasites attack about 1% of

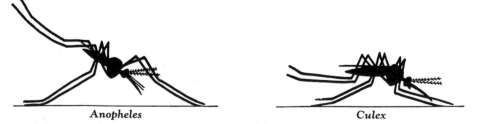

Anopheles   *Culex*

Malaria is transmitted to man only by mosquitoes of the genus *Anopheles*, whose characteristic attitude while biting is shown contrasted with that of *Culex*, one of our most common mosquitoes. *Culex* transmits the sporozoan that causes bird malaria.

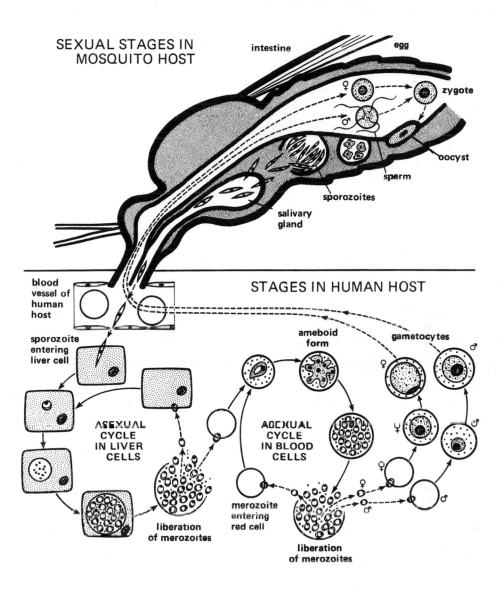

SEXUAL STAGES IN
MOSQUITO HOST

intestine

egg

♀

zygote

♂

oocyst

sperm

sporozoites

salivary
gland

blood
vessel of
human
host

STAGES IN HUMAN HOST

ameboid
form

gametocytes

♂

sporozoite
entering
liver cell

♀

ASEXUAL
CYCLE
IN LIVER
CELLS

ASEXUAL
CYCLE
IN BLOOD
CELLS

♀

♂

♀

♂

merozoite
entering
red cell

liberation
of merozoites

liberation
of merozoites

The **life cycle of the malarial parasite,** *Plasmodium vivax,* begins with the union of egg and sperm to form a **zygote** in the stomach of the mosquito. The zygote encysts in the stomach wall as an **oocyte** which undergoes multiple fission into large numbers of slender **sporozoites.** These invade all the tissues of the mosquito. Those that lodge in the salivary glands are carried, in the saliva of a biting mosquito, through the skin and into the blood stream of a human host. The sporozoites leave the human blood stream and enter tissue cells, such as the liver, where they undergo multiple fission. The resulting **merozoites** then enter new liver cells and repeat the foregoing history The second generation merozoites now invade the blood stream, enter red blood cells, and grow into ameboid forms that engulf portions of the cytoplasm of the red cell. The ameboid form undergoes multiple fission, the resulting merozoites enter new red cells, and repeat the cycle. Eventually some of the merozoites develop into sexual forms, male and female **gametocytes,** which produce either eggs or sperms in the stomach of a mosquito that takes a blood meal from the human host.

the blood corpuscles. The cycles of multiple fission take about 48 hours, so that chills and fever recur at 48-hour intervals. Since the merozoites of both tissues and red cells may reinfect tissue cells, *vivax* malaria is likely to recur after a time. The more malign disease caused by *Plasmodium falciparum* attacks 10% of the red cells, resulting in more severe paroxysms of chills and fever, and clumping of corpuscles that may result in coma or in death from cerebral obstruction.

Malaria was once the most widespread of human diseases, and supposedly responsible for the fall of the Greek and Roman empires. Worldwide there are still some 200 million cases with two million deaths each year. Screening patients, draining swamps, using drugs, and especially spraying houses with DDT has in recent decades practically eliminated malaria from the United States, Europe, and many parts of Asia. But DDT-resistant strains of anopheline mosquitoes are developing, and the disease may yet increase again.

THE OPALINIDS

The class **Opalinata** includes a small group of entirely parasitic protozoans which are evenly covered with rows of locomotory cilia. In this respect they resemble a paramecium or various other members of the main class of ciliated protozoans. But they differ sharply in two important respects. Their two or many nuclei are *all of the same kind*. And when they divide the plane of division is down the center of the animal, parallel with the rows of cilia. In this they resemble flagellates rather than other ciliated forms. *Opalina*, typical of the group, lives parasitically in the rectum of the frog, apparently without ill effects on its host. It is evenly covered with short cilia. It lacks a mouth and presumably obtains its food by diffusion of the intestinal contents of the host through the protozoan cell membrane. There is no contractile vacuole.

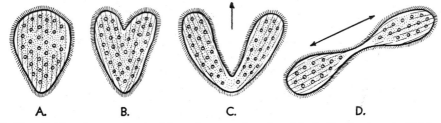

A.   B.   C.   D.

Opalina divides. **A,** normal animal has a large number of nuclei. **B, C, D,** stages in division of the cytoplasm. Division is longitudinal. Arrows show direction of swimming.

THE CILIATES

The class **Ciliophora** embraces the vast majority of ciliated protozoans, and is the most conspicuous of the protozoan groups. Almost any drop of fresh water or sea water put under a microscope lens will reveal a variety of ciliates crossing the field so rapidly as to defy easy interpretation of their mode of locomotion. The beating cilia by which they row themselves about may cover the whole body in evenly spaced rows or may be limited to certain areas. These differences in ciliation are accompanied by striking differences in feeding and other habits and form the basis of their classification into subclasses and their constituent orders. All have cilia during at least part of the life cycle, almost all have *two types of nuclei*, and most divide at right angles to the rows of cilia.

In the **holotrichs,** of which the paramecium is a member, the cilia are all short, fairly equal in length, and are evenly distributed over the surface in rows or restricted to certain regions. Most of the members of this order have trichocysts, but these structures are rare in other orders. Next to the paramecium, **Colpoda** is perhaps the most common of the fresh-water holotrichs. **Didinium** is a holotrich that works hard for a living; it eats almost nothing but paramecia. From the center of the front end of the animal projects a snout which is armed with structures resembling trichocysts. Didinium swims about at top speed, "trying" to pierce everything with which it comes into contact— plants, another Didinium, or even the glass walls of an aquarium. With these it has no success. When it chances to strike a paramecium, the snout penetrates and the prey is swallowed bodily.

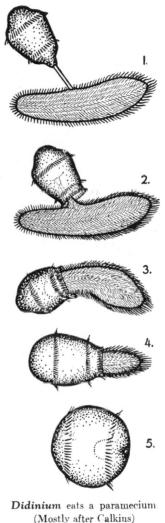

Didinium eats a paramecium
(Mostly after Calkins)

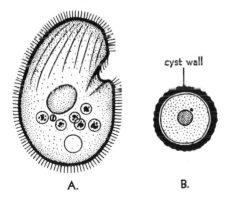

cyst wall

A.   B.

Colpoda. A, active animal. **B,** encysted animal. (**B,** after Kidder and Claff)

Balantidium coli is the largest protozoan that lives in humans and is the only ciliate that is at all common as a human parasite. It inhabits the large intestine, with consequences that are often fatal and something like those resulting from the activities of the dysentery ameba. *Balantidium* occurs not only in humans but also in pigs and monkeys. People probably become infected through too close association with other infected animals. Although parasitic protozoa generally lack contractile vacuoles, *Balantidium* has a conspicuous vacuole. Perhaps this is related to the fact that *Balantidium*, unlike many parasitic forms, feeds by means of a mouth.

The **membranelles** of *Stentor* are triangular plates formed by the fusion of many cilia. (Modified after Doflein)

The **spirotrichs** include a variety of forms, among which may be mentioned the two sub-orders most commonly seen in fresh water. The **heterotrichs** have, in addition to a covering of short cilia, a zone of large cilia around the mouth which increases the efficiency of the feeding currents. These large and powerful cilia, called **membranelles,** are really triangular plates formed by the fusion of a number of cilia. They usually occur in a row or circlet around the mouth end of the animal. Coincident with the development of membranelles, these animals frequently have sessile habits, that is, they tend to fasten themselves to the substratum. One of the most familiar of such forms is *Stentor* a trumpet-shaped animal which can swim about freely but, when feeding, attaches by its lower end to a water plant or similar object, stretches out to full length, and vibrates its circlet of membranelles so rapidly that they look like a swiftly rotating wheel. The water current thus produced sweeps small animals into the gullet. Stentor has a remarkable type of large nucleus, resembling a string of beads, and may have up to eighty small nuclei. Beneath the surface there are lengthwise contractile fibers which shorten the animal. The differentiation of cilia seen in the heterotrichs is carried to very curious extremes in the **hypotrichs,** which are flattened and have the cilia confined almost entirely to the lower surface. The upper

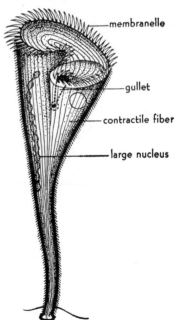

membranelle

gullet

contractile fiber

large nucleus

Stentor is a giant among protozoa; some specimens may be $\frac{1}{15}$ inch in length. An investigator who placed a hungry *Stentor* in a thick suspension of euglenas estimated that the flagellates were taken in at a rate of about 100 per minute.

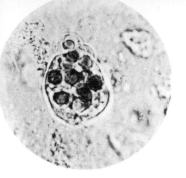

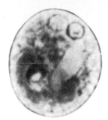

Dysentery ameba (*Entamoeba histolytica*). *Left*, living animal containing engulfed red blood cells, from a fresh stool of an amebic-dysentery patient. *Right*, cyst. This is the resistant form which is infectious, since it alone can survive the trip from person to person. (Photos, courtesy Army Medical Museum)

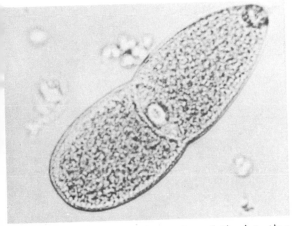

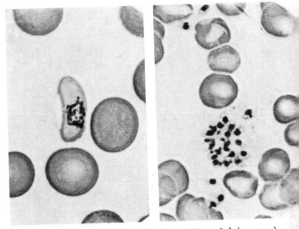

Gregarines are sporozoan parasites of the intestine, body cavity and blood of many animals. This one was found in the digestive gland of an acorn worm and was clinging to a host cell by the end shown here on the right. (Photo of living animal. Bermuda)

Malarial parasite. *Left* (*Plasmodium falciparum*), female sexual form among red blood cells. *Right* (*P. vivax*), red blood cell breaking up and liberating young spores which are about to enter now red blood cells. (Photo, courtesy Army Medical Museum)

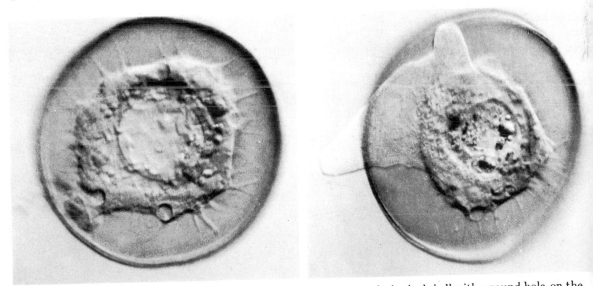

Arcella is a shelled ameba common in fresh water. It has a brown hemispherical shell with a round hole on the undersurface through which the pseudopods are extruded. Shown here are two views focused at different levels of the undersurface of the shell, (*right*), with large pseudopod extended and (*left*) with pseudopods withdrawn into the shell. View at left shows four contractile vacuoles and the two nuclei. (Photos by P. S. Tice)

Didinium is a holotrich, belonging to the same group of ciliates as does the better-known Paramecium. Instead of gathering in bacteria by means of ciliary currents, a didinium is a predaceous protozoan, and a very aggressive one. It has a protrusible snout that it fastens onto its prey. Then it retracts the snout, drawing the prey with it into the widely opened mouth. It feeds almost exclusively on paramecia and may eat as many as eight in a day. This is a notable feat, considering that a didinium is smaller than a paramecium. (Photomicrograph of living animals by Roemmert, courtesy *Nature Magazine*)

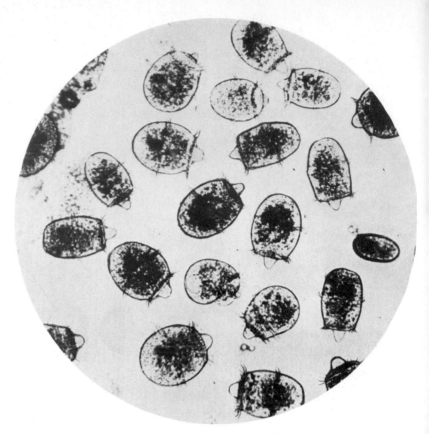

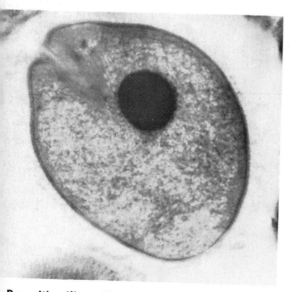

Parasitic ciliate (*Balantidium coli*), largest (1/300 inch) protozoan inhabiting the human intestine. It causes dysentery. The most common source of infection is the pig. (Photo, courtesy Army Med. Mus.)

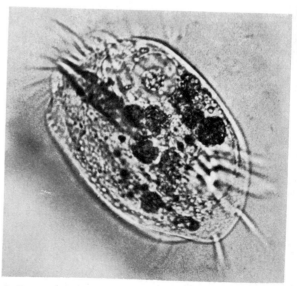

A hypotrich does not row itself about by means of numerous fine cilia but progresses by jerky movements of the large fused cilia on the lower surface of the animal. (Photo of living animal)

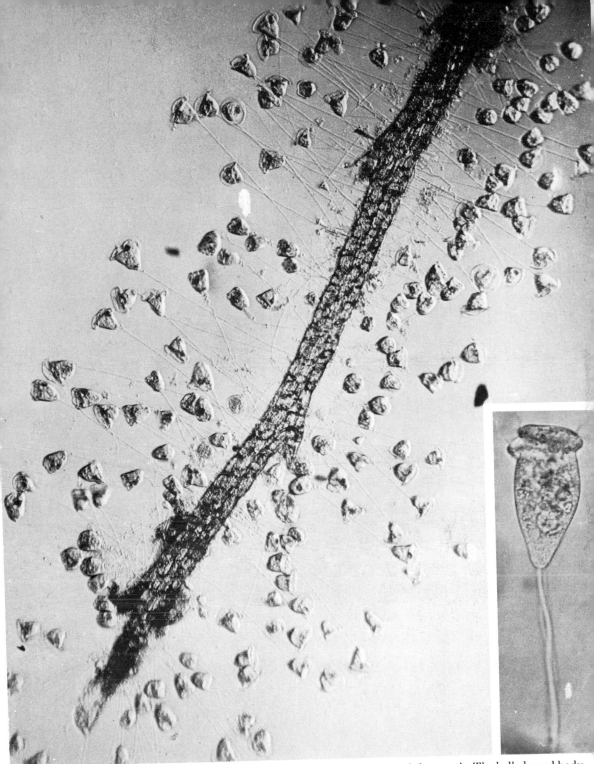

Vorticellas are peritrichs, ciliates in which the cilia are limited to a ring around the mouth. The bell-shaped body is firmly attached to some solid object in the water (in this case a plant stem) by a long slender stalk containing a spiral contractile fiber. At the least disturbance the stalk contracts like a coiled spring, and the bell contracts also, folding in its cilia. Several contracted individuals can be seen among the many extended ones. (Photomicrograph of living vorticellas by P. S. Tice). *Insert:* Closeup of living individual.

Stentors are large trumpet-shaped ciliates, which, with Balantidium, belong to the heterotrichs. *Stentor coeruleus*, shown here, is blue in color. It can swim about freely, as one of those shown is doing, but most of the time remains attached to some solid object. Note the row of large fused cilia around the top, the dark spiral gullet, the longitudinal stripes of which the white ones are underlain by contractile fibers, and the elongated large nucleus which looks like a string of beads. (Photomicrograph taken at 1/30,000 second)

surface may bear a few stiff, bristle-like cilia, but on the lower surface some of the cilia are fused in groups to form heavy **cirri.** These cirri do not beat rhythmically like ordinary cilia but are used like legs as the animal crawls about on vegetation. *Stylonichia* is one of the most common of the hypotrichs in fresh water. *Euplotes,* mentioned in chapter 3, is a hypotrich.

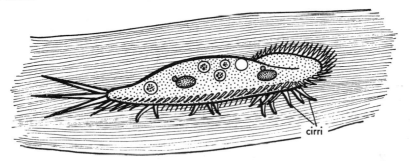

Stylonichia swims across the field of a microscope in rapid jerks, but it is easy to observe when it is crawling about on its cirri, as shown here.

In another order, the **peritrichs,** the cilia are usually limited to a ring around the mouth. Some of these, like *Vorticella,* live singly, while others form branching, treelike colonies. The vorticellas have lost nearly all their cilia except the circlets of cilia around the broad end of the bell-shaped body. The body is firmly attached to the substratum by a slender stalk containing a spiral contractile fiber, and in the surface layer of the bell there are contractile fibers like those in *Stentor.* When undisturbed, the bell is poised, fully expanded, on the end of the long, straight stalk, with the cilia in rapid action. The least disturbance causes the stalk to contract like a coiled spring, while the bell also contracts, folding its edge over the circlets of cilia. The vorticellas feed chiefly on bacteria brought into the gullet by the ciliary current made by the cilia. They reproduce by lengthwise division; one of the daughters retains the stalk,

Vorticella was so named because of the little vortex or whirlpool which it creates in the surrounding water by the action of the cilia about its mouth. (Based on Noland and Finley)

while the other develops a posterior circlet of cilia and swims away, later developing a stalk and attaching itself. Any bell can also develop such a ciliary girdle, break loose from its stalk, and swim away; and vorticellas exhibit this behavior when conditions become unsuitable. The vorticellas have one large sausage-shaped nucleus and one small nucleus. Conjugation occurs, but the two conjugating animals are of very different size.

THE SUCTORIANS

THE **suctorians** are a highly specialized group which are thought to be allied to the holotrich ciliates. Only the young suctorians have cilia. The adults lose the cilia and are characterized by the possession of long, hollow protoplasmic extensions, called "tentacles." through

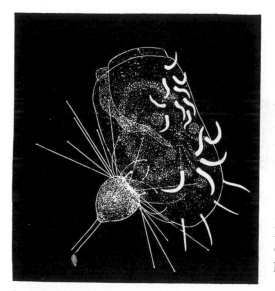

A **suctorian,** *Tokophrya*, feeds on a *Euplotes* many times its own size. It will take the suctorian about 15 minutes to suck its victim "dry," and by that time the suctorian will be stretched to several times its normal size. (From A. E. Noble)

which they suck up the protoplasm of their prey. *Tokophrya,* a typical suctorian, is attached to the substratum by a long noncontractile stalk and bears numerous tentacles which are held rigidly extended. When a ciliate happens to come in contact with the tentacles, it is seized. The tentacles are enlarged at their ends into sucking funnels, which in some way puncture or dissolve the outer protective covering of the prey and then suck up the protoplasm.

IN THIS chapter we have seen that the limitations of differentiation within a small protoplasmic mass have not prevented the development of an enormously varied and extremely successful group of animals.

CHAPTER 6

A SIDE ISSUE—SPONGES

SPONGES, or rather the skeletons of sponges, were commonly used by the ancient Greeks for bathing, for scrubbing tables and floors, and for padding helmets and leg armor. The Romans fashioned them into paintbrushes, tied them to the ends of wooden poles for use as mops, and made them serve, on occasion, as substitutes for drinking-cups. Today, sponges have an even wider variety of uses, and "sponge-fishing" is an industry which every year produces over one thousand tons of sponges. Bath sponges grow only in warm shallow seas; but many other kinds live in the ocean depths, and some are successful in fresh waters.

A living bath sponge looks more like a slimy piece of raw liver than like the familiar sponge of the bathroom. It grows attached to the substratum like a plant, and to the casual observer shows the same kind of unresponsive behavior. For a long time sponges were variously described as animals, plants, both animal and plant, and even as nonliving substances. secreted by the many animals that take shelter in the cavities of a sponge. In fact, it was not until about the mid 19th century that the last skeptics were finally convinced of the true animal nature of sponges. The question which they had been asking and which had previously not been satisfactorily answered was: "Since sponges do not move about and apparently do not respond rapidly to conditions about them, how can they capture food?"

This question is readily answered by adding a suspension of colored particles to the water near a sponge, thus disclosing a great deal of unsuspected activity. A steady jet of water is seen to issue from one or more large holes at the top of the animal. Closer inspection reveals that water is at all times entering through microscopic pores that riddle the entire surface. The sponge lives like an animated filter, straining out the minute organisms contained in the stream of water that passes constantly through its body. From the possession of the millions of pores, this phylum of animals has been called the **PORIFERA,** or "pore-bearers."

UP TO now, all the animals that we have considered were microscopic masses of protoplasm. Larger animals, like the sponges, are not merely larger masses, but their protoplasm is subdivided into microscopic units or cells. This many-celled structure is necessary for increase in the size of animals chiefly because the diffusion of oxygen and of metabolic substances is such a slow process that the interior of a large mass of protoplasm would not receive oxygen or dispose of wastes fast enough to support life. Cellular construction divides a large mass into a great number of small masses, or cells, making it possible to have spaces between them and thereby exposing an aggregate surface area many times that of the undivided bulk. This increases the surface through which substances can diffuse in and out of the protoplasm. Also, the distance that they must travel by diffusion is reduced because substances can be brought to nearly every cell in water currents. A glance at the diagram of a simple sponge shows that much more surface is exposed by this cellular construction than if the same amount of protoplasm were in a simple, solid mass.

Increase in size depends upon cellular construction for another reason. Since living protoplasm has a consistency much like that of raw egg white, a large amount of ordinary protoplasm could not maintain a constant shape or erect position. When the protoplasm is partitioned into microscopic cells, each inclosed in a membrane, it is more like beaten egg white, which has "cells" of air and albumin, instead of protoplasm. A cellular protoplasmic mass can take on almost any form, especially when supported by skeletal structures.

In one-celled animals specialization is that of different kinds of inter-mingling protoplasms. Consequently, a protozoan is limited not only in size but also in degree of specialization. In contrast with this protoplasmic level of organization, the most primitive of the many-celled animals—the sponges—may be said to be constructed on a **cellular level of organization.** No one cell must carry on all of the life-activities, but different cells may become specialized for different functions. The various kinds of cells are not rugged individualists like the protozoa but show definite socialistic tendencies. Only certain types feed; and these pass on some of the food to cells that specialize in protection, mechanical support, or reproduction. This division of labor among cells makes for greater efficiency and in-creases the possibility of exploiting sources of energy not available to simpler organisms. On the other hand, cells that specialize in certain jobs lose the ability to perform other functions and therefore become less able to lead an independent life.

The advantages of multicellularity are clear, but just how animals came to be composed of many cells is a question to which no definite answer can be given. One view is that as organisms grew larger and the amount of protoplasm increased, the nuclei divided, and then cell boundaries appeared. Another view is that the many-celled condition resulted from the failure of single protozoan-like cells to separate completely from each other after division. An example of this is seen in some of the colonial organisms such as *Volvox* (p. 40).

A SIMPLE sponge (like *Leucosolenia*) is a vase-shaped sac with a large **excurrent opening** at the top and microscopic **incurrent pores** per-forating the sides. The sac is covered and pro-tected on the outside by flattened **covering cells,** which fit together like the tiles in a mosaic.

The large internal cavity of the sponge is lined by a special kind of cell, called a **collar cell** because its free end is encircled by a deli-cate collar of protoplasm; this free end also bears a long flagellum whose base passes

A colony of simple sponges
(*Leucosolenia*), natural size.

through the collar. The beat of the flagella of the collar cells creates the water current which passes through the sponge, entering by the minute pores, passing through the main cavity, and leaving by way of the large hole at the top. As the water current passes through, the collar cells capture and ingest food organisms in the same way as do certain flagellate protozoa. The flagellum undulates from base to tip, causing a stream of water to flow away from its tip. Such a stream brings particles toward the base of the cell. The water passes through slits in the collar, which thus acts as a sieve. Particles stick to the outside of the collar and pass down its outer surface into the cytoplasm at the base, where they are engulfed.

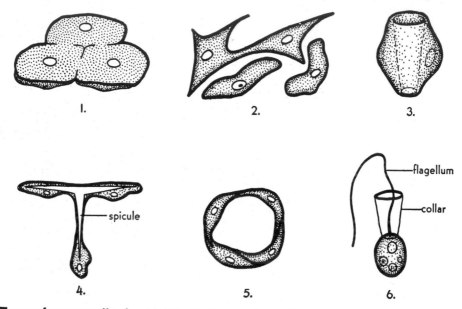

Types of sponge cells. 1, epithelial cells. **2,** mesenchyme cells. **3,** pore cell. **4,** three mesenchyme cells forming a spicule. **5,** contractile cells around a pore. **6,** collar cell.

The collar cells digest food in food vacuoles or pass it on to certain other cells that carry on digestion.

The collar cells look and behave almost exactly like the collar-flagellates (described on p. 43). For this reason it is believed that sponges evolved from the same group of ancestral protozoa that also gave rise to the modern collar-flagellates.

Between the outer covering cells and the collar cells is a nonliving jelly-like material containing living **mesenchyme cells** which move about in ameboid fashion and are more or less attached to each other by pseudopods. These are the least specialized cells and can develop into any of the more specialized types. They receive partly digested food particles from ' the collar cells, complete the digestion, and carry the digested food from one place to another. They probably also transport waste material to surfaces, from which it can be carried away by the outgoing current of water.

One of the chief functions of some of the mesenchyme cells is to secrete needles of calcium carbonate called **spicules.** The spicules form a skeletal framework which supports the soft cellular mass, keeps the canals from collapsing, and enables the sponge to grow to considerable size.

Spicules from **calcareous** sponges.

The spicules of sponges vary considerably in form. One sponge may have spicules of several shapes. But, as they are fairly constant for any particular species, they serve as an important criterion in identification.

The simple sponge, *Leucosolenia*, belongs to the class Calcarea, composed of small chalky or **calcareous sponges** with spicules of calcium carbonate. The **glass sponges** are all deep-water sponges known mostly from their dried but beautiful glasslike skeletons. They comprise the class Hexactinellida, in which all the members have six-rayed (hexactine) silicious spicules, and most have, in addition, a latticework of fused silicious spicules. The third class of sponges, or Demospongiae, is by far the largest class. It includes all the so-called **silicious sponges** in which there are silicious spicules (not six-rayed) or in which such spicules are combined with spongin fibers. Most of the sponges in temperate marine waters, and all the fresh-water sponges, are silicious sponges. The Demospongiae also include the **horny sponges,** which have no spicules at all but have a skeleton made of a horny elastic substance called "spongin," chemically related to silk and horn. The familiar sponges of commerce are all horny sponges.

Spicules from **silicious** sponges.

Another type of mesenchyme cell is the **pore cell,** shaped like an old-fashioned napkin ring. In a simple sponge the pore cells lie with their outer ends opening among the covering cells and with their inner ends opening among the collar cells, so that they form the pores through which water is drawn into the sponge.

Fibers of a **horny** sponge.

In some sponges there are special elongated **contractile** or **muscle cells,** which produce movement by becoming shorter (and thicker), thus drawing closer together the structures to which they are attached. They are

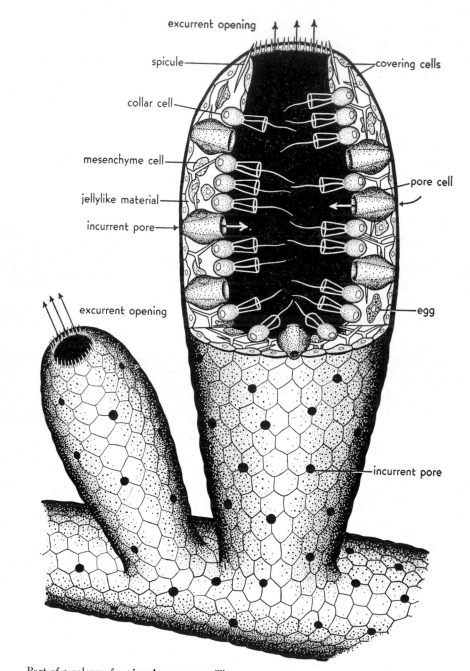

excurrent opening

spicule

covering cells

collar cell

mesenchyme cell

jellylike material

incurrent pore

pore cell

excurrent opening

egg

incurrent pore

Part of a colony of a **simple sponge.** The upper part is cut away to show the structure.

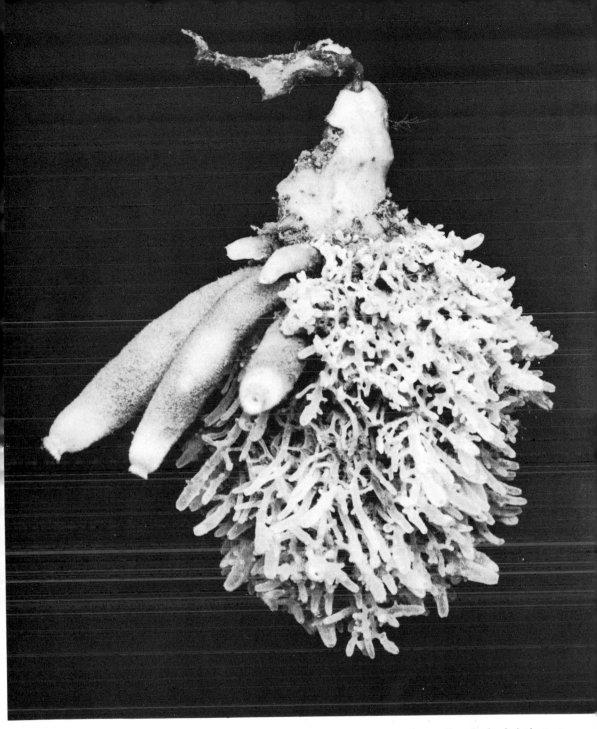

Calcareous sponges common in temperate shallow marine waters are mostly small and of a bristly texture. Shown here growing together in a cluster, which was attached to a pile of a harbor wharf, are three species of sponges. The main mass is the simple, much-branched, *Leucosolenia*, described in the text. At the left protrude three large and two small vase-shaped individuals of *Sycon*, which has calcareous spicules as in Leucosolenia, but a more complex internal structure. At the top of the picture, and surrounding the stalk by which Leucosolenia was attached, is a flat encrusting sponge described on the next page. (Photo of living sponges by D. P. Wilson, Plymouth, England)

All sponges are sessile as adults and with few exceptions live permanently attached to rocks or other solid objects. Sponges are abundant in shallow marine waters throughout the world, and there are some deep-water kinds. Only one family of sponges live in fresh water. Most sponges grow irregularly, forming rounded masses, sheets, or branching plantlike growths. But some have a simple vaselike shape with definite radial symmetry, like these, shown growing in the waters around the Bahama Islands. (Underwater photo by Johnson, courtesy Mechanical Improvements Corp.)

Encrusting sponge, *Halichondria panicea,* is sometimes called the "bread-crumb sponge" because of the way in which it crumbles when broken. It grows as a flat adhering crust on the underside of overhanging rocky ledges (as shown here) or attached to any hard surface where it is protected against direct sunlight when the tide is out. The colony has numerous excurrent openings, each at the tip of a conical projection. The soft flesh is supported by interlacing monaxon silicious spicules. (Photo of living colony by D. P. Wilson. Plymouth, England)

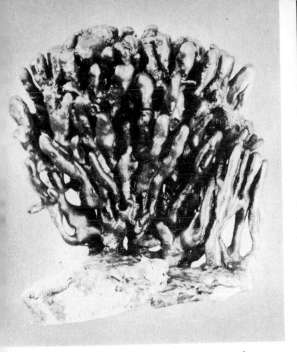

ilicious sponge (*Microciona*). A bright red sponge much used in regeneration experiments. It lives in hallow water and grows up to 6 inches high. Common on our East Coast. (Photo, courtesy American Museum of Natural History)

Horny sponge, the elephant's-ear sponge, from the Mediterranean Sea, has numerous excurrent openings on the inner wall of the cup. Its texture and flatness (when cut up) make it valuable to potters. (Photo of dried skeleton)

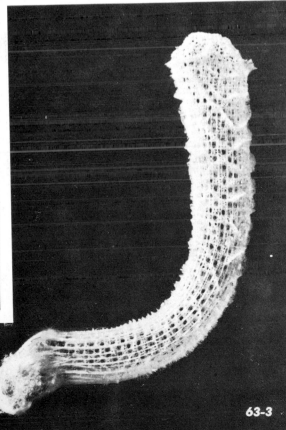

Glass sponges grow only in deep marine waters. *Staurocalyptus* (*above*) and *Euplectella* (*right*) (often called "Venus' flower basket") have skeletons consisting of separate silicious spicules plus interlacing lattice framework of spicules. (Photos of dried skeletons)

Fresh-water sponges grow in streams, lakes, and ponds as irregular encrusting masses on sticks and stones. They grow up to the size of one's hand and are usually yellow or brown in color but, when growing in strong sunlight, may be green from the presence of unicellular algae living in their tissues. *Spongilla*, shown here, has a horny skeleton combined with silicious spicules. (Photo, courtesy American Museum Natural History)

Another fresh-water sponge (*Ephydatia*). This specimen, collected from a Wisconsin lake, was dark gray-green in color and only about 1 inch long. (Photo, courtesy J. R. Neidhoefer and F. A. Bautsch)

arranged around openings, and, when irritating substances are present in the water, the contraction of the muscle cells narrows the openings.

There are no sense cells that receive stimuli from the environment and no nerve cells to transmit them to other parts of the sponge. Contractile cells are stimulated directly or by adjacent cells. Any sponge cell is irritable and will react if directly affected, but the animal cannot respond as a unified whole. A strong touch or even an actual cut is apparently not transmitted beyond a short distance. Sponges are very insensi-

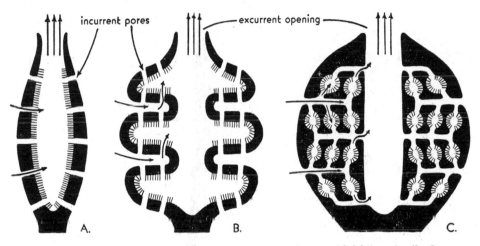

Types of sponges. **A, simple sponge. B, more advanced type** with folding of wall. **C, complex sponge,** like the bath sponge, with elaborate system of canals and flagellated chambers. (Modified after various sources)

tive to conditions about them, and, when they do react, as in decreasing the size of their openings, it is only with extreme slowness.

This lack of a specialized co-ordinating system accounts for the low grade of **individuality** of the sponge. Many sponges are of irregular size and shape and consist of numerous individuals inextricably fused into a single large mass or colony. An individual can only roughly be distinguished as a part of the sponge colony served by one of the numerous large openings.

In addition to bringing a constant supply of food, the continuous current of water passing through a sponge furnishes an ample supply of oxygen for all the cells and carries away carbon dioxide and waste nitrogenous substances. Consequently, even very large sponges do not have any spe-

cial mechanisms to aid in respiration or excretion, but simply have a larger
and more efficient system of canals.

The evolution from small, simple sponges to large, complex sponges revolves chiefly
about the problem of increasing the surface in proportion to the volume. If a simple sponge
like Leucosolenia were to enlarge indefinitely without any modification in structure, it
would soon reach a point at which the internal surface available for the location of collar
cells would not be large enough to bear the number of collar cells necessary to take care of
the food demands of all the other cells composing the large bulk. In some sponges this problem has been solved by a simple folding of the body wall, which increases the surface available for the location of the collar cells and strengthens the body wall. In most sponges,
like the bath sponges, there has been a further increase in the folding of the walls, resulting
in very intricate systems of canals with innumerable chambers lined with flagellated cells.

Complexity serves to increase the efficiency of the sponge in another way. A sessile
animal cannot get up and leave when the food or oxygen supplies of its environment run
low. The water which leaves a sponge has been filtered of much of its food and oxygen and
is loaded with poisonous wastes resulting from metabolism. If this water is not thrown
sufficiently far away it will enter the animal over and over again. The evolution of the
structure of sponges, then, has been in improving the rate of flow of water through the
sponge and in separating as completely as possible the outgoing from the incoming water.

Sponges may **reproduce sexually.** Some cells of the mesenchyme enlarge greatly with reserve food and become female sex cells or **eggs;** other

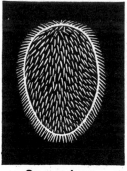

Sponge larva.

mesenchyme cells divide to form male sex cells or
sperms. In some sponges both kinds of sex cells
(gametes) may arise in one individual. In others
they occur in different individuals, in which case
the sperms are brought into the female sponge in
its water current. The fertilized egg develops into
a flagellated **larva,** which finally escapes from the
parent body and swims about. The term "larva"
is applied to young free-living stages in the development of animals. After swimming about for a
short time, the sponge larva settles down, becomes
firmly attached, and grows into a young sponge. Through their larvas
the sessile sponges are able to spread geographically and to send some of
their offspring far enough away from home so that they do not set up business in direct competition with their parents.

Sponges **grow** by budding and branching, somewhat like plants. A
simple sponge (*Leucosolenia*) sprouts horizontal branches which extend
over the rocks and give rise to an extensive mass of vase-shaped uprights.
Also, **asexual reproduction** may occur when buds break off and grow into
new individuals. Some sponges grow as irregular incrustations and in-

crease their mass indefinitely over the surface of rocks, vegetation, wharf pilings, or even on the backs of crabs. Thus, it is not surprising that a complete individual will grow from almost any piece which has been broken from a living sponge.

Many sponges produce asexual reproductive units known as **gemmules,** which consist of a mass of food-filled mesenchyme cells surrounded by a heavy protective coat strengthened with spicules. The gemmule survives drying and freezing and carries the sponge over

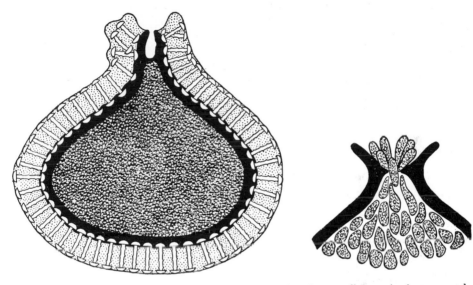

Gemmule of a fresh-water sponge. (Modified after Evans)   Sponge cells emerging from gemmule. (Modified after Wierzejski)

the winter season or a period of drought. Under favorable conditions the sponge cells emerge through a thin spot in the gemmule coat, aggregate in a small mass, and grow into a new sponge.

All animals—but particularly the less specialized ones have some capacity to replace lost or injured parts, a process called **regeneration.** Some sponges are noteworthy in this respect, and under a certain kind of treatment their cells display a behavior which reminds us strongly of the absurd cartoons which show a dog being ground to bits in a meat-grinder, only to emerge as small, compact sausages which still retain some of the behavior of the original dog. When certain sponges are pressed through very fine silk bolting-cloth, such as is used for sifting flour in flour-mills, the cells are separated from one another and come through singly or in small groups. In a dish of sea water these separated cells creep about on

the bottom in ameboid fashion. When they happen to come in contact
with each other, they stick together; and after some time most of the cells

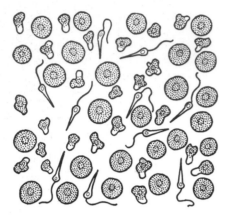

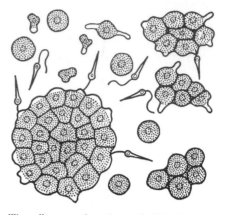

Cells separated by pressing a living sponge
through fine silk bolting-cloth.

The cells move about in ameboid fashion and
aggregate in small masses which grow into
small sponges. (After H. V. Wilson)

are found to have united into one or more small masses. Finally, these
masses of aggregated cells grow up into new sponges.

THE sponge body plan is unique. No other many-celled animals use
the principal opening as an exhalant opening instead of a mouth, or
have the peculiar collar cells, or show so low a degree of co-ordination between the various cells. Hence, it is thought that the sponges have evolved
from a group of protozoa different from the ones that gave rise to all the
other many-celled animals. And the phylum PORIFERA has sometimes
been set aside as a separate subkingdom of animals. There is no evidence
that the sponges have ever given rise to any higher group. This does not
mean that the sponges have been a failure, for they are an abundant and
widespread phylum. The "sponge plan" is of interest to us because it
illustrates the cellular level in animal structure (cell differentiation without much cell co-ordination), a stage which can no longer be found among
the other many-celled animals. But in the general trend of animal evolution the sponges are little more than a side issue.

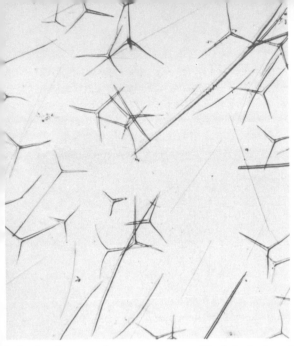

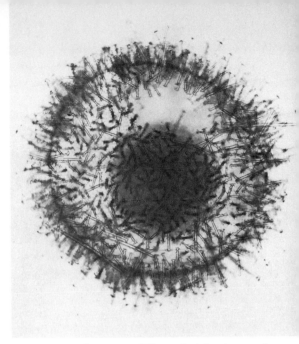

Spicules from a calcareous sponge. The interlacing of these triradiate and single needles of calcium carbonate provides support for the soft walls of the sponge and keeps the canals from collapsing. (Photo, courtesy General Biological Supply House)

This gemmule, from a Wisconsin lake, has anchor-shaped silicious spicules, which help protect the cells within. The gemmule withstands the winter; in the spring the cells emerge and grow into a new sponge. (Photo, courtesy J. R. Neidhoefer and F. A. Bautsch)

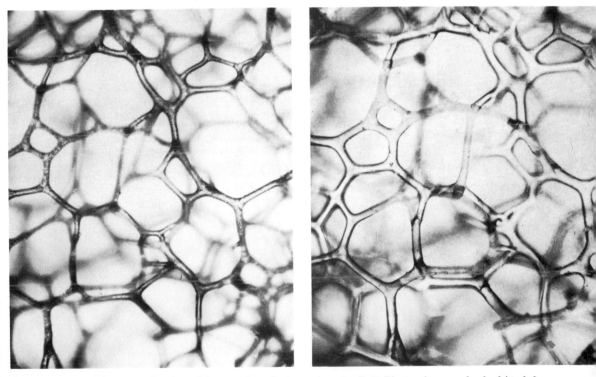

Horny sponge fibers are elastic and form an interlacing network. *Left:* Photomicrograph of a bit of *dry* sponge skeleton. *Right:* Water has been added and is taken up by the fibers as well as into the spaces between fibers.

Sponge-fishing in the United States has its center at Tarpon Springs, Florida, where the sponge fishermen bring in their "catch." The sponges are brought up by divers and after preliminary cleaning are hung on the rigging, where the rest of the protoplasm decays. (Photos from a motion picture made by James Rhodes Co., courtesy L. E. Diamond)

Hooking sponges is a method used in shallow water. While one man rows, the other scans the bottom through a glass-bottomed bucket (which cuts off surface reflections) and pulls up the sponges with a two-pronged hook on the end of a long pole.

Living bath sponge, (*right*), pulled from its attachment on the sea bottom, is covered with a tough, dark, leathery membrane. *Below:* The same sponge cut open looks like a slimy piece of raw liver. In this gross aspect the supporting framework of spongin fibers is not distinguishable from the mass of protoplasm. Like all horny sponges, it has no spicules. The larger horny sponges, and all those of commercial value, live in tropical and subtropical waters. (Photos of living sponge. Batábano, Cuba)

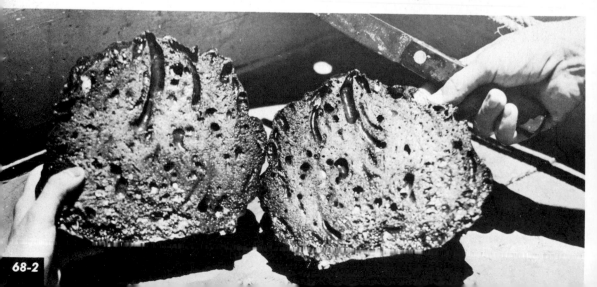

Preparing sponges for the market. At the sponge exchange, where fishermen bring their dried sponges, the sponges are washed, sorted, and strung up in convenient bundles (*left*). Before they can be used, they must be cleaned to remove further the cellular debris, pounded with a mallet to break up the shells or skeletons of various invertebrates that lived in the cavities of the living sponge, and then trimmed to a regular shape with a shears, as shown at *right*. (Photos taken at Tarpon Springs, Florida)

Commercial sponges used for taking up and expelling water are highly elastic. The compressibility of a load of such sponges is used to advantage in shipping them. *Left:* A baling machine compressing sponges. *Right;* A basketload of sponges shown beside a bale compressed from a basketload of the same size. (Photos taken at Batábano, Cuba)

Millions of sponges are used every year for washing automobiles, for washing walls, floors, and ceilings, and for various cleaning and polishing purposes in industry. In the home sponges are used mostly for bathing and household cleaning and also for applying face cream and shoe polish. The tough elastic fibers of natural sponges stand hard wear much better than those of rubber or other synthetic "sponges." Of the thousands of different kinds of sponges in nature, only a few species of horny sponges are soft enough to be of much commercial value. Our Florida sponges are good enough for most cleaning purposes, but the finest ones come from the Mediterranean Sea.

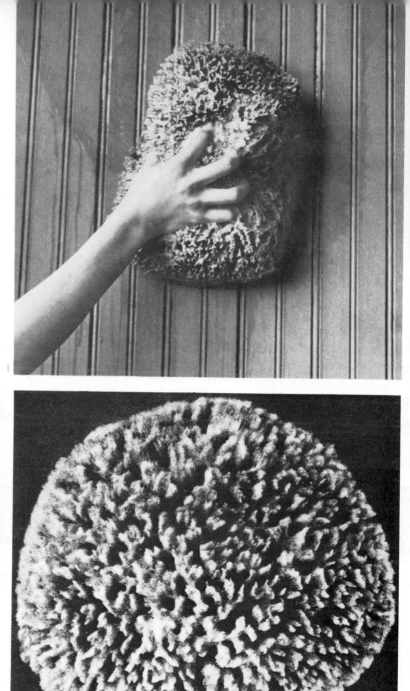

Sponge culture has been tried with some success for restocking grounds depleted by excess sponge fishing. Live sponges were cut into small pieces (about the size of the one on the *left*), affixed to tiles, and lowered to the bottom. Under favorable conditions such pieces grew from 2 cubic inches to 12 cubic inches in two months. The sponge on the *right* (actual size 6 inches in diameter) was "harvested" after three years of growth in experimental "gardens" off the coast of Florida. (From *Bulletin, U.S. Bureau of Fisheries*, Vol. 98)

CHAPTER 7

TWO LAYERS OF CELLS

THE hydra lives in ponds, lakes, and streams, attached to rocks or water plants by a sticky secretion from its disklike base. The body, with its tentacles encircling the mouth, looks like a half-inch of string with the unattached end frayed out into several strands. Because of its small size, transparency, and habit of contracting down into a little knob when disturbed, the animal is readily overlooked. Yet, hydras are abundant and are the only really successful ones among the few members of their phylum that have invaded fresh water. The marine relatives of the hydras—the jelly fishes, sea anemones, and corals—are the more conspicuous members of the large and varied phylum **COELENTERATA**. The name of the phylum is derived from "coel," meaning "hollow," and "enteron," meaning "gut," and refers to the fact that the main cavity of the body is the digestive cavity. Beginning with the coelenterates, all higher animals have a digestive cavity which connects with the outside through a mouth. For this reason, among others, it is believed that modern coelenterates, unlike sponges, evolved from the same stock that gave rise to all the higher phyla.

ONE might suppose that in the course of evolution a great many different types of cells would arise; but when animals are examined microscopically, it is found that their **cells may be classified into a few main**

types—about five: *epithelial, mesenchyme* or *connective, muscular, nervous,* and *reproductive.* All but nerve cells were already present in the

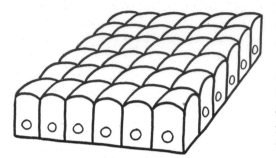

most primitive many-celled animals, the sponges. Nerve cells are added by the coelenterates.

An association of cells of the same kind which work together to perform a common function is called a **tissue.** Thus, a mass of mesenchyme cells or some other type of connective cells is known as a "connective tissue," a bundle of muscle cells is spoken of as

An **epithelium** is a group of cells covering a surface.

"muscular tissue," and a group of nerve cells as "nervous tissue." Sponges were presented as animals organized on a cellular basis, but they have some beginnings of tissue formation. For instance, the flattened cells covering the exterior and lining some of the chambers were fitted closely together to form a covering membrane. Such cells, clothing a free surface, are called "epithelial cells"; and the tissue they form is called an **epithelium.**

The higher animals—man, for example—have many more different kinds of cells than a hydra, but they are all modifications of these same basic cell types. An epithelium covers the exterior surface of man, lines his mouth and digestive tract, lines his heart and blood vessels, and is folded in various places to form glands. Liver and thyroid cells are epithelial; blood and bone cells are mesenchyme or connective-tissue types.

The organization of cells into tissues is a distinct advance in structure, since various cell functions can be performed better by a group of cells of the same kind acting together than by separate cells. Scattered epithelial cells would be a poor protection for the surface of an animal, and epithelial cells almost always occur close together in a sheet as epithelial tissue. Similarly, a single muscle cell does not have enough strength to produce much movement, while a bundle of muscle cells contracting together can lift a heavy weight. Because their cells act together in a more co-ordinated fashion than do the cells of sponges, the coelenterates may be said to have reached the **tissue level of organization.**

THE hydra consists of **two layers of cells.** The outer layer, or **ectoderm** (literally "outside skin"), is a protective epithelium, as in sponges; but it contains several other kinds of cells. The "inner skin," or **endoderm,** lines the internal cavity and is primarily a digestive epithelium. The cells of both layers differ from ordinary epithelial cells in that their bases are drawn out into long contractile muscle fibers. The muscle fibers of the ectoderm run lengthwise. When they contract equally on all sides, the body is shortened. When they contract more on one side than on the other, the body is bent in the direction of greatest contraction. The muscle fibers in the endoderm cells run circularly; and when they contract, the body becomes narrower and longer. There is no separate muscular tissue in the hydra; the muscle fibers occur only in the bases of epithelial cells,

which also perform other functions. Since they cannot be strictly classed as either epithelial or muscular, we name these cells according to function and call those of the ectoderm the **protective-muscular cells** and those of the endoderm the **nutritive-muscular cells.**

Between the two layers of cells is a thin layer of jellylike material produced by both ectoderm and endoderm. Among the bases of the epithelial cells of both layers, but particularly in the ectoderm, are small **mesenchyme cells.** The mesenchyme cells of a hydra, as in sponges, are the least specialized cells and are capable of developing into some of the

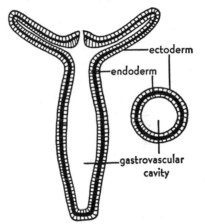

A hydra consists of **two layers of cells.** *Left,* a hydra in longitudinal section; *right,* in cross-section. Between the two layers is a jellylike material.

kinds of cells which will be described in connection with the hydra's activities.

In **feeding,** the hydra does not chase its prey, but remains attached to the substratum, with the almost motionless tentacles trailing in the water. When a small crustacean or worm brushes one of the tentacles in passing, the unlucky victim is suddenly riddled with a shower of poisonous, numbing threads shot out from certain of the **thread capsules,** effective weapons with which the body, and particularly the tentacles, are heavily armed. There are four kinds of thread capsules, all produced within specialized mesenchyme cells, known as "thread cells." Each consists of a fluid-filled capsule containing a long spirally coiled, hollow thread. The largest and most conspicuous type (distinguished by large barbs at the base of the thread) is called a **stinging capsule** and lies in a cell in the ectoderm, which has a fine protoplasmic extension, or trigger, projecting

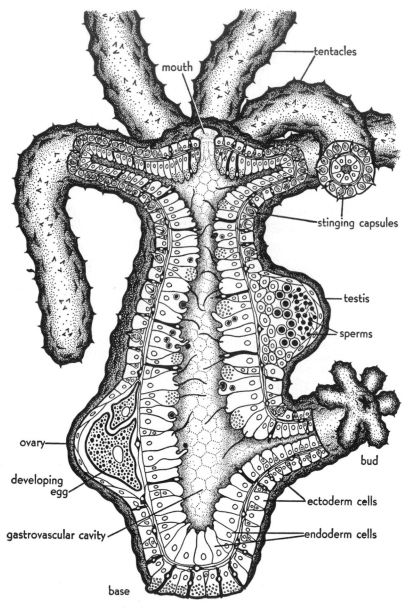

mouth

tentacles

stinging capsules

testis

sperms

bud

ectoderm cells

endoderm cells

ovary

developing
egg

gastrovascular cavity

base

A **hydra,** cut away to show structure.

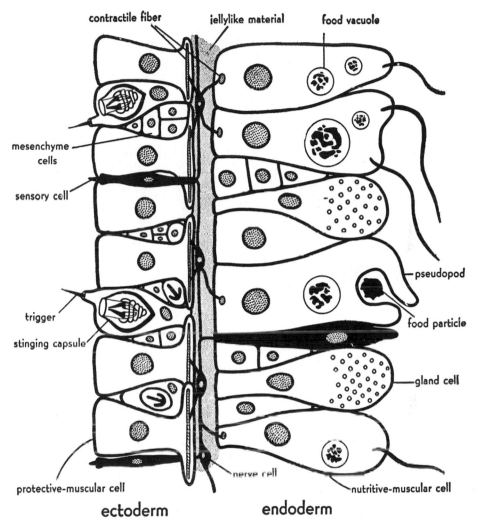

contractile fiber   jellylike material   food vacuole

mesenchyme cells

sensory cell

trigger

stinging capsule

protective-muscular cell

pseudopod

food particle

gland cell

nerve cell

nutritive-muscular cell

ectoderm   endoderm

Portion of body of a hydra enlarged to show **cell types.** The contractile fibers in the endoderm run circularly and are shown in section as small ovals.

Although ectoderm and endoderm layers may appear superficially similar, and both consist mostly of the same two types of cells, slender sensory cells and columnar epithelial cells with muscle fibrils in their bases, it has been shown that during development they have undergone a basic differentiation which cannot be reversed merely by changing their position in the hydra. In one experiment about sixty hydras were turned inside out (by manipulation with needles). About a third of the hydras did not adjust but underwent a period of "depression" followed by death. And the rest were able to return themselves right side out.

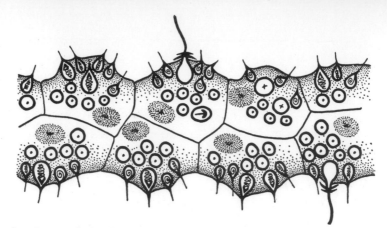

Small portion of a tentacle showing the prominent **batteries of thread capsules.** Two of the large stinging capsules are discharged. Circles containing a central dot are capsules in top view. (Modified after Pauly)

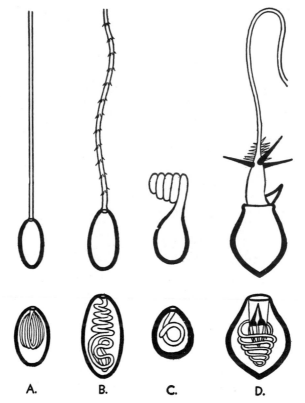

A.   B.   C.   D.

Thread capsules of the hydra. *Bottom row,* undischarged; *top row,* discharged. **A,** adherent capsule, anchors tentacles in locomotion. **B,** defensive capsule, kills or repels animals other than prey. **C,** volvent type, aids in holding prey by winding about bristles of prey. **D,** stinging capsule, pierces body of prey and injects a poison. (After various sources)

A thread capsule is also called a **nematocyst** or a **cnida.** From this last name comes the phylum name **Cnidaria,** now often preferred to Coelenterata.

from the surface. When the trigger is stimulated, a physiological change
occurs, such that the pressure within the capsule is suddenly increased
and the coiled hollow thread is
turned inside out. This has been
compared to the way in which one
everts the finger of a rubber glove
by blowing into the glove. The
thread is discharged with such
explosive force that it pierces the
body of the prey and injects a
poisonous substance contained in
the capsule. After the poison from
a number of these stinging capsules has paralyzed a small animal, the tentacles are wrapped

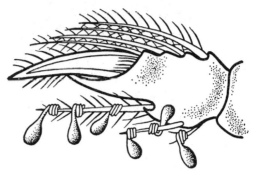

Several thread capsules of the type which wind about
small projections are seen clinging to the bristles of a
small crustacean. (After Toppe)

around the prey and contract, drawing the prey toward the mouth, which
opens widely to receive it. The victim is swallowed by means of muscular
contractions of the body wall, aided by a slimy secretion from gland cells
lining the inside of the mouth region.

The mechanism by which the hydrostatic pressure within the thread capsule is suddenly
raised is not definitely known. The most prevalent theory assumes an increase in pressure
due to a rapid intake of water, which causes a swelling of the capsule contents. Another
theory attributes the increase in pressure to the force exerted by contractile fibers surrounding the capsule.

A thread capsule (or **nematocyst**) can be discharged only once. Used ones are discarded
and are replaced by new ones, produced in specialized mesenchyme cells.

Digestion takes place in the interior cavity. Gland cells in the endoderm secrete enzymes, chiefly of the protein- and fat digesting types, which
reduce the digestible parts of the prey to a thick suspension containing
many small fragments. This material is then engulfed by the pseudopodal activity of the nutritive-muscular cells. The process of digestion is
completed within food vacuoles in these cells, for the hydra has retained,
in part, the protozoan method of food ingestion and digestion. Since preliminary digestion takes place in the large digestive cavity, where enzymes
poured out by many cells act together to disintegrate a food organism, a
hydra can eat animals which are very large as compared with those that
can be taken by a sponge. (In the sponge, as in protozoa, the prey must
be of a size that can immediately be engulfed by a single cell.)

The indigestible remnants left in the central cavity are **eliminated** through the mouth, which serves as both entrance and exit. The digested food is passed by diffusion from cell to cell. Currents, set up by muscular movements of the body and by the beating of long flagella on the endoderm cells, circulate the food throughout the cavity of the body and of the

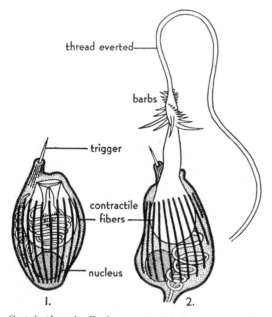

thread everted

barbs

trigger

contractile
fibers

nucleus

1.   2.

Certain thread cells show contractile fibers surrounding the capsule. **1**, capsule undischarged. **2**, capsule discharged. (After Will and P. Schultz)

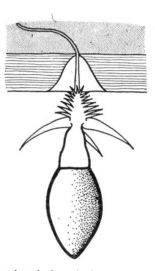

The thread of a stinging capsule has penetrated the hard chitinous covering of the prey and lies imbedded in the soft tissue (stippled). The clear area in the chitin represents the part that has been mechanically injured by the barbs and chemically dissolved by some substance from the capsule. (Modified after Toppe)

hollow tentacles. The cavity thus has the double function of digestion and circulation. For this reason it is called the **gastrovascular cavity,** which means "stomach-circulatory" cavity.

Respiration and excretion take place by diffusion, as in protozoa, for the hydra is still a relatively small animal. Because of the thinness of the walls and the circulation in the gastrovascular cavity, most of the cells, inside and out, are freely exposed to the surrounding water.

When well fed and healthy, hydras **reproduce asexually by budding.** The buds occur about one-third the length of the body up from the base. Here both layers hump up, forming a projection which elongates and soon sprouts tentacles at its outer end. In two or three days the bud looks like a little hydra, complete with stalk, tentacles, and mouth. At its base the gastrovascular cavity of the bud is continuous with that of the parent, and in this way it receives a supply of nourishment. Shortly after this, it constricts off from the parent and takes up the serious responsibilities of an independent life.

Regeneration would seem to be an easy matter for an animal that always has a reserve supply of unspecialized mesenchyme cells. And the hydra does, in fact, show a marked capacity for replacing tentacles or speedily repairing the more serious injuries likely to be incurred by so delicate an animal. Even if the hydra is cut into a number of pieces, most of the pieces will grow the missing parts and will become complete and independent hydras (see chap. 12).

The ability of these animals to replace lost or injured parts won for them the name of "hydra." An early naturalist saw in this habit a resemblance to the mythical monster, Hydra, which was finally slain by Hercules. Hydra had nine heads; and when Hercules cut one off, two grew in its place.

Hydras **reproduce sexually** at certain times of the year, generally in the fall or winter. In some species both male and female sex cells occur in the same individual, which is then known as a **hermaphrodite.** In other species, the two sexes are always separate, and male and female individuals can be distinguished. The sex cells come from the mesenchyme cells in the ectoderm. In certain regions these cells suddenly start to grow rapidly, causing the body wall to bulge locally. Such bulges are known as **testes** when filled with sperm-forming cells and as **ovaries** when filled with egg-forming cells. In each testis the mesenchyme cells first enlarge and then divide a number of times to form many sperms. An ovary also contains many mesenchyme cells at the start; but several of these fuse, all the nuclei but one degenerate, and the result is a single, large ameboid egg. The ameboid egg finally incorporates the remaining yolk-filled mesenchyme cells and becomes the spherical ripe egg, packed with food reserves that will later nourish the developing embryo.

Just how the formation of sex cells is initiated is not definitely known. That one of the factors may be low temperature is suggested by the fact that some species of hydras will produce testes and ovaries if kept in a refrigerator for two or three weeks. Sometimes abundant food appears to stimulate the development of sexual maturity.

The ripe egg breaks through the covering ectoderm and projects with its outer surface freely exposed to the water. Sperms, discharged from a testis, swim through the water and surround the egg; one enters and effects

fertilization. If not fertilized within a short time after it is first exposed, the egg dies and disintegrates.

Development begins, as in sponges and in nearly all other many-celled animals, with the division of the fertilized egg or **zygote** into two cells. These promptly divide, forming four cells. Continued division results in a

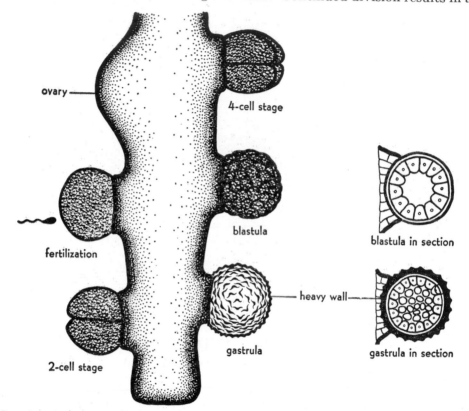

ovary

4-cell stage

fertilization

blastula

blastula in section

heavy wall

2-cell stage

gastrula

gastrula in section

Development of a hydra. A hydra frequently has several eggs developing at the same time. (Based on Tannreuther and other sources)

one-cell-layered hollow ball known as the **blastula.** In the hydra the cells composing the single layer now divide, and some of the surface cells migrate inward so that cells accumulate in the interior cavity. In this two-layered stage, known as a **gastrula,** the outer layer of cells produces a heavy covering membrane or shell, and the gastrula usually drops from the parent and becomes fastened, by a sticky secretion, to the substratum. Under favorable circumstances the young hydra may hatch from the shell after a week or more. In winter the egg may lie dormant until the following spring. When development is resumed, the outer layer, or ectoderm, becomes differentiated into a protective epithelium. The inner mass of cells, or endoderm, becomes hollowed out and differentiates as a digestive epithelium. Tentacles develop, a mouth breaks through, and the young hydra hatches out by rupture of the shell.

The **behavior** of the hydra is far more varied and complex than that of the unco-ordinated sponge but perhaps not much more so than that of the complex protozoa which have a system of co-ordinating fibers. The many-celled hydra has a network of nerve cells extending through the entire animal. This **nerve net** is slightly more concentrated around the mouth than elsewhere, but there is very little evidence of a specialized controlling group of nerve cells or brain which characterizes the nervous systems of more complex animals. The nerve net is thought to be composed of separate nerve cells, so that a traveling impulse, in passing from one nerve cell to another, must pass across definite breaks at the junctions between nerve endings. Such

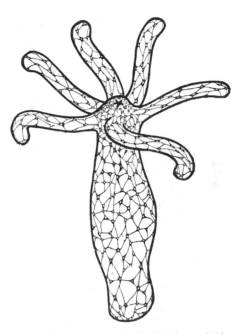

The **nerve net** of the hydra lies beneath the ectoderm. It is more concentrated around the mouth and base than elsewhere.

junctions or synapses are characteristic of the nervous systems of higher animals, but in the latter the junctions are so constituted that impulses can pass across them in only one direction, making for definite specialized pathways in the nervous system. In the nerve net of the hydra impulses can cross the synapses in either direction, and there are *no definite pathways* for the impulses. An impulse picked up at any place in the network can spread from one nerve cell to another in all directions. A very strong stimulus applied to the tip of one tentacle may cause the whole animal to contract. This is not a very efficient kind of nervous mechanism, and the impulses travel very much more slowly than in higher types; but apparently it is adequate for the limited activities of the sedentary hydra.

The response of the whole body of the hydra to a strong local stimulus applied to a part of one tentacle illustrates another primitive feature of the coelenterate nerve net—the poor development of simple reflexes. Such reflexes, which are specific localized responses to local stimuli, enable man, for example, to adapt to many ordinary external stimuli by reflexes such as blinking the eyes or withdrawing one hand without having to contract all the muscles in the body, as does the hydra.

All the *essentials of a simple nervous mechanism* in a many-celled animal are present in the hydra. Stimuli are received by **sensory cells** which are peculiarly sensitive to touch or to chemical substances in the water. These are slim, pointed cells, scattered among the cells of both layers, and lying with their pointed ends projecting to the outside if in the ectoderm, or into the digestive cavity if in the endoderm. From the sensory cells the impulses are picked up by **nerve cells** which lie at the bases of the ectoderm cells and connect with each other to form a network extending throughout the animal. They transmit the nervous impulses to muscle cells which contract or gland cells which secrete.

Even nonliving things can respond to stimuli. If one pushes a rock, it may "respond" by rolling over; but in doing so, it may roll down a hill and break as it strikes some other rock. The response of a hydra, or of any living organism, however, is usually *adaptive*, that is, the response results in a favorable adjustment of the animal to its environment. If a hydra is touched, the sensory cells receive the stimulus. The change started in them by the touch affects, in some way, the nerve cells of the network, and causes them to transmit a stimulus to the lengthwise muscle fibers. These contract, shortening the hydra and getting it out of the way of the "offending" object.

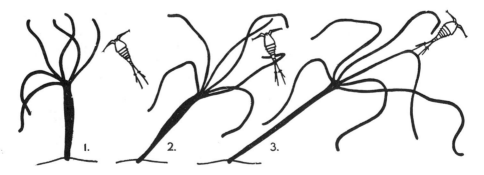

A hydra catches . . . .

The nerve net not only transmits impulses but also **co-ordinates the hydra's activities.** When a small animal touches one tentacle, the other tentacles will finally come to grasp the prey and will work together to cram it into the mouth, which has already opened in response to chemicals leaking from the food. The nerve net likewise co-ordinates the muscular contractions involved in swallowing food or in forcibly removing indigestible particles. Thus it enables an animal composed of many thousands of cells to react as one integrated individual.

The importance of co-ordinated activity is emphasized by what happens when it fails. Sometimes the hydra swallows its food so rapidly that it takes in one or more of its own tentacles, and the hydra has even been observed to swallow its own base along with the prey. Fortunately, it does not digest its own cells; and after a time the swallowed parts emerge, apparently uninjured.

The simplest method of **locomotion** in a hydra is a *gliding* on the base due to a creeping ameboid movement of the basal cells. The most rapid method is a kind of *somersaulting*. The animal bends over, attaches its tentacles to the bottom by means of the adhesive thread capsules, loosens its base, swings the base over the mouth, and attaches it to the bottom; then it loosens the tentacles and repeats. In species of hydras in which the tentacles are two to five times longer than the body, the animals can move by throwing out the extended tentacles and catching hold of some object, then loosening the base, and contracting the tentacles until the body is pulled up to the object—very much like an athlete "chinning" himself.

Hydras react to a variety of stimuli, usually by a kind of trial-and-error procedure, but almost always with a result likely to lead to the continued existence of the animal. A hydra will move away from a region in

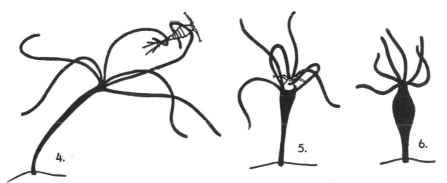

4. 5. 6.

. . . . and eats a cyclops (microscopic crustacean)

which the temperature rises above 25° C., or will move from the bottom of a dish to the top if carbon dioxide or products of decay accumulate. Many species react to light and tend to move toward the lighted side of their container, where there are usually more food organisms.

A hydra somersaulting. This is its most rapid method of locomotion.

The behavior changes with the physiological condition of the animal. A well-fed hydra usually remains attached, with the tentacles quietly extended. At intervals the body and tentacles suddenly contract and then slowly extend in a new direction. Presumably, this increases the amount of territory controlled by the animal. If a food organism does not appear after some time, the tentacles begin to wave "restlessly," and the body contracts and expands in a new direction more frequently. If food is still not forthcoming, the hydra may move off to new hunting-grounds.

Even the food reaction varies greatly according to the state of the animal. A very well-fed hydra will not react to food when it is presented. A very "hungry" hydra exhibits the behavior of the feeding reaction when only the chemical stimulus—glutathione added to the water—is present. Also, while a light touch on the tentacles may cause the food-taking reaction, a stronger touch on the tentacles or the body, as well as shaking or jarring, causes contraction of the animal in varying degrees, depending upon the strength of the stimulus. Unless the hydra has been injured by the stimulus, the body and tentacles will soon again be extended slowly, and the "patient" life of trapping and digesting will be resumed.

OBELIA

THE nearest marine relatives of the hydras are the branching colonial coelenterates (known as hydroids) which are usually seen as delicate plantlike growths on kelps, rocks, and wharf pilings along the seacoasts. One of the commonest of these is the **obelia,** colonies of which are about an inch to several inches in height. The colony arises by budding from a single hydra-like individual. The buds fail to separate; and after repeated budding, there results a treelike growth, permanently fastened to some object and consisting of numerous members united by stems. Because the activities of the members are subordinate to the colony as a whole, they are sometimes called **subindividuals.** More often a member of a colony is referred to as a **polyp,** a name applied to any tubular coelenterate which bears a whorl of tentacles around the free end of the body and is attached at the other end.

Obelia, natural size.

The hydra is also called a "polyp," a name which means "many feet." Polyps use their "many feet," or tentacles, chiefly for feeding and only occasionally for moving about. Hence, the name is not particularly appropriate but comes to coelenterates by an indirect route. It was derived from *poulpe,* the French word for octopus, because an early French naturalist thought that coelenterate tentacles resembled the "feet" of an octopus.

Polyps and stems are protected and held erect by a **horny covering,** secreted by the ectoderm, which incloses all the stems and extends around each polyp as a transparent cup, shaped like a goblet. When irritated, the polyp can withdraw into this cup; and the rapid contraction and slow expansion of the polyps are about the only movements that can be seen in an obelia colony. The stems are unable to move because of the rigidity of the covering; but at certain points the covering is arranged in rings, which allow for flexibility as the stems are swayed by water currents.

An obelia polyp is built on the same plan as a hydra, and consists of the same two cell layers, **endoderm** and **ectoderm.** These are composed of cell types similar to those of the hydra.

The obelia **feeds** in the same way as the hydra, capturing small prey by means of tentacles armed with stinging capsules. The tentacles are not hollow, as in the hydra, but are solid, having a central core of large endodermal cells. The polyps and stems are hollow, and the **gastrovascular cavity** of every polyp is continuous with that of every other polyp in the colony. The food is partly digested in the cavity of the polyp, and the resulting fluid is circulated about through the stems by the beating of the flagella of the digestive epithelium and by the muscular contractions of the polyps. Thus, food is distributed throughout the colony in thoroughly cooperative fashion, and digestion is completed in food vacuoles within the cells lining the gastrovascular cavity.

Asexual reproduction by budding is the usual method of increasing the number of polyps. In addition to the continuous budding on any vertical axis, rootlike horizontal stems from the base grow over the substratum and give rise to a series of upright colonies, so that the entire colony may, after a time, consist of hundreds of subindividuals.

Sexual reproduction does not occur in the polyp colony. We could search in vain throughout the year for any signs of testes or ovaries, for the obelia polyps, unlike hydras, never have any sex cells.

If we examine older obelia colonies carefully, we see that the polyps are not all alike. Those we have already described have tentacles with which they catch prey, and may be called the **feeding polyps.** In some of the angles where the feeding polyps branch from a stem, there occur **reproductive polyps.** These have lost their tentacles and the capacity to feed, and are nourished through the activities of the feeding members of the colony. They are specialized for **asexual reproduction** of a special type. Each is inclosed by a transparent horny vase-shaped covering and consists of a stalk on which are borne little saucer-like buds, the largest and most completely developed near the top, the smallest and least developed near the base. If live obelia colonies are kept in a dish of sea water, it is easy to observe that the top-most "saucer" escapes through an opening at the upper end of the vase-shaped covering and swims about as a tiny animal called a **medusa,** a name applied to any free-swimming "jellyfish" type of coelenterate. (The name is derived from a fancied resemblance of the waving tentacles of jellyfishes to the snaky tresses of the Gorgon Medusa.)

The **medusa** of the obelia looks like a tiny bell-shaped piece of clear jelly. From the middle of the under surface, where one expects to find the clapper of a bell, hangs a tube which bears, at its free end, the **mouth.** The mouth leads up this hollow tube into the **gastrovascular cavity,** which branches out into canals that carry food to all parts of the medusa. From the margin of the bell hangs a circlet of **tentacles,** well armed with **stinging capsules.** The tiny animal swims by alternately contracting and relaxing the muscle cells of the bell. As it swims or drifts with the current, the trailing tentacles catch small organisms.

The primary function of the medusa is **sexual reproduction.** From the under side of the bell hang four sex organs. In female medusas they are the ovaries and produce eggs; in male medusas they are testes and produce sperms. Eggs and sperms are shed into the sea water, where fertilization takes place. The zygote develops into a hollow blastula and then into a two-layered gastrula, with ectoderm and endoderm. The outer

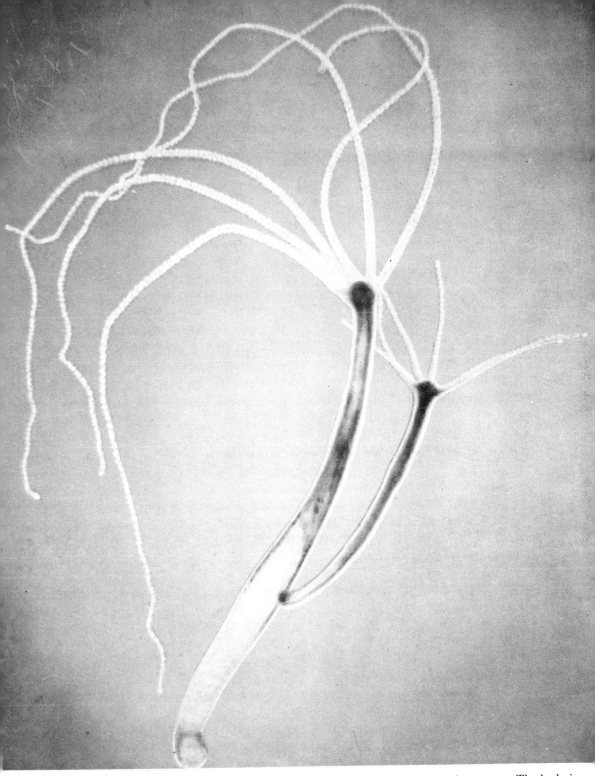

Hydra with bud. This particular species is probably *Hydra littoralis*, common in running water. The body is about ½ inch long. When well extended, the tentacles are 1½ times the length of the body. Other species vary in size and shape of body, number and length of tentacles (which may be shorter than, or 3 or 4 times longer than, the body), shape of stinging capsules, color, habitat, etc. (Photo of living animal by P. S. Tice)

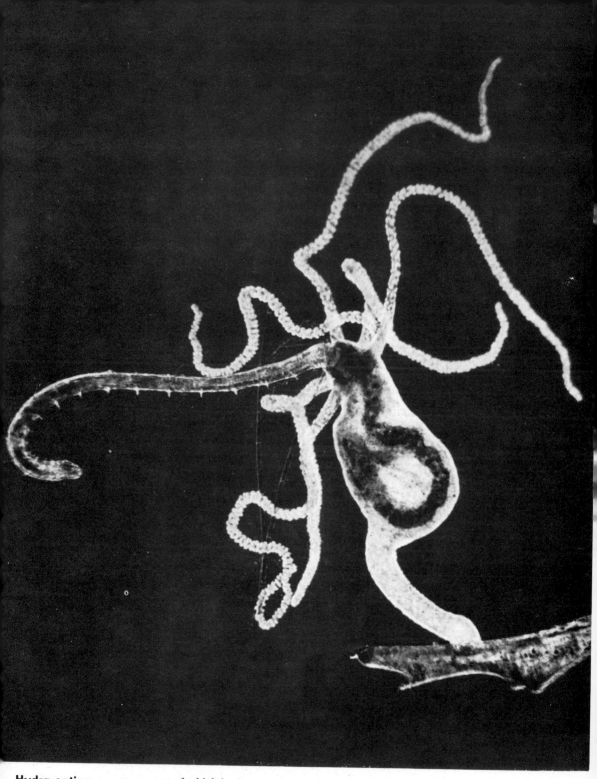

Hydra eating a worm, part of which is already in the gastrovascular cavity and can be seen through the thin body wall. Besides these fresh-water aquatic relatives of the earthworm (oligochetes), the hydra eats small crustaceans, very young fish, and other small animals which come within reach of the long tentacles; the bumps on the tentacles are batteries of stinging capsules. (Photo of living animal by P. S. Tice)

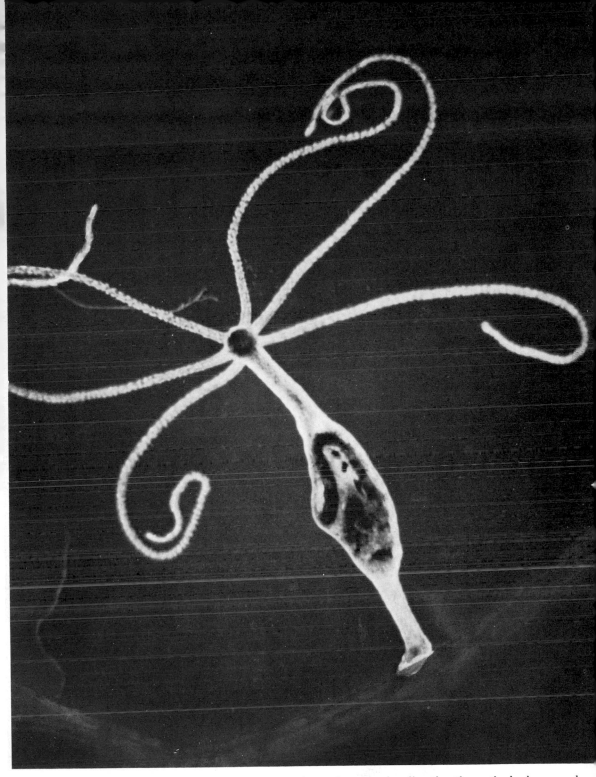

The worm safely tucked away in the gastrovascular cavity and undergoing digestion there, the hydra spreads its tentacles again and awaits a new victim. The long tentacles radiating out from a central mouth, and controlling the surrounding territory in all directions, are admirably suited to the needs of an animal that spends most of its time in one place waiting for prey to approach. (Photo of living animal by P. S. Tice)

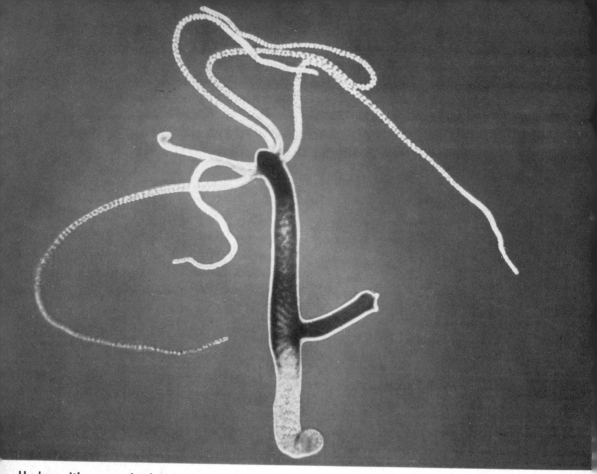

Hydra with young bud. The cavity of the bud is in direct communication with that of its parent, and the food can be seen washing back and forth. The lower third of the body contains no food and serves chiefly as a stalk which is more sharply set off from the trunk region in some hydras. (Photo of living animal by P. S. Tice)

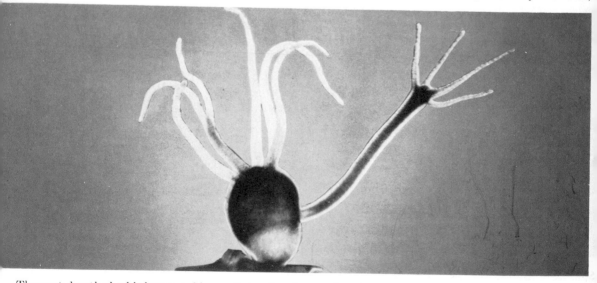

The next day the bud is larger and has well-developed tentacles. Its behavior is relatively independent of that of the parent. Here the parent is contracted, while the bud is extended. (Photo of living animal by P. S. Tice)

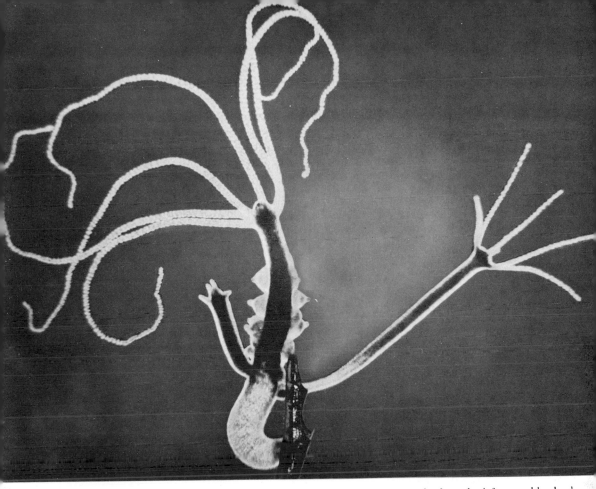

Hydras reproduce sexually as well as asexually. This male hydra has a young bud on the left, an older bud on the right, and rows of testes along the sides of the body. (Photo of living animal by P. S. Tice)

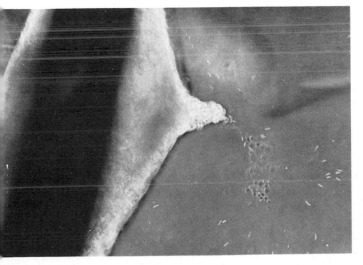

Sperms being discharged from a testis. The ectoderm is ruptured, liberating sperms which swim through the water to female hydras bearing ripe eggs. (Photo of living animal by P. S. Tice)

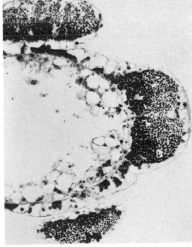

Three testes show in this cross-section. (Photo of stained preparation, courtesy Gen. Biol. Supply House)

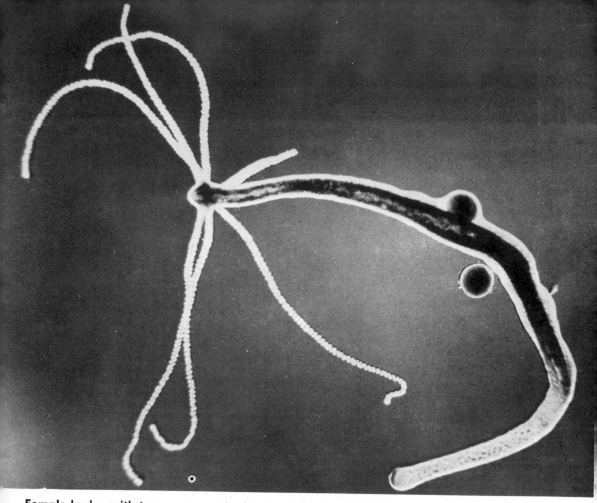

Female hydra with two eggs, one in the ovary and covered by the ectoderm, and one extruded. In this species the sexes are separate, but some hydras are hermaphroditic. (Photo of living animal by P. S. Tice)

An immature egg lying between ectoderm and endoderm is shown in this cross-section through a female hydra. As in the stained preparation on the preceding page, only about one-half of the complete section through the animal is shown. (Photo of stained preparation, courtesy General Biological Supply House)

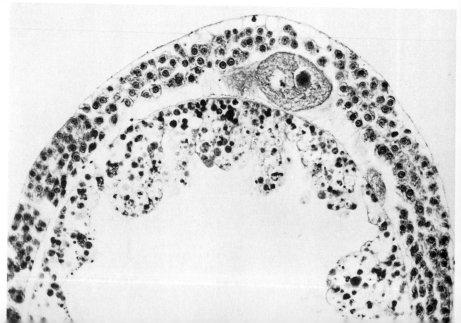

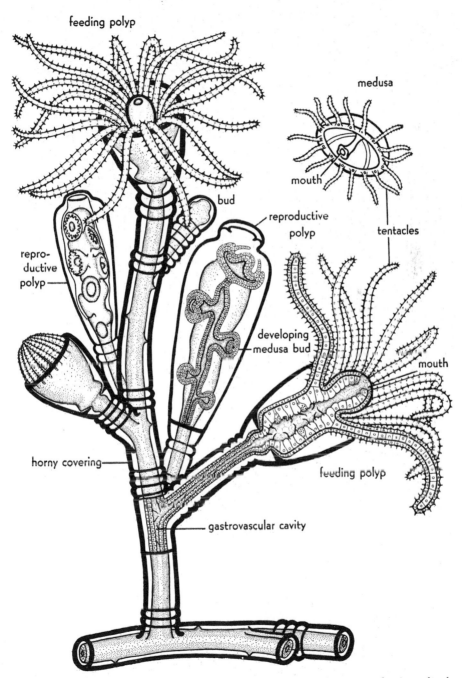

feeding polyp

medusa

mouth

tentacles

bud

reproductive
polyp

repro-
ductive
polyp

developing
medusa bud

mouth

horny covering

feeding polyp

gastrovascular cavity

Obelia, portion of a **colony,** and **medusa.** One feeding polyp and one reproductive polyp have
been drawn in section, showing the two layers of cells characteristic of coelenterates.

cells bear cilia, whose beating propels the little animal through the water. While the developing hydra was called an "embryo" because it developed first attached to the parent and later within a shell, a free-swimming young stage, like that of the obelia, is called a **larva.** The larva swims about for a time, finally settles on a rock or on a piece of kelp, becomes fastened at one end, develops tentacles and a mouth at the other end, and grows into a polyp which, by asexual budding, produces a new colony of sessile

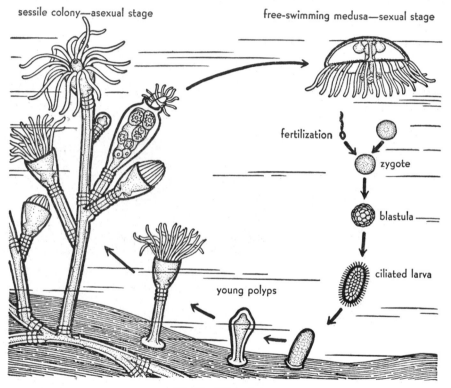

sessile colony—asexual stage   free-swimming medusa—sexual stage

fertilization   zygote

blastula

ciliated larva

young polyps

Obelia, life-cycle.

polyps. The larva and the free-swimming medusa both serve as a means of spreading the obelia to new localities.

Although a medusa seems very different from a polyp in general appearance, the two forms are really similar in construction. Both are composed of ectoderm and endoderm, and both consist of similar types of cells. In the medusa, ectoderm covers the entire surface of the bell and the tentacles, while the endoderm lines the various parts of the gastrovascular cavity. The chief difference from the polyp is the great thickness of the jelly between the two layers of cells. In the absence of supporting structures like the spicules of sponges or the connective tissue of higher animals, the jelly gives a firm consistency to the otherwise fragile body and adds to its buoyancy. Correlated with its greater locomotor activity, the medusa has a much more highly developed **nerve net** with a *marginal ring* of nerve cells acting as a controlling center. There are also more specialized sensory cells. The polyp and medusa may be regarded as having adapted the same general pattern to two different ways of life—attached and free-swimming.

The occurrence of coelenterates in two forms, the medusa and the polyp, is a phenomenon termed **polymorphism** (meaning "many forms"). Polymorphism is not unique among coelenterates, for many other groups,

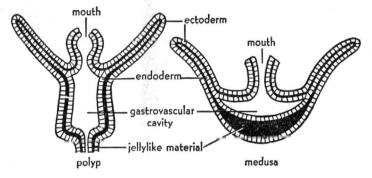

The polyp and medusa are constructed on the same general plan.

notably the social bees and ants, show structural differentiation of individuals which fits them for different roles in the life of the species. The ordinary polyps of the obelia colony capture and digest food and are capable only of asexual reproduction; they have been called feeding polyps. The stalks which bud off medusas are specialized reproductive polyps. The medusa is the sexually mature stage of the obelia and also serves to spread the species. Therefore, the obelia may be said to consist of three kinds of individuals: the *feeding polyps*, the (asexual) *reproductive polyps*, and the (sexual) *reproductive medusas*, each with its own functions. The work that in most other animals is done by every individual of the species is performed in these coelenterates by different kinds of individuals. In some colonial coelenterates, still other kinds of individuals are found, such as protective polyps which do not feed or reproduce but are heavily armed with stinging capsules.

THE life-history of coelenterates may be interpreted as follows: The ancestral coelenterate was a medusa-like form which produced other medusas directly from eggs and sperms, by way of a larval stage resembling a polyp. Eventually this larval polyp-like stage became more and more important in the life-history and took on an independent existence. At first the polyps were incapable of sexual reproduction and would grow up into medusas which produced eggs and sperms. But some coelenterate polyps eventually developed the capacity for forming sex cells and dropped out the medusa stage altogether, as has happened in hydras.

The alternation of a polyp and a medusa stage has been called "alternation of generations" or "metagenesis." However, it seems better not to make a distinct principle of this phenomenon. The polyp colony is simply a juvenile stage, the medusa the fully adult form, just as a caterpillar is a juvenile form capable of carrying on all activities except sexual reproduction, and the butterfly is the adult sexual form. If the caterpillar could develop mature sex organs, the butterfly stage could be dropped out of the life-history. Such a process of sexual maturity of juvenile forms is actually known to occur in quite a variety of animals besides the coelenterates, even in animals quite high up in the scale of life, such as salamanders.

THE coelenterates have only two layers of cells, the ectoderm and the endoderm. However, they show several advances in structure over the sponges because they have a network of nerve cells for co-ordinating cell activities and because they are organized on the tissue level of construction. Their most characteristic structures are the thread capsules, which are produced in no other phylum of animals. The variations among coelenterates result chiefly from a shifting emphasis in the evolution of the several groups—first upon the medusa stage, and then upon the polyp stage of the life-cycle. Some, like the obelia, have polyp and medusa almost equally developed; in others the medusa stage is more or less degenerate or wholly suppressed, as is believed to be the case with the hydra; and in others, large and well-developed medusas predominate while the fixed polyp stage is greatly reduced or has disappeared. Some of these variations are described in the next chapter.

CHAPTER 8

POLYPS AND MEDUSAS

THOSE who have never seen the ocean can hardly be impressed with the coelenterates as a phylum. They have seen only the diminutive hydra and probably not at all the relatively rare fresh-water medusas. When one first walks down to a rocky ocean shore and looks in the shallow pools left by the outgoing tide, he is amazed at the display of anemones—huge, brightly colored polyps that look more like flowers than like animals; from the algae and from rock surfaces hang delicate fringes of white, pink, or violet hydroids, like the obelia; in the open water just off shore large jellyfish drift by. But it is in warm shallow seas that the coelenterates really come into their own. There, towering growths of colonial polyps and massive banks of reef-building corals occupy almost every available square foot of the bottom, replacing the plants and dominating the lives of the other invertebrates and even of the fishes, as the trees in a forest dominate the other plants and the animals.

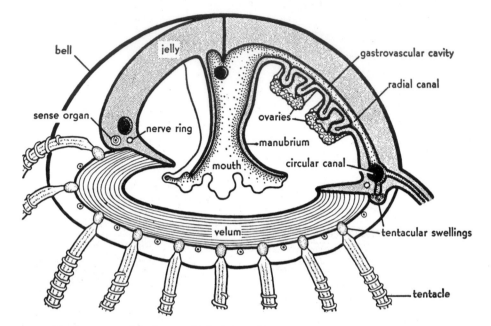

Gonionemus, with about a third of the bell removed to show internal structures.

Coelenterates occur in two forms: polyps and medusas. The obelia, described in the last chapter, shows both forms but is more conspicuous in the polyp phase. Yet it was suggested that the primitive coelenterate was

Gonionemus, natural size.

probably a kind of medusa and that the polyp was at first only a young transitory stage in the development of the medusa. Now we come to *Gonionemus*, a jellyfish that is thought to be close to the ancestral coelenterate. It has a well-developed medusa and a simple inconspicuous polyp.

The gelatinous bowl-shaped bell of Gonionemus has a convex outer surface and a concave under surface. From the center of the concave surface hangs a tube, the **manubrium**, with the **mouth** at its tip. The other end of the manubrium leads into four **radial canals** that traverse the jelly to the margin of the bell. There they join a **circular canal** which runs around the margin and connects with the cavities of the hollow tentacles. This continuous cavity through manubrium, radial canals, circular canal, and tentacles is the **gastrovascular cavity;** it distributes partly digested food to all regions of the body.

The medusa **swims** slowly by rhythmic pulsations of the bell. Under the ectoderm are specialized muscle fibers which do not have a protective function and serve only for contraction. The muscular part of each individual muscle cell is very much elongated; and the epithelial part, which in hydra was so prominent, is inconspicuous here. From the margin of the bell, a muscular shelf, the **velum,** projects inward. The velum contracts strong-

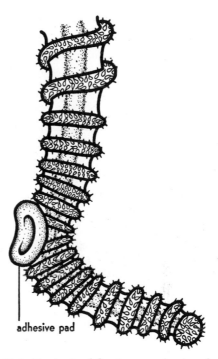

adhesive pad

End of **tentacle of** *Gonionemus* showing adhesive pad and batteries of stinging capsules.

ly; and this, together with the contraction of muscle fibers in the bell, forces water out from the concavity of the bell and drives the animal in the direction opposite to that in which the water is expelled.

Gonionemus **feeds** while actively swimming about or by a kind of "fishing" technique. The medusa swims upward, turns over on reaching the surface of the water, and then floats slowly downward with the bell inverted and the tentacles extended horizontally in a wide snare, from which passing worms, shrimps, or small fish seldom escape. When at rest, the medusa attaches to the bottom or to vegetation by the **adhesive pads** near the tips of the tentacles.

The free-swimming habit of Gonionemus requires greater activity and a more elaborate nervous mechanism than that of sedentary polyps like the hydra and the obelia. The **nerve network** which runs beneath the ectoderm of the bell surface is concentrated around the margin of the bell into a **nerve ring,** which controls and co-ordinates the animal's behavior. Specialized **sense organs** occur around the bell margin, imbedded in the jelly between the bases of the tentacles. Each sense organ is a small sac containing a hard mass. As the medusa swims about, the movements of the mass within the sac probably act as stimuli which control and direct the swimming movements. In addition, the prominent swellings at the bases of the tentacles are abundantly supplied with sensory cells. These tentacular swellings are hollow and communicate with the ring canal. They are lined with pigment, which may be related to a special light-receptive function, although the whole epithelium of the lower surface is generally sensitive to light. These swellings probably serve chiefly as a site for the formation of stinging capsules, which from there migrate out along the tentacles and take their places in the stinging batteries.

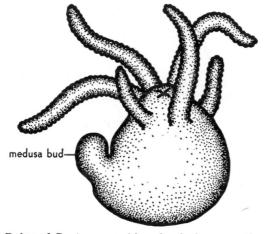

medusa bud—

Polyp of *Gonionemus* with medusa bud, very greatly enlarged. (Based on Joseph)

The **ovaries** or **testes** occur on separate female or male individuals and appear as folded ribbons which hang from beneath the four radial canals. The eggs or sperms break through the surface ectoderm and are shed directly into the water, where fertilization occurs. The zygote divides many times to form a ciliated larva, called a **planula,** a name applied to any ciliated coelenterate larva. The planula has an outer layer of ectoderm and an inner mass of endoderm, and thus corresponds to the gastrula stage in development. It swims about for a time, then finally settles down, loses its cilia, and develops an internal cavity. A mouth breaks through at the unattached end, tentacles push out around the mouth, and the young Gonionemus soon resembles a minute, squat hydra. The polyp feeds and buds off little larvas, which also become feeding polyps. Finally, the polyps produce medusa buds which detach and grow directly into adult medusas.

The life-history of Gonionemus resembles that of the obelia, with the emphasis shifted to the medusa. Many jellyfish closely related to Gonionemus have no attached polyp at all. One of these, *Liriope*, has a larva that sprouts its tentacles before the bell develops, and it looks like a free-swimming polyp. If this larva were to settle down and become attached, it would resemble the minute polyp stage of Gonionemus. In some coelen-

terates this juvenile, fixed polyp stage has been
elaborated into a relatively large and flourishing colony, while the medusa is very small, as in
the obelia. In other hydroids the medusa stage
has been reduced still further. The medusa buds
of *Hydractinia*, for example, begin to develop
in the usual way but never have any medusa-
like features, and fail to detach from the colony. They are degenerate saclike structures
that shed eggs and sperms into the water. The
final step in this direction is the complete elimination of the medusa—a condition illustrated
by the hydra.

In the obelia we saw division of labor not only
in the life-history but also in the composition of
the polyp colony. *Hydractinia* carries this **polymorphism** a step farther. The colony has *feed-*

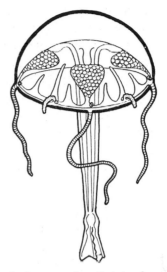

Liriope, a medusa that develops
directly from a zygote without a
fixed polyp stage. (After Mayor)

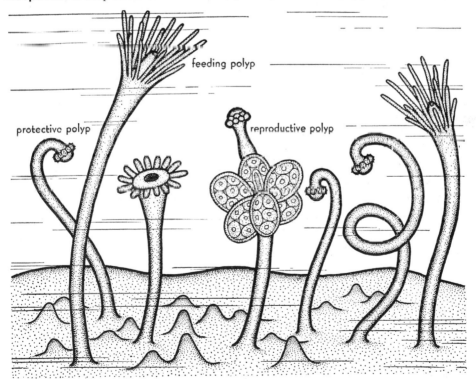

feeding polyp

protective polyp

reproductive polyp

Hydractinia, a polyp colony that shows polymorphism and degenerate medusa buds. (Modified
after Allman)

ing polyps, with mouths and long tentacles; *reproductive polyps*, which have degenerate tentacles but bear the saclike medusa buds; and *protective polyps*, which have short knoblike tentacles and cannot feed but are richly supplied with stinging capsules that serve to protect the colony or to aid in paralyzing prey.

It is among the **siphonophores,** however, that we find the extremes to which colonial organization can be carried. These complex floating colonies have not only more than one kind of polyp but also more than one kind of medusa. In addition to the sexual medusas (either free or attached) they may have numerous modified medusas, called "swimming-bells," which cannot feed or reproduce but serve only to propel the colony. There are also leaflike structures that hang as protective flaps over the various members, and often a gas-containing float which can be filled or emptied to adjust the level of the colony in the water. *Physalia*, often called the "Portuguese man-of-war," has no swimming-bells and is driven about by the action of the wind on its crested float. From the under side of the float there hang down into the water several kinds of specialized polyps, clusters of attached medusas, and tangles of long tentacles that may reach a length of 60 feet and are armed with especially large stinging capsules that can readily paralyze a large fish. The vivid blue float is a familiar and a very beautiful sight on the surface of warm seas all over the world—but it is not a very welcome one to swimmers, who know that the trailing tentacles can inflict serious and sometimes fatal injury on man.

Velella ("little sail") is a common **chondrophore** and whole "fleets" of them may be driven into a bay by a strong wind. The velella looks like a single flat oval polyp with an erect sail-like projection from the upper surface. Around the margin is a single row of stinging protective tentacles. The under surface is covered by numerous reproductive stalks (each with a mouth) which bud off free-swimming medusas; in the very center hangs a single, large central mouth.

In many animal groups there are marked differences between male and female individuals, and in termite colonies between individuals specialized for feeding, reproduction, and defense. These are all separate organisms—discrete physiological units. In the siphonophores it is frequently difficult to tell where one polyp ends and the next begins; and they are all closely knit into one complex organism, almost like the various organs in the body of man.

T HE phylum COELENTERATA is divided into three classes. All of the animals mentioned thus far belong to the class **Hydrozoa** ("hydra-like animals"). Most hydrozoans are polyp colonies that give rise to free medusas (*Obelia*) or to degenerate attached medusas (*Hydractinia*). But there are exceptions, like the hydra and *Gonionemus*. Further, both polyps and medusas are small, delicate, and much simpler in structure than the members of the other two classes.

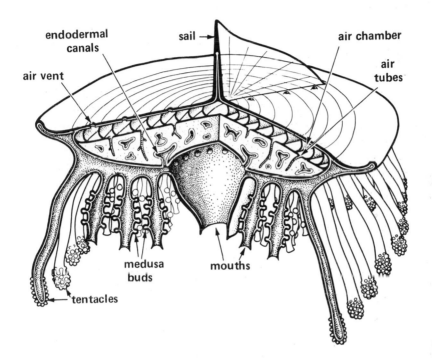

Velella, a large polyp (often up to 3 inches in its widest diameter), floats at the surface. The front portion has been cut away to reveal the gastrovascular cavity, budding stalks, and endodermal canals. The large central mouth and smaller mouths at the tips of the budding stalks all ingest prey. (Modified after Delage and Herouard)

THE second class of coelenterates, the **Scyphozoa** ("cup animals"), includes the larger jellyfish. All are marine medusas and can be roughly distinguished from the hydrozoan jellyfish by their large size and by the absence of a velum. Moreover, the polyp stage either is lacking altogether or is very small.

Aurelia is one of the commonest of the scyphozoan jellyfish and occurs all over the world. From ocean liners one often sees large shoals of them drifting along together or swimming slowly by rhythmic contractions of the shallow, almost saucer-shaped bell. They range in size from less than

3 inches to about 12 inches across the bell. Exceptional individuals may reach a diameter of 2 feet.

At the end of a very short manubrium is a square mouth, the corners of which are drawn out into four trailing **mouth lobes.** Each lobe has a ciliated groove. Stinging capsules in the lobes paralyze and entangle small

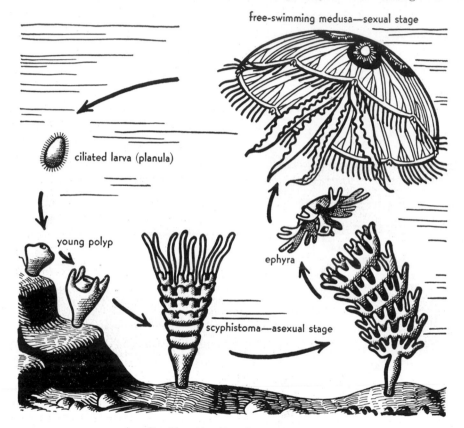

Aurelia, life-cycle. (Based on various sources)

animals, which are then swept up the grooves, through the mouth, into a spacious cavity in the center of the bell, and from there through eight radial canals to the margin of the bell. Flagella lining the entire gastrovascular cavity maintain a steady current of water, which brings a constant supply of food and oxygen to, and removes wastes from, the internal parts of this large animal.

The central cavity extends into four pouches, in which there are tentacle-like projections of the endoderm, called **gastric filaments.** These are covered with stinging capsules which paralyze prey that arrives in the pouches still alive and struggling. The presence of these filaments is one of the characters that distinguishes a scyphozoan from a hydrozoan medusa.

The margin of the bell bears a fringe of short and very numerous tentacles, set closely together except where they are interrupted by eight equally spaced notches. In each notch lies a **sense organ,** consisting of a pigmented eyespot, sensitive to light; a hollow sac, containing hard particles whose movements set up stimuli that direct the swimming movements; and two pits, lined with cells that are thought to be sensitive to food or to other chemicals in the water.

The four horseshoe-shaped colored bodies by which Aurelia is usually recognized are the **testes** or **ovaries,** which occur (in separate individuals) on the floor of the large central part of the gastrovascular cavity. In a male medusa the sperms are discharged into the cavity and are shed to the outside through the mouth. In the female the eggs are shed into the cavity and are fertilized there by sperms which enter with the food current. The fertilized eggs go out of the mouth and lodge in the folds of the mouth lobes, where they continue to develop. The ciliated larva escapes and eventually settles down on a rock or on seaweed. There it grows into a small polyp with long tentacles. The polyp feeds and stores food, and may survive in this stage for many months, meanwhile budding off other small polyps like itself. At certain seasons, usually fall and winter, it develops a series of horizontal constrictions which gradually deepen until the polyp resembles a pile of saucers. One by one the "saucers" pinch off from the parent and swim away as little eight-lobed medusas, which gradually develop into adult aurelias.

This type of development in which the fixed polyp stage (hydratuba) becomes an elongated and deeply constricted polyp (scyphistoma), which successively splits off young eight-lobed medusas (ephyras), is characteristic of Scyphozoa and does not occur in the Hydrozoa.

The stinging capsules of Aurelia do not readily penetrate the human skin; but even a small *Cyanea* can raise huge weals on the arms or legs, and the monster orange and blue cyaneas of the North Atlantic are a real danger to swimmers. The largest one on record had a disk 12 feet in diameter and trailing tentacles over 100 feet long. Such huge masses of jelly are among the largest of the animals without backbones. Their long tentacles are probably responsible for some of the "sea-serpent" stories.

THE third class of coelenterates, the class **Anthozoa** ("flower-ani-
mals"), consists of marine polyps which have no medusa stage. Anthozoa are technically distinguished from hydrozoan polyps by the fact that
the gastrovascular cavity is divided up by a series of vertical partitions,
and the surface ectoderm turns in at the mouth to line the gullet. But
superficially there is no difficulty in telling the large fleshy sea anemones

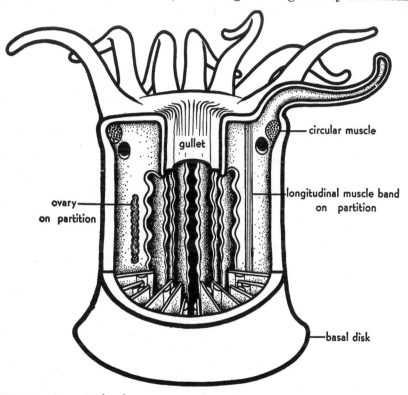

Sea anemone cut away to show large gastrovascular cavity and the many partitions. The free edges
of the partitions are thickened and bear gland cells.

or the limestone-secreting corals from most of the small fragile hydrozoan
polyps.

The **sea anemone** has a stout muscular body, the **column,** expanded
at its upper end into an **oral disk** having a central **mouth** surrounded
by several circlets of **hollow tentacles.** The basal end forms a smooth,
muscular, slimy **basal disk** on which the anemone can slide about very
slowly and by which it holds to rocks so tenaciously that one is likely to
tear the animal in trying to pry it loose.

From the mouth a muscular gullet hangs down into the gastrovascular cavity and is connected with the body wall by a series of vertical **partitions.** Between these primary partitions is a secondary set of incomplete ones, which extend only part way from the body wall to the gullet; and between these are still less-complete tertiary and sometimes quarternary sets. The chambers between the primary partitions are in open communication with each other below the gullet; but above the point at which they attach to the gullet, they communicate only through one or more holes in the wall of each partition (these are shown but not labeled in the diagram of the anemone).

The partitions are double sheets of endoderm supported by a central layer of jelly, and they serve to *increase the digestive surface of the cavity,* making it possible for an anemone to digest a relatively large animal, like a fish or a crab. The free edges of the partitions are expanded into convoluted thickenings or **digestive filaments** which bear the gland cells that secrete digestive juices. Digestion is completed, as in the hydra, by pseudopodal endoderm cells lining the gastrovascular cavity.

The gullet is not cylindrical but is flattened, and at one or both ends of its long diameter is a longitudinal groove lined with flagella that are much longer than the ones lining the rest of the gullet. The flagella in these **gullet-grooves** beat downward, drawing a current of water into the gastrovascular cavity and providing the internal parts of the anemone with a steady supply of oxygen. At the same time, the flagella lining the gullet proper beat upward, creating an outgoing current of water that takes with it carbon dioxide and other wastes. When small animals touch the tentacles, the flagella of the gullet proper reverse their beat, and the food is swept down the gullet and into the digestive cavity.

Anemones are among the most highly specialized of the polyp types of coelenterates. They have a well-developed **nerve net,** mesenchyme cells between ectoderm and endoderm, and several sets of specialized **muscles.** A layer of circular muscles serves to narrow, and therefore to extend, the body. The longitudinal muscles are concentrated into prominent bands which run, one on each partition, from mouth to basal disk. Their contraction pulls the mouth disk, with its tentacles, completely inside. A strong circular muscle just below the mouth disk then closes over the opening, much as a pouch is drawn closed by a string. In the contracted condition anemones resist drying or mechanical injury during low tide.

Anemones sometimes **reproduce asexually** by pulling apart into two halves. In certain species, owing either to injury or simply to poor co-ordination, pieces of the body are left behind as the animals slide about. These fragments regenerate into tiny anemones. In **sexual reproduction** the eggs or sperms form on the partitions of the gastrovascular cavity and are ejected through the mouth, or the eggs may be fertilized and brooded internally. The fertilized egg develops into a planula, which finally settles down in some rocky crevice and grows into a single anemone.

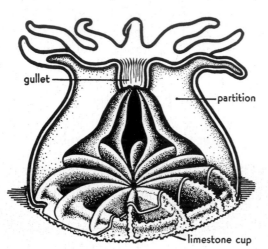

Stony coral polyp. The animal has been cut away to show the gastrovascular cavity and, beneath the polyp, the beginning of the formation of the stony cup. (After Pfurtscheller)

THE **stony corals** are like small anemones but are usually **colonial** and secrete protective **limestone cups** into which the small delicate polyps can retract. From the wall of the cup a series of radially arranged vertical plates project inward and alternate with the digestive partitions. The stony cup and its plates are, of course, outside the polyp and merely in contact with the ectoderm which secretes them.

Many kinds of small solitary cup corals grow in the temperate waters along our coasts. Colonies of five to thirty individuals of *Astrangia* incrust shells and rocks as far north as Cape Cod. And even the cold, deep waters of the Norwegian fjords support great banks of a colonial branching coral, *Lophohelia*. But the reef-building cor-

A colony of **Astrangia** from the Cape Cod region.

als build reefs only in tropical or subtropical waters, where the temperature never falls below 70° F. Where the reef is exposed to strong wave action, the corals grow mostly as flat or rounded masses or with short branches. In the sheltered waters behind the reef front are found the taller corals with longer, more slender branches.

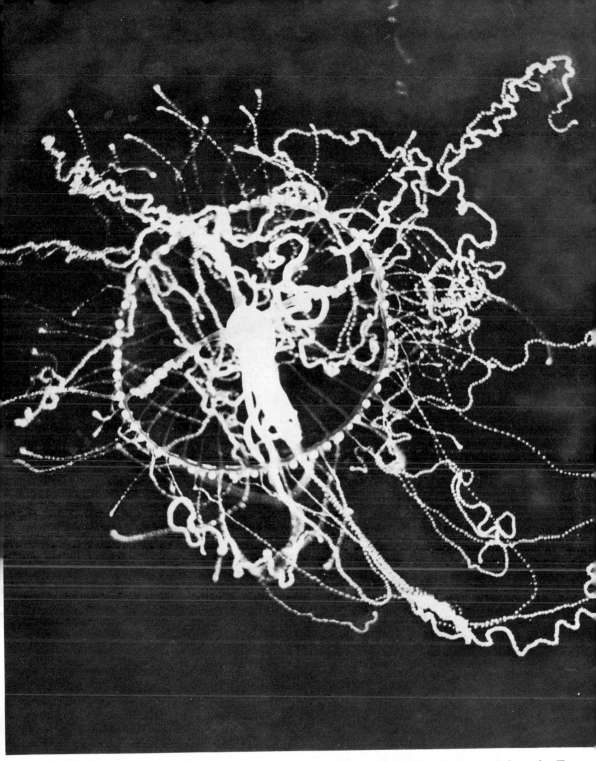

The most primitive living coelenterates are thought to be the small hydrozoan medusas of the order Trachylina, to which *Olindias*, shown here, belongs as do Gonionemus and Liriope, presented earlier. *Olindias* resembles Gonionemus very closely but differs in having two sets of tentacles, one short and one long. The medusa has just swallowed a small fish, the eye of which can be seen. (Photo of living animal, $2 \times$ natural size. From shallow water, Bermuda)

Polyorchis, a hydrozoan jellyfish common in bays along the West Coast. The mouth is at the end of the long trumpet-shaped stalk. The stringlike structures around the mouth stalk are testes or ovaries. At the base of each tentacle is a sense organ. (Photo of living animals, about natural size. Monterey Bay, California)

Fresh-water jellyfish *Craspedacusta*, from a lake in Ohio. It has the habit of swimming to the surface of the water (*above*) and then coasting down (*below*) with tentacles outspread and forming a trap. (Photos of living animal about natural size from a motion picture, courtesy James A. Miller)

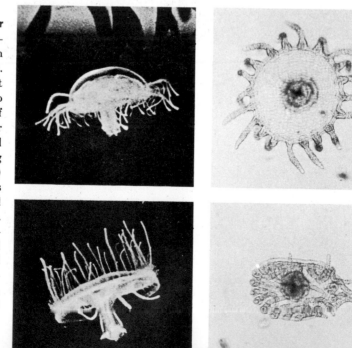

Obelia jellyfish is only about 1/25 inch in diameter. This one was set free when a bit of an obelia colony was placed in a drop of sea water on a slide. The number of tentacles increases as it grows. *Above:* bell expanded; *below:* bell contracted. (Photos of living animal. Pacific Grove, California)

Hydroids are of two main types. Tubularian hydroids have tubular horny coverings encasing only the stems, from the free ends of which emerge the naked polyps bearing tentacles. *Tubularia larynx*, (*above*), lives attached to piles or floating buoys, to rocks below low-water mark, and to mussel shells. The campanularian hydroids shown in both photos (*below*), have a more complete horny covering which extends up around the polyps as bell-shaped cups into which the polyps can withdraw. (Photo of living colony, 1.5 × natural size, by D. P. Wilson. Plymouth, England)

Ostrich-plume hydroid (*Aglaophenia*) is commonly found on beaches, where it is cast up along with sea-weeds. About ½ natural size. (Pacific Grove, California)

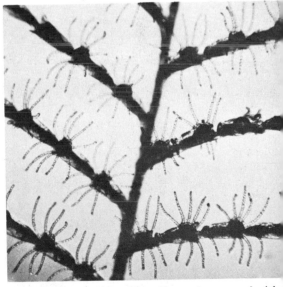

Portion of hydroid colony. Tentacles covered with nematocysts give small animals little chance of escape. (Photomicrograph of living colony. Bermuda)

Physalia, often called the "Portuguese man-of-war," is a member of the Siphonophora, an order of colonial coelenterates. The most prominent of the subindividuals is the modified medusa that forms the large, crested, gas-filled float. The action of the wind on this float carries the colony along, and the trailing tentacles, armed with stinging capsules, paralyze and entangle any small animals they touch. The colony shown here has just caught a fish. About ½ natural size. (Photo, courtesy New York Zoölogical Society)

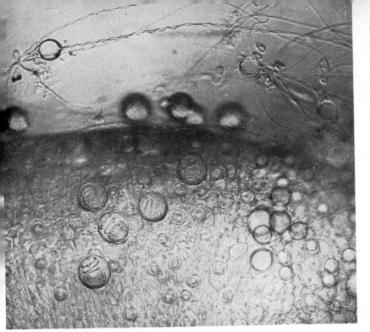

Portion of a tentacle of *Physalia* seen through a microscope and enlarged 180 times. The bottom half of the microscope field is occupied by the edge of the tentacle, whose surface can be seen to be almost solid with undischarged, and a few discharged, thread capsules of several kinds. The largest ones clearly show the hollow coiled thread within. Above are a number of discharged stinging capsules with long hollow threads extending out of the field. The unusual size of these capsules, and the great length of their threads, suggests why Physalia is the most dangerous to man of all coelenterates. (Photomicrograph by Douglas P. Wilson. Plymouth, England)

The feeding polyps of a *Physalia* can be seen stretching down, with their thin transparent lips spread over the surface of the 4-inch fish and meeting edge to edge so as to inclose the fish completely. From their mouths digestive fluids are poured onto the fish, disintegrating its substance so that it can be sucked up into the feeding polyps and later distributed to all members of the colony. (Photo of living Physalia by D. P. Wilson. Plymouth, England)

A jellyfish swims by alternately relaxing and contracting the bell. In this photograph the bell is contracted, forcibly expelling the water from its concavity and so pushing the animal in the direction opposite to that in which the water is expelled. On the page opposite, the bell is relaxed to admit water again.

Chrysaora hysocella, usually called the "compass jellyfish" because of its markings, is one of the commonest of scyphozoan medusas. It occurs in great numbers, toward the end of summer, along the Atlantic coast of Europe. Related species are found on our coasts. (Photos of living animal by F. Schensky. Helgoland)

Aurelia is one of the commonest and most widespread of all scyphozoan jellyfishes. It occurs in the coastal waters of all seas. In colder waters it has been seen in shoals so dense that the water appeared to be almost solid with them. It is easily recognized by the shallow, saucer-shaped bell, the fringe of numerous small tentacles, and the four large horseshoe-shaped gonads. Here viewed from the oral surface, one can see the four highly transparent mouth lobes whose cilia maintain a steady food-bearing current of minute organisms into the centrally located mouth. (Photo of living animal, natural size, by D. P. Wilson. Plymouth, England.)

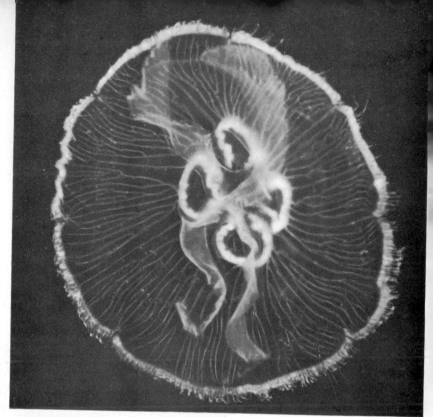

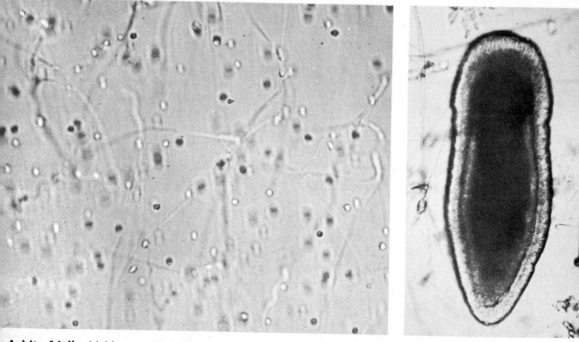

A bit of jelly, highly magnified, from the bell of an *Aurelia*. The streaks are fibers that strengthen the jelly. The dots are ameboid cells, whose presence in large numbers in the jelly distinguishes a scyphozoan from a hydrozoan medusa. The jelly is approximately 95% water, 4% salts, and 1% organic matter. (Photomicrograph of living jelly. Bermuda)

A planula is the ciliated free-swimming larva of most marine coelenterates, including Aurelia. (Photomicrograph of living animal. Pacific Grove, California)

Fixed stages of Aurelia growing on a hollowed stone are typical of scyphozoan development. The planula settles on a rock and develops into a small polyp with long tentacles. All summer it feeds and grows and buds off small polyps. In winter and spring it elongates, loses its tentacles, and undergoes a series of successive constrictions. At first these are shallow grooves, but they gradually deepen until the constricted polyp, or scyphistoma, resembles a pile of saucers, each one a developing jellyfish. One by one, the young jellyfishes break away. They feed and grow all spring and summer and by August are adult male and female aurelias. Then the cycle starts over again. *Above:* Several small polyps and many scyphistomas in various stages of constriction, or strobilation. An ephyra (young jellyfish) released only a few seconds before is swimming away ($2\frac{1}{2} \times$ natural size). *Below:* Three scyphistomas in successive stages of strobilation ($2 \times$ natural size). (Photos of living animals by D. P. Wilson. Plymouth, England)

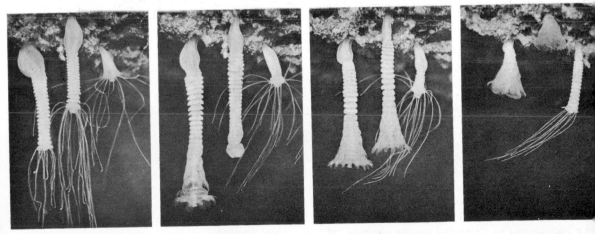

February 24   **March 3**   **March 10**   **March 20**

Jellyfishes are often washed up on the beach in great numbers during a storm. A large jellyfish is surprisingly stiff; one can jump on such an animal without crushing it. (Photo by W. K. Fisher. Pacific Grove, California)

Cassiopeia is a scyphozoan jellyfish in which there are no tentacles and the mouth lobes are divided and their grooves narrowed or fused. The original mouth opening is small or absent, and food organisms are taken in through numerous small mouths on the mouth lobes. Cassiopeia has the lazy habit of lying in shallow water with the mouth side up. This exposes the algae, which live in the mouth lobes, to the sunlight; the relationship is supposedly one of mutual benefit. Natural size. (Photo of preserved specimens from Florida, courtesy A. Novak and E. M. Miller)

Sea anemones, so called from their resemblance to flowers, are among the most familiar coelenterates because they are easily seen in tide pools on rocky shores. The one on the right has just captured a small fish, which is held by thread capsules on the tentacles. (Photo of living animals by F. Schensky. Helgoland)

Sea anemones fill every crevice in the tide pools. Animals like these have been observed to occupy the same spot for over thirty years. (Photo of living animals by W. K. Fisher. Pacific Grove, California)

Anemone dividing (*Metridium*). After a time the constriction will extend to the base, and the two anemones, asexually produced, will then separate. (Photo of living animal by D. P. Wilson. Plymouth, Eng.)

Metridium has extremely numerous delicate tentacles with which it catches minute organisms instead of larger animals like fish, on which other anemones, like the one on the lower right, feed. A large *Metridium* may attain a height of more than 8 inches. The animal does not remain always in one spot but can glide on the slimy disk. One was seen to move 1½ feet in 24 hours. (Photo of living animals by F. Schensky. Helgoland)

Anemones expanded (*Metridium*) in a darkened aquarium. Most anemones are negative to light and in nature expand fully and feed only in partial or complete darkness.

Same anemones contracted after being illuminated. When contracted, they can resist mechanical injury or drying during low tide. (Photos of living animals. Mount Desert Island, Maine)

Oral disk of *Metridium* shows the elongated mouth and the gullet groove (light-colored) at one end. Both are modifications of the basic radial symmetry of coelenterates (see chapter 11).

Basal disk of *Metridium* (seen through glass wall of aquarium) emphasizes the primary radial symmetry of coelenterates. (Photos of living animal. Mount Desert Island, Maine)

Solitary corals. *Left:* The polyp, *Balanophyllia*, has a stony cup about ¼ inch in diameter. The spots on the tentacles are batteries of stinging capsules. (Photo of living animal. Pacific Grove, California.) *Right:* Radial symmetry of coelenterates is strikingly shown by the empty cup of a small solitary coral seen from above. In the living animal the stony plates support delicate partitions which divide up the gastrovascular cavity.

A coral colony (*Astrangia*) arises by budding from a single individual. Each polyp is ⅜ inch high. *Astrangia* can grow in the cold waters off Cape Cod. (Woods Hole, Mass., courtesy American Museum of Natural History)

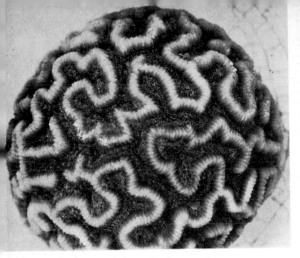

Live brain coral with polyps retracted as it appears in the daytime. Most corals remain contracted during the day and only expand and feed at might. A closeup of this same coral colony at night is shown below. (Bermuda)

Cleaned skeleton of a brain coral shows the sinuous grooves, which the live polyps formerly occupied, and the delicate stony partitions. Even such a small coral as this is the accumulated deposit of many generations of polyps. (Slightly less than natural size)

Live brain coral, with partly expanded polyps, at night. The polyps do not occupy separate cups but are continuous with each other, except for separate mouths which lie at intervals along the bottom of the groove. The tentacles lining the sinuous grooves were outspread until the light was turned on to make the exposure. (Bermuda)

Coral blocks for building houses in Bermuda are cut from the hilltops with a saw. The material consists of sand, mollusk shells, and coral skeletons ground fine by wave action, deposited on the beach, and then cemented together to make the soft "coral rock" which forms the surface layers of the island.

Great Barrier Reef of Australia at low tide. The reefs, which form an underwater barrier 1,260 miles long, many miles wide, and at least 180 feet high, are a serious hazard to ships. Most of this limestone has been deposited by countless small, delicate polyps. This view is unusual in that it consists almost entirely of a single type, the stag's-horn coral (*Madrepora*). (Photo from W. Saville-Kent)

Coral skeletons form the Great Barrier Reef. In life they differ even more because the polyps show a striking variety in form and in their brilliant colors. These dried and bleached skeletons are beautiful and are used as ornaments. But they give about as good an impression of the exquisite beauty of living, expanded corals as one would get of the beauty of a woman from her whitened bones. (Photo from W. Saville-Kent)

Coral reef exposed at low tide. Black-and-white photographs give a poor impression of the beauty of such reefs. The corals are all colors, from delicate blues and pinks to yellow-green, bright green, violet, or brown. (Photo taken on Great Barrier Reef by Mrs. C. M. Yonge)

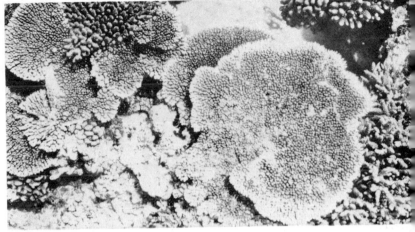

Close-up of coral reef seen above. Only the skeletons show; the delicate polyps are all retracted. They usually expand at night, spreading their tentacles to trap the small animals on which they feed. (Photo by Mrs. C. M. Yonge)

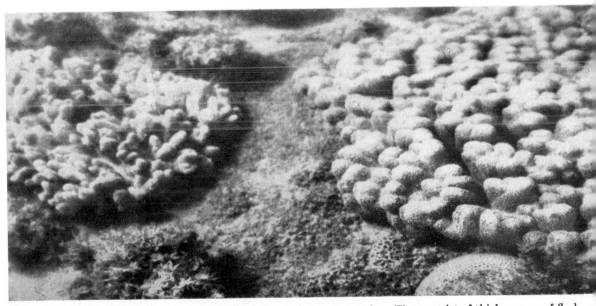

Soft corals are alcyonarians, coelenterates with eight feathery tentacles. They consist of thick masses of flesh, toughened by the particles of limestone imbedded in them, and bear small delicate polyps. When the tide is out and the polyps are retracted, they look like flabby masses of leathery seaweed. (Photo taken looking down through shallow water on the Great Barrier Reef by Otho Webb)

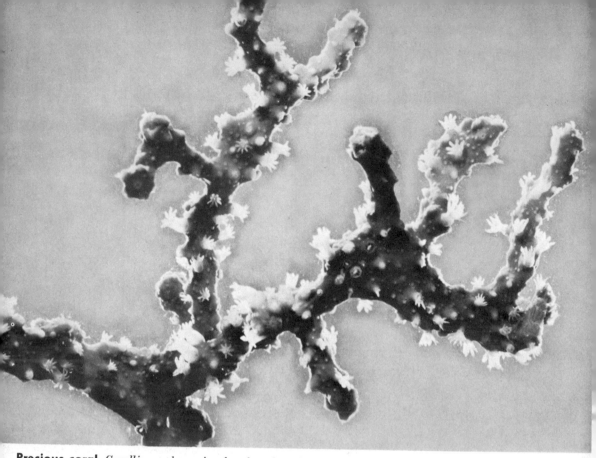

Precious coral, *Corallium rubrum,* is related to the soft corals, the gorgonians, and the organ-pipe coral, as can be seen in the closeup of a single polyp (*below, left*) which shows the eight feathery tentacles characteristic of alcyonarians. The tissues of the precious coral are stiffened by limestone particles which are fused in the center of the colony into a solid core of red limestone, often used in making jewelry. (Photo by P. S. Tice)

Polyp of precious coral, greatly enlarged, shows the eight feathery tentacles surrounding the elongated mouth.

Sea fan, *Eunicella verrucosa,* has branches all in one plane. $\frac{1}{5}$ natural size. (Photo by D. P. Wilson. Plymouth, England)

Portion of sea fan (*Eunicella verrucosa*) with fully expanded polyps, 2 × natural size. As in the precious coral, the polyps have eight feathery tentacles and share the food they catch with the other members of the colony. The whole colony appears on the opposite page. (Photo of living colony by D. P. Wilson. Plymouth, England)

Gorgonians, often called "sea whips," spring like shrubbery from among the rounded masses of corals on the ocean floor. (Photo taken off the Bahama Islands by Johnson, courtesy Mechanical Improvements Corp.)

Making an underwater photograph. The camera is inclosed in a watertight box held by the man in the diving helmet. (Photo by Johnson, courtesy Mechanical Improvements Corp.)

Sea whip brought up from 20 feet of water. The dark spots represent the positions of retracted polyps. The skeleton is not rigid as in stony corals but is quite flexible. About ⅓ natural size. (Photo of living colony. Bermuda)

Three main types of coral reefs are recognized. **Fringing reefs** grow in shallow water and border the coasts closely or are separated from them at the most by a narrow stretch of water that can be waded across when the tide is out. **Barrier reefs** also parallel coasts but are separated from them by a channel deep enough to accommodate large ships. Captain Cook sailed within the Great Barrier Reef of Australia for over 600 miles without even suspecting its presence until the channel narrowed and he was wrecked on the reefs. **Atolls** are ring-shaped coral islands inclosing central lagoons, and thousands of them dot the

South Pacific. Atolls are hundreds or thousands of miles from the nearest land, and their steep outer sides slope off into the depths of the ocean.

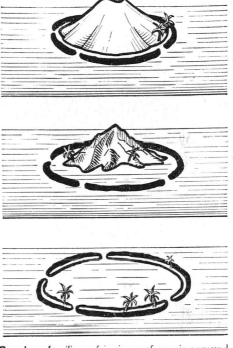

Darwin noticed that all the known coral reefs were in regions where a sinking of the land was known to have taken place or where there were evidences that it probably occurred. He reasoned that, if an island, surrounded by a fringing reef, were to subside very slowly, so slowly that the reef could grow upward at about the same rate, the island would grow smaller and smaller, and the fringing reef would become separated from it by a wide, deep channel, finally becoming converted into a barrier reef. If this process were to continue, the island would finally disappear entirely beneath the surface of the water, and the rising barrier reef would become a ring-shaped island, or atoll. This theory is still the most widely accepted one, though there are others almost as well in accord with known facts. In some cases an atoll may have been formed directly, without going through a fringing reef and a barrier stage, upon a submarine platform built up close to the surface by volcanic activity.

Coral reefs. *Top,* a fringing reef growing around an oceanic island. *Middle,* a small barrier reef widely separated from the island. *Bottom,* an atoll. (Based on various sources)

THE sea anemones and corals are zoantharians, anthozoans in which the tentacles and internal partitions are numerous and often arranged in multiples of six. There is another large group of anthozoans, the alcyonarians, in which the polyps always have **eight branched tentacles** and eight internal partitions. Almost all members of this group are colonial, and the body cavities of the polyps are in communication with each other through endodermal canals which penetrate the whole colony. The polyps are remarkably similar, but the skeletons produced by the various colonies are strikingly diverse. The skeletons are made of minute particles, either of a horny substance or of limestone, which lie loosely in the soft tissues or grow together to form a hard, compact support, or do both in the same colony. Since any coelenterate which secretes a compact skeleton of limestone is called a "coral," there are a number of well-known corals among alcyonarians.

The polyps of the **organ-pipe coral** live in separate vertical limestone tubes which are joined at intervals by horizontal platforms in which run the endodermal canals. The well-known **precious coral** has a firm outer tissue, stiffened by numerous scattered limestone particles, and containing openings into which the delicate polyps can be retracted. In the center of the colony the particles are fused into a solid, hard axial core of red limestone, which is sold as "precious coral."

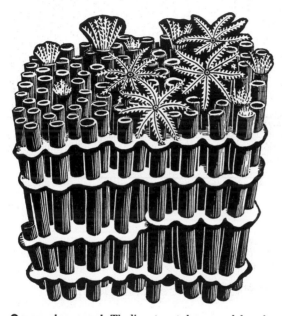

Organ-pipe coral. The limestone tubes are red, but the polyps are a bright green. In some places this coral is one of the important reef-builders. (Modified after Haeckel)

The **gorgonians** (sea fans, sea plumes, and sea whips) are branching, treelike colonies built on a plan similar to that of the precious coral. Their central supporting axis is made of a flexible horny material, so that they sway gracefully with the currents and form the most conspicuous and attractive feature of the coral reefs of Florida, Bermuda, and the West Indies.

In the warm, clear waters off Bermuda it is easy to descend 15 or 20 feet in a simple diving helmet and walk about on the white coral sand among tall purple gorgonians that tower overhead and low "bushy" ones that spring from all sides like blossoming shrubbery. Among them are massive dome-shaped heads of greenish-yellow "brain corals," huge pink or orange anemones spreading their flower disks, and many other kinds of colorful invertebrates and fishes taking refuge in this coelenterate jungle.

CHAPTER 9

COMB JELLIES

COMB jellies are transparent gelatinous animals which float in the surface waters of the sea, mostly near shores. Being feeble swimmers, they are carried about by currents and tides, so that they often accumulate in vast numbers in some bay where winds have driven them. During a storm their fragile bodies are swept ashore by the high waves and are strewn about on the beaches. They are, therefore, not an unfamiliar sight to people who live on the seacoasts; and they have been given many common names, such as "sea gooseberries" and "sea walnuts." These names describe the shape and size of two of the most common types; but they give no suggestion of the unique character from which has been derived the common name, "comb jelly" and the technical name, phylum **CTENOPHORA.** The ctenophores ("comb-bearers") swim about in the water by means of eight rows of **ciliary combs.** Each row consists of a succession of little plates formed of large cilia fused together at their attached ends like the teeth of a comb. The rows radiate over the surface of the animal from the upper pole to the lower pole, like the lines of longitude on a globe.

The combs are lifted rapidly in the direction of the upper pole, then slowly lowered to their relaxed position. Those in each row beat one after the other from the upper toward the lower pole. All eight rows beat in unison, and the animal is slowly propelled through the water with the

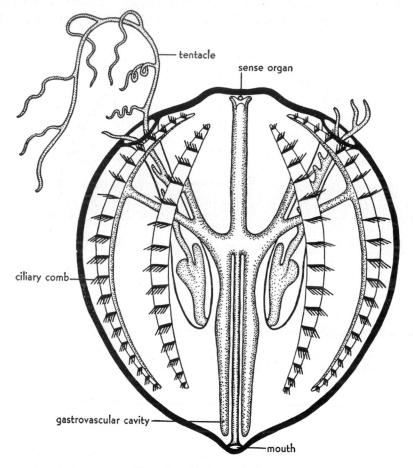

Pleurobrachia is a typical ctenophore.

lower pole (mouth end) in front. The rapidly beating combs refract light and produce a constant play of changing colors. Comb jellies are noted for the beauty of their daytime iridescence, but this is certainly matched at night by those comb jellies that are luminescent. When the animals are disturbed as they move slowly through the dark water, they flash along the eight rows of combs.

At the upper pole of the animal is an area composed of nerve cells and sensory cells. In the center of the area is a covered pit containing a **sense organ,** which consists of a little **mass** of limestone particles supported on four tufts of cilia connected with sense cells. It is thought to act as a sort of balancing device or "steering-gear." Any turning of the body causes the limestone mass to bear more heavily upon the ciliary tuft of one side or

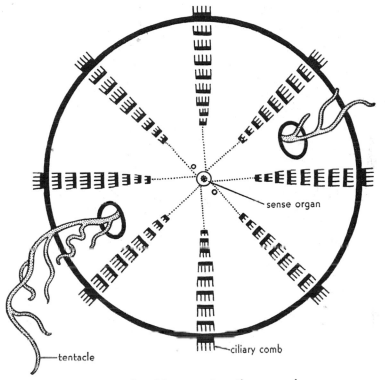

Pleurobrachia as seen from the upper pole.

another. Presumably, this stimulates the sensory cells, and the stimulus is carried by nerve cells to the swimming combs, causing them to beat faster on one side, thus righting the animal. From this polar area a **nerve net** extends all over the body and is concentrated into **eight nerve cords,** one under each row of combs. This system regulates and co-ordinates the activity of the eight ciliary rows, for, if the polar area is removed, the combs become disorganized. And if any row is cut across, the swimming combs below the cut get out of step with those above.

The general **body plan** of a comb jelly resembles that of a coelenterate medusa. There is an epithelium of ectoderm covering the outer surface, an epithelium of endoderm lining the gastrovascular cavity, and a thick jelly between. The jelly contains ameboid mesenchyme cells and long, delicate muscle cells which run from one part of the body to another.

The more primitive comb jellies have globular or pear-shaped bodies with a branched muscular **tentacle** on each side which can be withdrawn into a pouch. These tentacles have no stinging capsules, but the branches are covered with special **adhesive cells** (not to be confused with the adhesive thread capsules of coelenterates) which stick to, and entangle, the

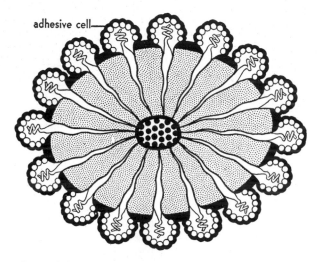

Cross-section through one of the branches of a **tentacle.** The outer surface is covered by the sticky heads of the **adhesive cells.** Each cell has a coiled thread (attached to the central muscular axis of the tentacle branch) which acts as a spring to prevent the cell from being wrenched off by the struggling prey. (Combined from several sources)

prey. Such ctenophores catch small shrimps or fishes by extending their tentacles full length and then curving and looping through the water, with the sticky tentacles sweeping a wide area. Other kinds of ctenophores have the tentacles reduced to very short filaments; they feed mostly on larvas and other small organisms which are caught by the ciliated grooves and swept toward the mouth.

The mouth is situated in the center of the lower pole and leads into a branched **gastrovascular cavity** which extends through the jelly, eventually giving off eight branches, one below each row of combs. Undigested food comes out through the mouth, though the gastrovascular cavity does open to the outside by two minute pores situated near the sense organ and shown (but not labeled) in the diagram of the upper pole of *Pleurobrachia.*

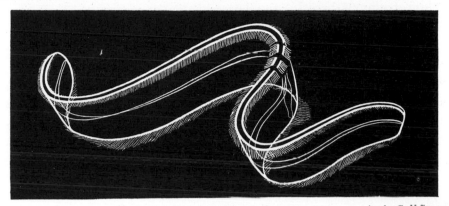

Cestum lives in warm seas but is sometimes carried north along our eastern coast in the Gulf Stream. Because of the elongated shape and the beauty of the transparent body, which shimmers with blue or green in the sunlight, it is commonly called "Venus' girdle." (After Chun)

All ctenophores are **hermaphroditic.** Both ovaries and testes occur on the walls of each of the gastrovascular branches that run below the rows of combs. The eggs and sperms are shed to the out side through pores in the ectoderm. The free-swimming larva develops directly, through rather complicated changes, into the adult.

A number of bizarre forms occur among the comb jellies. *Cestum* is flattened from side to side and reaches a length of over 3 feet. It swims by muscular undulations of the ribbon-like body as well as by the beating of the elongated swimming-plates.

Coeloplana is flattened in the other direction, so that the two poles, bearing mouth and sense organ, are brought close together. It has a typical comb-jelly structure with two tentacles, but has lost its combs. By virtue of extensive development of muscle fibers it is able to creep about like a worm. Such an animal illustrates how round, free-swimming organisms can become flat-

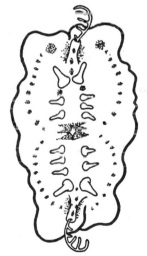

Coeloplana, as viewed from above. This animal is common on the coasts of Japan, where it creeps about on the sea-weeds. (Modified after Krempf)

tened, bottom-creeping forms. As we shall see, the next phylum is characterized by the flattened, creeping type of animal. And it is difficult to resist the temptation to build a theory which would derive the next higher phylum of animals directly from flattened ctenophores such as Coeloplana. For, although ctenophores are essentially, like coelenterates, animals of

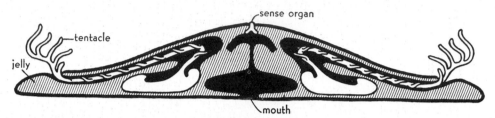

Coeloplana, as seen in diagrammatic section. The body is flattened so that the mouth and sense organ are relatively close together as compared with a more typical ctenophore like Pleurobrachia. (Modified after Komai)

the tissue level of construction, Coeloplana does seem to point in a number of different ways to the shape of things to come in the animal world. It even has special ducts leading from the testes to pores on the surface of the body, a foreshadowing of the organization of more than one kind of tissue into specialized structures or organs, which we see in all higher phyla. Unfortunately, the preponderance of facts do not support the attractive theory that ctenophores like Coeloplana are the connectinglink between coelenterates and flatworms. They show rather that both ctenophores and flatworms are probably descended from a primitive coelenterate stock from which higher coelenterates also have come.

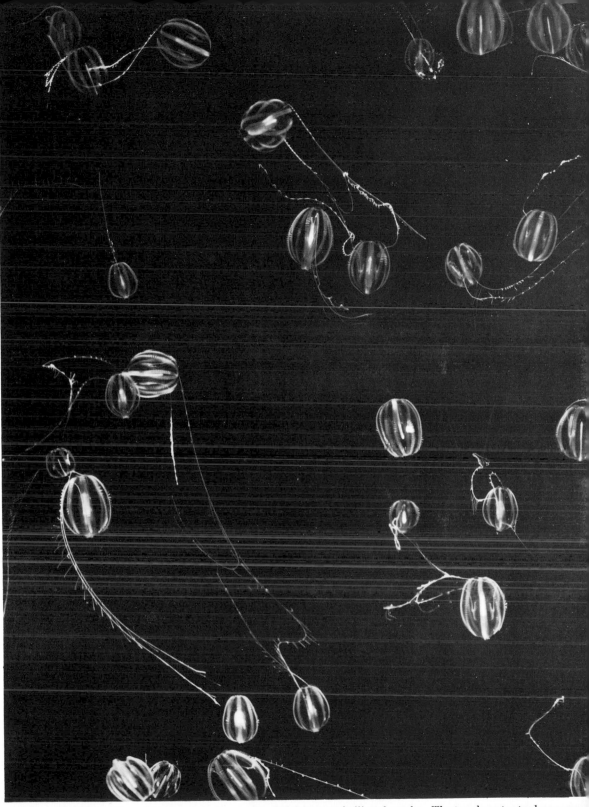

Sea gooseberries (*Pleurobrachia*) swim by means of eight rows of ciliated combs. The two long tentacles sweep the water for small food organisms. About natural size. (Photo of living animals by F. Schensky. Helgoland)

Mnemiopsis is a ctenophore or "sea walnut" common on our East Coast, where it is often seen in large swarms at the surface, swimming by means of eight rows of ciliated plates, nearly all of which show in the photograph. Rows of cilia, around the mouth end (toward the right in this photo), direct minute particles of food into the mouth. The animals are luminescent, and when disturbed at night, as by the passing of a boat, they light up along the rows of swimming plates. About twice natural size. (Photo of living animals, Woods Hole, Mass.)

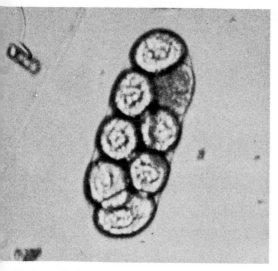

Eight-celled stage of a developing ctenophore already shows the biradial symmetry character- istic of adult ctenophores (see chap. 11). (Photo of living embryo. Pacific Grove, Calif.)

Small ctenophores, often called "sea gooseberries," are frequently cast upon ocean beaches. These are *Pleuro- brachia,* shown in a previous photograph with tentacles ex- tended. (San Francisco, California)

CHAPTER 10

THREE LAYERS OF CELLS

P UT a piece of raw meat into a small stream or spring and after a
few hours you may find it covered with hundreds of black worms
that are feeding upon it. These worms, each about half an inch
long, are called **planarias.** When not attracted into the open by food, they
live inconspicuously under stones and on the vegetation.

Planarias belong to the phylum **PLATYHELMINTHES** ("platy,"
flat; "helminthes," worm), which also includes many free-living marine
species and two important groups of parasites, the **flukes** and **tapeworms.**
There are many kinds (species) of planarias, just as there are many kinds
of amebas and hydras.

The planaria differs from the hydra in that one end of the body has a
definite **head,** with eyes and other sense organs. The head is always directed forward in locomotion; and the body is clearly differentiated into
front, or **anterior,** and rear, or **posterior,** ends. The planaria has an
elongated flattened body; and if we watch it move, we see that one surface
of the body always remains upward while the other is kept against the
bottom. The upper surface is termed **dorsal** (meaning back), and the
lower surface, **ventral** (meaning belly). We also notice that the various
parts, such as the eyes, are symmetrically arranged on the two sides, as in
ourselves.

The planaria **moves** about in a characteristic slow, gliding fashion, with
the head bending from side to side as though it were "testing" the environ-

ment. If we prod the animal, it hurries away by marked muscular waves. All these movements result from two mechanisms. The gliding is **ciliary** and **muscular;** the other movements are **muscular.** The surface of the planaria consists of a protective epithelium, as in the hydra; but this is ciliated, particularly on the ventral side. Numerous gland cells, which secrete a mucous material, open on the surface of the body and pour out mucus, upon which the worm moves. The cilia obtain traction on this bed of mucus, and as they beat backward, they help to move the animal forward. Planarias do not swim freely through the water, but move only in contact with a solid object or on the underside of the surface film. When a worm leaves the surface, it glides down attached to a thread of mucus. Just underneath the epithelium are layers of muscle cells. The outer layer runs in a circular direction, and the inner layer in a longitudinal direction. Muscles also run between dorsal and ventral surfaces and help to make possible all sorts of agile bending and twisting movements. The muscles are not part of the epithelial cells, as in the hydra, but are independent muscle cells specialized for contraction. Also, they are not developed from the ectoderm or endoderm but arise in a different way.

Beginning with flatworms, all higher animals have a mass of cells between the ectoderm and the endoderm, appropriately called the **mesoderm** ("middle skin"). This layer gives rise to muscles and to other structures which make possible an increasing complexity and efficiency in animal activities. Like almost all characters of animals, the mesoderm does not appear suddenly in fully developed form. Its early beginnings are perhaps represented in the ameboid mesenchyme cells of the hydra. But we do not consider mesenchyme to be true mesoderm until, as in the flatworms and in all higher animals, it is more massive than either ectoderm or endoderm and gives rise to definite structures, such as the reproductive organs.

Two-layered animals, like the hydra and the obelia, are usually small and fragile, because ectoderm and endoderm are essentially single layers of epithelium. The largest two-layered animals are certain jellyfishes; these have attained great size and a certain amount of body firmness by the secretion of great quantities of a viscous, nonliving jellylike material between ectoderm and endoderm. In three-layered animals the mesenchyme has been increased from a scattered group of wandering ameboid cells to a many-layered tissue that gives firmness and bulk to the body.

The coelenterates were animals organized on the tissue level. From flatworms to man, animals are constructed on a still higher level of organization. Not only do cells work together to form tissues, but tis-

sues of various kinds are closely associated to form one structure, called
an **organ,** adapted for the efficient performance of some one function.
Thus the human stomach is an organ composed of epithelium, connective tissue, muscle layers, and nervous tissue. The epithelium lines the
cavity and contains the gland cells which secrete the gastric juice; the
muscle layers cause the stomach contractions; the nervous tissue co-ordinates the muscle contractions and relates stomach activity to the whole
body; and the connective tissue binds the various layers together. An
organ usually co-operates with other organs or parts in the performance
of some life-activity, and such a group of structures devoted to one activity is termed an **organ-system.** Thus, the stomach is part of the digestive
system; and all the other parts of this system, such as the esophagus, liver,
and intestine, are necessary for the proper performance of digestion in

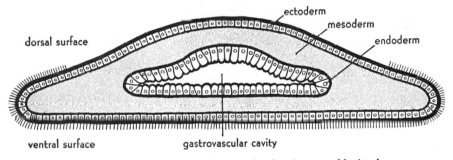

Diagrammatic cross-section of a planaria showing the general body plan.

man. The higher animals are made up of a number of such organ-systems,
as the digestive system, excretory system, circulatory system, nervous
system, and so on. Flatworms do not have all of these, and the ones they
do possess are not all very well developed; but they are the lowest phylum
of animals built on the **organ-system level of construction.**

In the **digestive system** of planaria the mouth, curiously enough, is
not on the head but near the middle of the ventral surface. It opens into a
cavity which contains a tubular muscular organ, the **pharynx,** attached
only at its anterior end. The pharynx contains complex muscular layers
and many gland cells. By means of the muscles, the pharynx can be greatly lengthened and then protruded from the mouth for some distance; it
behaves in this way during feeding. Planarias feed on small live animals
or on the dead bodies of larger animals. They can sense the presence of
food from a considerable distance by means of sensory cells on the head.
They move toward their food, mount upon it, and press it against the

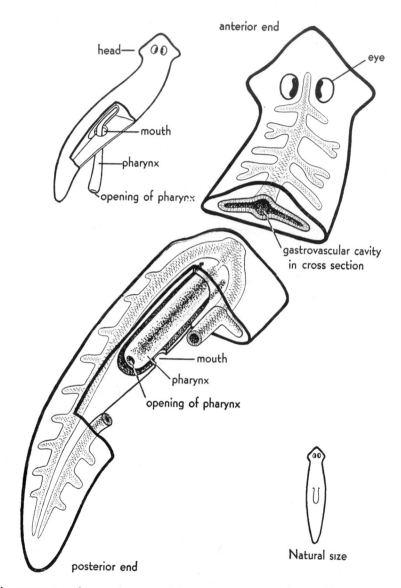

anterior end

head — eye

mouth

pharynx

opening of pharynx

gastrovascular cavity
in cross section

mouth

pharynx

opening of pharynx

posterior end

Natural size

Planaria, cut open to show construction of **digestive system,** pharynx withdrawn. Small drawing in upper left shows animal with part of body cut away and pharynx extended.

bottom by means of their muscular bodies. Struggling prey can be successfully held in this way, especially after they have become entangled in the slimy secretion from the worm. The pharynx is protruded through the mouth and inserted into the prey. It secretes enzymes that soften the prey tissue, and its sucking movements tear the tissue into microscopic bits which are swallowed along with the juices of the prey.

From the anterior attached end of the pharynx the rest of the digestive system extends throughout the interior of the animal. It consists of one anterior branch which runs forward and two posterior branches which pass backward, one on either side of the pharynx, to the posterior end. All three branches of this **gastrovascular cavity** have numerous and fairly regularly spaced side branches, thus providing for the distribution of the food to all parts of the body. The epithelium of the gastrovascular cavity consists simply of the endoderm and corresponds to the endoderm of the hydra.

There is some **digestion** of food in the gastrovascular cavity of the planaria, such as occurs in coelenterates, but the food is broken up into small particles before it enters the cavity, and is mostly ready to be taken up by the epithelial cells in ameboid fashion and formed into **food vacuoles.** The digested food is absorbed and passes by diffusion throughout the tissues of the body. There is only one opening to the gastrovascular cavity; indigestible particles are eliminated through the mouth, as in the hydra.

Experiments on a common species of planaria which was fed on liver showed that, after a meal, all the ingested liver was taken into the epithelial cells in about eight hours, and that three to five days were required for the complete digestion of the food vacuoles so formed. Much of the food was found to be converted into fat, which was stored in the digestive epithelium.

Practically all animals can store food reserves upon which they draw in time of need. A small animal like the ameba stores very little and, unless it goes into the inactive encysted state, will die after about two weeks without food. Hydras survive much longer periods of starvation. But planarias are peculiarly adapted to go for many months without food while remaining active. During the starvation period they use the food stored in the digestive epithelium, whole cells breaking down. Later they begin to digest other tissues, the reproductive organs usually going first. Externally one can observe only that the worms grow steadily smaller though retaining the same general appearance. A worm starved for six months may shrink from a length of 20 mm. to one of 3 mm. Because of their ability to go for months without food, planarias make ideal household pets for people who are too busy or too absent-minded to keep an "exacting" animal like a canary.

The region between the outer protective epithelium (ectoderm) and the inner digestive epithelium (endoderm) is filled with various organs surrounded by **mesenchyme** in the form of ameboid cells, most of which are united into a network, although some are free and move about. The mus-cle layers already mentioned are imbedded in this mesenchyme, and it also contains many gland cells which open to the surface and secrete slime or sticky substances. The gland cells are largely derived from the ectoderm; but the organs, muscles, and mesenchyme are meso-dermal.

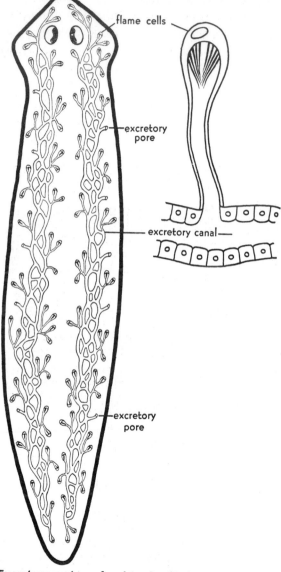

A new system which was not found in any of the forms already studied is the **excretory system.** This lies in the mesenchyme and consists of a network of fine tubes which run the length of the animal on each side and open to the surface by several minute pores. Side branches from these tubes terminate in the mesenchyme in tiny enlargements known as **flame cells.** Each flame cell has a hollow center in which beats a tuft of cilia simulating a flickering flame. The hollow center is continuous with the cavity of the tubules of the system, and the ciliary beat causes a current of fluid to move along the tubules to the pores. Since metabolic

Excretory system of a planaria. On the right is shown a single flame cell attached to a portion of the excretory canal.

wastes seem to be excreted to the outside largely by way of the endodermal epithelium and mouth, the flame-cell system (like the contractile vacuoles of protozoa) appears to function primarily for the regulation of the water content of the tissues.

The planaria has a highly complicated **reproductive system** for sexual reproduction. We saw that in sponges single mesenchyme cells became eggs and sperms, and that in coelenterates mesenchyme cells aggregate into simple ovaries and testes, which discharge their contents, the eggs and sperms, directly to the exterior. In planarias, ovaries and testes arise in the mesenchyme; but there is a system of tubules and chambers in which fertilization occurs, and there are complicated sex organs for the transfer of sperms. The animals are hermaphrodites, forming both male

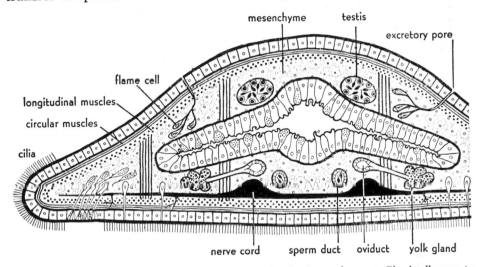

Cross-section through a **sexually mature planaria,** showing internal organs. Gland cells open to the surface.

and female sex organs in every individual; but exchange of sperms takes place so that cross-fertilization is effected. After the reproductive season the reproductive system degenerates and is regenerated anew at the beginning of the next sexual period.

When sexually mature, each individual has a pair of **ovaries** close behind the eyes. From each ovary a tube, the **oviduct,** runs backward near the ventral surface. Multiple **yolk glands,** consisting of clusters of **yolk cells,** lie along the oviduct, into which they open. There are numerous **testes** along the sides of the body. From each testis leads a delicate tube, and all these tubes unite on each side to form a prominent **sperm duct,** which runs backward near the oviduct. The sperm ducts, packed with sperms during the time of sexual maturity, connect with a muscular, protrusible organ called the **penis,** which is used for transfer of the sperms to another planaria. The penis projects into a chamber, the **genital chamber,** into which there also open the oviducts and a long-stalked sac called the

copulatory sac. The genital chamber opens to the exterior by the **genital pore** on the ventral surface behind the mouth.

Although each planaria contains a complete male and female sexual apparatus, self-fertilization does not occur; instead, two worms come together and oppose their ventral

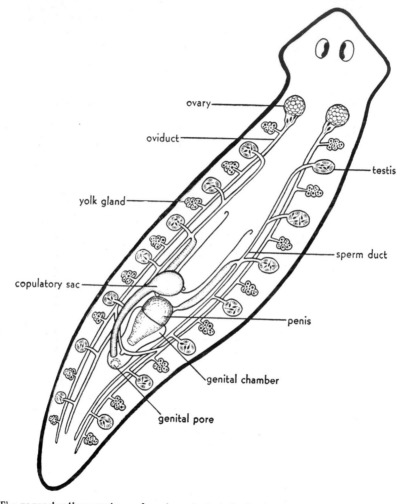

ovary

oviduct

testis

yolk gland

sperm duct

copulatory sac

penis

genital chamber

genital pore

The **reproductive system of a planaria** includes both male and female sex organs.

surfaces. The penis of each is protruded through the genital pore and deposits sperms in the copulatory sac of its partner. After copulation, the worms separate. The sperms soon leave the copulatory sac and travel up the oviducts until they reach the ovaries, where they fertilize the ripe eggs as they are discharged. The fertilized eggs pass down the oviducts, and at the same time yolk cells are discharged from the yolk glands into the oviducts. When eggs and yolk cells reach the genital chamber, they become surrounded by a shell to form an egg capsule. The eggs of flatworms are peculiar in that food reserves do not occur in the eggs themselves but are kept in the yolk cells, which accompany the eggs. The capsules (each containing less than ten eggs and thousands of yolk cells) are passed out through the genital pore and are often fastened to objects in the water. They hatch in two or three weeks to minute worms, like their parents except that they lack a reproductive system.

Asexual division. *Left,* just before division. *Right,* just after. The rear piece will soon develop a head, pharynx, and other structures.

Many planarias have no method of reproduction other than the sexual, but some multiply **asexually.** In this process the worm, without any evident preliminary change, constricts at a region behind the pharynx, and the posterior piece begins to behave as though it were "revolting" against the domination of the anterior piece. When the whole animal is gliding quietly along, the posterior part may suddenly grip the bottom and hold on, while the anterior head-piece struggles to move forward. After several hours of this "tug of war" the anterior piece finally breaks loose and moves off by itself. Both pieces regenerate the missing parts and become complete worms. Species which have this habit often go for long periods without sexual reproduction, and, in fact, some of them rarely develop sex organs.

In flatworms we see the first appearance of a **central nervous system,** the kind of nervous system possessed and further centralized by all higher

brain

nerve cord

Nervous system of a planaria.

animals. In the planaria there is in the head a concentration of nervous tissue into a bilobed mass called the **brain.** From the brain two strandlike concentrations of nerve cells, the **nerve cords,** run backward through the mesenchyme near the ventral surface. From these ventral nerve cords numerous side branches are given off to the body margins, and the two cords are connected with each other by many cross-strands like the rungs in a ladder. Because of its resemblance to a ladder, this type of system has been called the "ladder type" of nervous system. The brain and the two cords constitute the central nervous system, a kind of "main high-way" for nervous impulses going from one end of the body to the other. The brain is not necessary for the muscular co-ordination involved in loco-motion, for a planaria deprived of its brain will still move along in co-ordinated fashion. It serves chiefly as a sensory relay that receives stimuli from the sense organs and sends them on to the rest of the body. The re-sult is a much more closely knit behavior than is possible with the diffuse,

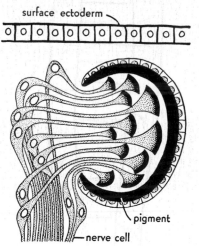

Section through **eye of a planaria.** The eye is sensitive only to light coming toward the open end of the pigment cup. The light-sensitive nerve cells run to the brain. (Modified after Hesse)

noncentralized nerve net of the hydra, which lacked definite pathways and a co-ordinating center.

The nerve net does not disappear in the planaria but persists in addition to the central nervous system. Nerve nets also occur locally in the tis-sues of almost all higher animals. In man, for ex-ample, a well-developed nerve net (connected with the central nervous system) occurs in the wall of the intestine.

Conditions in the external world are conveyed to the nervous system by **sen-sory cells,** slender, elongated cells that lie, with their pointed ends projecting from the body surface, between the epi-thelial cells. Probably different ones are specialized to receive touch, temperature, and chemical stimuli. Sensory cells are distributed all over the body surface, as in the hydra, but in addition are concentrated in the head to form sense organs. The **sensory lobes,** pointed projections on each side of the head, are known to be especially sensitive to touch and to water currents, and probably also to food and other chemicals. The two **eyes** are sense organs special-ized for light reception. Each consists of a bowl of black pigment filled

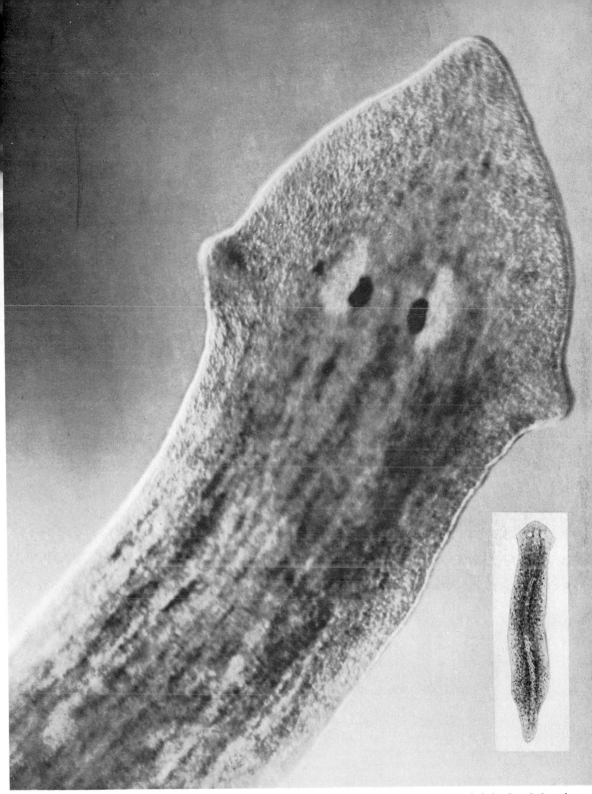

The planaria belongs to the lowest phylum of animals all the members of which possess definite heads bearing a concentration of sense organs, and have symmetry like that of man. (Photo of living animal by P. S. Tice) *Insert* shows whole worm (*Dugesia tigrina*) about 2×natural size. (Photo courtesy Libbie Hyman)

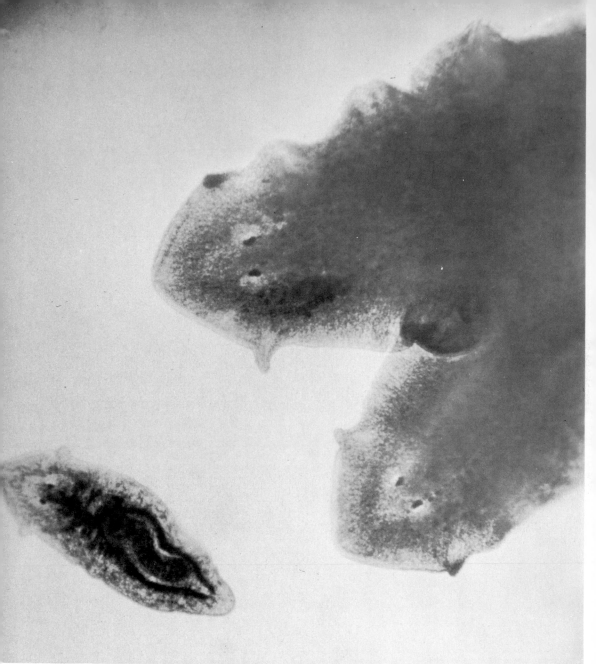

Two-headed planaria can be produced easily by making a longitudinal cut through the anterior end of a normal worm and then renewing the wound on several successive days so that the cut edges cannot grow together again. Each half of the head will regenerate the missing parts, and the result will be a two-headed worm in which both heads are of equal dominance over the rest of the body and often appear to try to go off in opposite directions. Some monsters do split apart, and each part regenerates into a normal worm. If the original cut extends far enough back into the body, each head will induce the formation of its own pharynx (see chap. 12 for more about regeneration in planarias). The very tiny planaria is of the same species as the two-headed worm (*Dugesia dorotocephala*). The small size and disproportionally wide body are a result of regeneration from a small piece of a worm that had been cut up by a series of transverse slices. In regenerated or in very young worms the pigmentation is light and the internal organs are more easily visible. 90 × natural size. (Photo of living animals. Illinois)

with special sensory cells whose ends continue as nerves which enter the brain. The pigment shades the sensory cells from light in all directions but one, and so enables the animal to respond to the direction of the light. Unlike other regions of the ectoderm, that which is immediately above the eye is unpigmented, and thus allows light to pass through to the sensory cells. Planarias whose eyes have been removed still react to light,

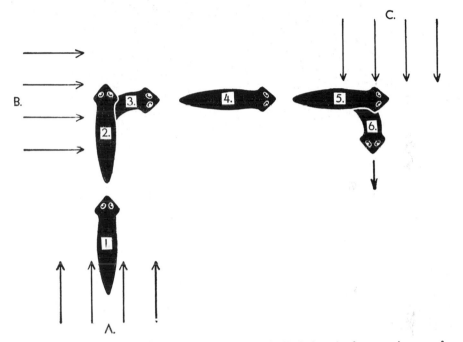

Orientation to light. 1–6, successive positions of a planaria. In **1**, the animal was moving away from light coming from source **A**. When it reached position **2**, the light was turned off at **A** and on at **B**. The worm turned and moved away from the light. At position **5**, light **B** was turned off and **C** turned on, the worm again oriented away from the light. (Based on Taliaferro)

but more slowly and less exactly than normal worms. This indicates that there must be some light-sensitive cells over the general body surface.

By virtue of abundant sensory cells, specialized sense organs, and a centralized nervous system the planaria shows a more varied **behavior** and much more rapid responses than does the hydra. Planarias avoid light and are generally found in dark places, under stones or leaves of water plants. If placed in a dish exposed to light, they immediately turn and move toward the darkest part of the dish. They are highly positive to contact and tend to keep the under surface of the body in contact with other objects. They respond to chemical substances in the water and quickly react to the presence of food by turning and moving directly toward it. That is why a piece of raw meat placed in a spring inhabited by planarias attracts hordes of worms, which glide upstream toward the food, guided by the meat juices in the current of water. Planarias react to water currents, and some species regularly move upstream against a current.

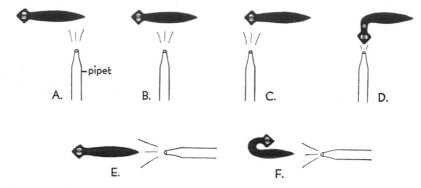

Reactions to water currents produced by a pipet. **A, B,** there is no response as the current strikes the middle or rear of the body. **C, D,** the current strikes the sensory lobe on the side of the head and the worm turns toward the current. **E, F,** the current from the rear passes along the sides of the body to the sensory lobes and the worm turns around toward the current. In nature these reactions orient the worm upstream. (Modified from I. Doflein)

They also respond to the agitation of the water produced by the animals upon which they prey.

THE flatworms, as illustrated by the planarias, are advanced over two-layered animals in a number of important characters which are possessed by all higher animals. The flatworms are the first animals to have specialized anterior and posterior ends and dorsal and ventral surfaces. They are the first to have a definite head with a concentration of sense organs and the development of a central nervous system. And they are the first to make extensive use of a third layer of cells, the mesoderm, which, either by itself or in combination with ectoderm or endoderm, gives rise to organs and organ systems.

CHAPTER 11

THE SHAPES OF ANIMALS

ALTHOUGH animals range in size from microscopic protozoa to massive whales, there are only three basic styles in animal shapes: spherical, radial, and bilateral. A spherical form is assumed by any isolated small quantity of liquid because of the physical forces acting upon it. Small bits of protoplasm or single cells assume this spherical form unless they have a stiff surface layer or skeleton which enables them to maintain some other shape. The ameba becomes spherical when at rest and must expend energy in extending pseudopods. This type of shape, called **spherical symmetry,** is characterized by the arrangement of structures with reference to the center of a sphere. Since all radii are alike, a spherical animal can be divided into two identical pieces by a cut in any direction through the center. There is no front or rear, no top or bottom, no right or left sides—at least no permanent ones. This is a disadvantage of the spherical form since such an animal can show only a very indefinite kind of locomotion—and, in fact, most spherical animals are free-floating. On the other hand, this type of symmetry is admirably suited to the needs of floating animals which do not move under their own power; they cannot swim toward food or away from enemies but must respond to these on any side from which they may approach. Spherical symmetry is found among adult animals only in protozoa, where it is best seen in certain shelled protozoa (radiolarians, one of which is shown in the drawing at the head of the chapter) which float near the surface of the ocean and feed by means of pseudopods that radiate out in all directions through openings in the shell.

If we imagine a sphere developing a mouth, the surface will no longer be everywhere the same but will be differentiated into the part bearing the mouth and the part not bearing the mouth. If the food-capturing devices of the animal, such as tentacles, are arranged in a circle around the mouth, this body plan just fits a polyp like the hydra. Both the polyps and the medusas of coelenterates exhibit **radial symmetry,** in which all radii are alike at any particular level, but there is a differentiation between levels along an axis from the mouth end to the end opposite the mouth. A radial animal may be cut into two identical pieces by a *lengthwise* cut (but not a crosswise cut) through the center in any direction, for the animal is alike all around the circumference: it has no differentiated sides. This lack of a definite side to go first in locomotion renders moving about somewhat ineffective. Radial animals either drift with the water currents most of the time or live a sessile life. A sessile animal has nothing to "fear" from below, and the basal end is specialized only for attachment. The circle of tentacles extending from the exposed mouth end are prepared to meet the environment from above and from all sides. Radial symmetry is seen in some protozoa and sponges but is most characteristic of coelenterates.

In many of the higher coelenterates such as the anemones and corals the radial symmetry is modified by an elongation of the mouth and some internal structures in a particular plane, so that all the radii are no longer alike. The animal can now be divided into two identical halves only by cutting it across in two particular planes, one running through the long axis of the mouth and the other at right angles to this axis. Such a modification of a basically radial symmetry is called **biradial symmetry.** Only one whole phylum of animals, the phylum Ctenophora, is characterized by biradial symmetry. In ctenophores the biradiality begins as early as the eight-celled embryo stage and continues throughout development.

For efficient locomotion it is essential that one end should go first. In the planaria there is a definite front or anterior end, which bears the sense organs and which always ventures first into a new environment; and a rear or posterior end, which merely follows along. Such an animal would seem to be open to attack from the rear and from the sides, as compared with a hydra (or medusa), which can detect enemies and ward off their attacks on all sides. However, the concentration of sense organs in the front end enables the animal to detect danger ahead and so better to avoid it. Also, specialization of anterior and posterior ends is related to efficient locomotion, and such animals can better escape from their enemies than can radial animals.

The specialization of the head end is accompanied by a differentiation of the upper and lower surfaces of the body. The undersurface of the planaria bears the mouth and most of the cilia and is quite different from the upper, exposed surface. This type of body form in which there is a difference between front and rear and between upper and lower surfaces is called **bilateral symmetry.** The term "bilateral" means *two sides* and refers not to these surfaces but to the fact that in these animals the body structures are arranged symmetrically on the two sides with reference to a central plane which runs from the middle of the head end to the middle of the tail end. The paired eyes and sensory lobes of planaria (and the paired structures of man) occur at equal distances to either side of this plane. Single organs are generally located in the mid-line and are bisected by the plane. Bilateral animals have right and left sides, while the hydra, for instance, has no defined sides. A bilateral animal can be cut into two similar pieces by only one particular cut—along the plane which runs down the middle of the body from head to tail and from back to belly. The two resulting pieces are not identical but are mirror-images of each other.

The bilateral symmetry is often imperfect in some degree, owing to the specialization of one side or part over the other. Thus, in man the right arm is usually larger and stronger than the left, and in the brain there is a speech center on the left side but none on the right. Much more marked asymmetries occur in other bilateral animals, such as the coiling of the snail shell into a spiral (because of unequal growth of the two sides).

When bilateral animals become sedentary, they tend to evolve a modified symmetry which appears superficially like radial symmetry. An example is the starfish, which is said to be secondarily radial or to display secondary radial symmetry.

Although typical bilateral symmetry is not found below the flatworms, we see approaches to it independently arrived at in lower phyla. For example, in many anemones the elongated mouth and gullet have a groove at only one end. This differentiates the two ends of a plane that passes through the long axis of the mouth, and such an animal can be divided into two identical halves by only one particular cut—that passing lengthwise through the long axis of the mouth.

The bilateral body form lends itself readily to "streamlining" and, with a head to direct movements, gives rise to fast-moving and therefore very successful animals. Beginning with the flatworms, all animals (unless secondarily modified) are bilaterally symmetrical.

A form which has no definite symmetry is seen in various protozoa and in most sponges (such animals are said to be *asymmetrical*). However, some kind of symmetry is characteristic of animals in general,

CHAPTER 12

NEW PARTS FROM OLD

REGENERATION, or the ability to repair damage, is almost universal among animals. In man, wounds of considerable size heal and broken bones grow together again, but a lost finger or toe cannot be replaced. Among the invertebrates, the power of repair is much greater, and, in general, the lower the degree of organization of an animal, the greater is its ability to replace lost parts. Such regeneration depends upon the ability of the uninjured cells to produce the kinds of cells destroyed by the injury. Consequently, the more specialized the cells of an animal have become, the less able are they to produce cells different from themselves and so replace the missing parts.

Protozoa have a notable capacity for regeneration. Any piece that is not too small will re-form a complete and perfect animal if it contains a nucleus; nonnucleated pieces fail to regenerate. This is not surprising when we recall that in the normal method of reproduction in protozoa the nucleus divides in two and the cell breaks into two parts, each containing half of the nucleus and each capable of growing into a complete individual. In many ciliates there is more than a simple replacement of missing parts. Injury to the basal part of a single large cilium may cause a complete breakdown and resorption of all the cell structures except the nucleus, followed by the differentiation of a new set of parts. Furthermore, even the most complex protozoa lose a certain amount of their specialization at each cell division, and then undergo complete differentiation again before the next division. Thus, their life-histories are fundamentally different from, and their powers of regeneration greater than, those of the many-celled animals whose development involves the progressive specialization of cells through countless cell divisions.

Among the loosely organized **sponges** we saw a very marked ability to regenerate; cells from finely macerated sponges fuse in small masses and develop into complete sponges, as discussed in chapter 6.

Coelenterates also regenerate very well. Hydroids, like the obelia, have been pressed through fine gauze, after which the separated cells have clumped together to regenerate new polyps. Pieces of the hydra body grow into small but complete hydras.

Similarly, some **planarias** will regenerate complete worms from almost any piece. The parasitic flatworms,

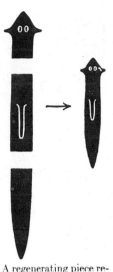

A regenerating piece retains its polarity—a head grows from the anterior end and a tail grows from the posterior end. (Based on Child)

on the other hand, have no ability to replace parts removed; and this statement applies to parasitic animals in general.

Although an earthworm can replace its head, a starfish its arms, and a lobster a leg or an antenna, the **higher invertebrates** show, in general, an increased specialization and a corresponding decrease in capacity for regeneration.

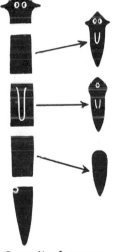

Capacity for regeneration decreases from the anterior to the posterior end. (Based on Child)

Consequently, regeneration has been studied mostly in the protozoa, sponges, hydras, hydroids, and planarias. Planarias, in particular, have been the subject of extensive researches, and certain general facts have been ascertained which apply to all the lower animals.

In the first place, any piece of such animals usually retains the same **polarity** it had while in the whole animal, that is, the regenerated head grows out of the cut end of the piece which faced the anterior end in the whole animal, and the regenerated tail grows out of the cut end which faced the posterior end. This *antero-*

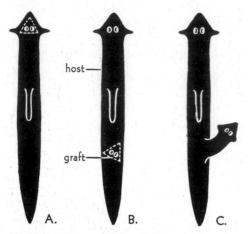

Grafting. A, a small piece, indicated by broken lines, is cut out of the head of the donor. **B,** the graft is placed in a wound made in the body of the host. **C,** the graft has grown into a small head. (Based on Santos)

bigger and more normal heads than pieces from posterior regions, and there is a gradual change in these respects along the anteroposterior axis. In some planarias only the pieces from anterior regions are able to form a head, while those farther back effect repair but do not regenerate a head.

The head of a planaria is dominant over the rest of the body, and, in general, any level controls the level posterior to it. One way in which this **dominance** can be demonstrated is by means of grafting. If a small bit of the head region of a planaria is grafted into a more posterior level, it will not only grow out into a head but will influence the adjacent tissues to co-operate with it so that a new pharynx, for example,

posterior differentiation is operative throughout the entire animal down to small portions.

Another generalization drawn from these experiments is that the capacity for regeneration is greatest near the anterior end and decreases toward the posterior end. Pieces from the anterior regions of a planaria regenerate faster and form

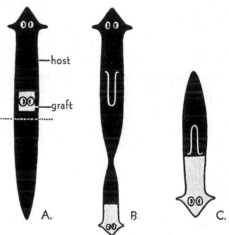

Reversal of anteroposterior axis by a graft. In this experiment host and graft were of two different species, and the tissues of each remained distinguishable as growth occurred. **A,** the host pharnyx was removed and in its place was grafted a piece of the donor, including the eyes, part of the brain, and adjacent mesenchyme. One week later, the host tail was cut off at the level of the dotted line. **B,** a few weeks later, the graft has grown out as a small head. The host has developed a new pharynx and is about to divide asexually as indicated by the constriction. **C,** 74 days after grafting. A pharynx has formed in the host tissue, but it is oriented in a direction opposite that of the old one. A tail has developed at the anterior end of the old host tissue where a head would be expected to develop if the graft were not present. The direction of beat of the host cilia has been reversed. (Based on J. A. Miller)

may be formed in the body near this grafted head. If a head piece is grafted into a planaria and then the host's head cut off, the grafted head may influence the anterior cut surface (which ordinarily would regenerate into a head) to form a tail. In other words, grafts of head pieces reorganize the adjacent tissues into a whole worm in relation to themselves. Grafts from tail regions do not have these effects but are generally absorbed.

The dominance of the head over the rest of the body is limited by distance. If the animal grows to a suf-

If the head of a planaria is cut down the middle, each half will regenerate the missing parts. (Based on Child)

ficient length, its rear part may get beyond the range of dominance of the head. This happens in asexual reproduction of the planaria when the rear part starts to act as if it were "physiologically isolated" and then finally constricts off as a separate animal. A diminishing of the control of the head over the body is the most important factor in asexual division, as shown by the fact that separation of the rear part can be induced by

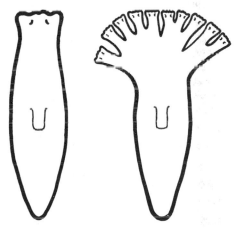

The head of this flatworm was cut repeatedly, and the cut edges not allowed to grow together again. The result is a monster with ten complete heads. (After Lus)

cutting off the head. And conversely, if the worms are kept in a vaselined dish, so that the posterior subindividual cannot get a good hold on the substratum, the worm is prevented from dividing.

Until the moment when a planaria starts to constrict, there is no external evidence of the physiological isolation of a posterior subindividual. Yet its presence can be shown experimentally. Pieces of worms taken from a region just behind the pharynx usually do not develop heads. But shortly behind this is a zone that almost always produces normal heads. This is the region where the worm constricts in asexual division. Apparently this zone represents the developing head of the posterior subindividual. In a longer worm it can be shown that there are three or even four subindividuals, one behind the other, and each indicated by a region which shows an increased ability to produce worms with normal heads.

If the anterior end of a planaria is cut down the middle, and the two halves prevented from growing together again into a single head by renewing the wound several times, then each half will regenerate the missing parts and a two-headed planaria will result, with dominance divided equally between the two heads. If the cut goes far enough back, each head will influence the formation of its own pharynx.

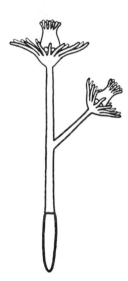

All these facts indicate that there is some sort of gradation in essential processes along the anteroposterior axis of a planaria, and we refer to this as the **anteroposterior gradient.** Similar experiments on hydras and hydroids have yielded similar results, and in these radially symmetrical animals the gradient is highest at the mouth end and decreases gradually toward the base, resulting in a **mouth-base gradient.**

Hydroid (*Corymorpha*) **with graft-induced polyp** growing from side of host stem. The side polyp becomes larger if the host polyp is removed. The stem of the side polyp grows longer the farther away it is from the dominant host polyp. (After Child)

One of the hydroids most frequently experimented upon is *Corymorpha*, in which it is easy to demonstrate a gradient that is highest at the polyp end and decreases toward the base of the long stalk. If a bit of hydroid stalk is removed from one individual and grafted into the stalk of another individual, a new polyp will develop, as shown in the figure. As we might expect, from what we already know of planarias, a higher percentage of complete polyps will develop from those fragments of stalk that were taken from near the polyp end of the donor than from fragments cut from more basal regions of the donor. If the host polyp is removed before the fragments are grafted into the host stem, the percentage of complete polyps formed is greater than if the host polyp is allowed to remain. Apparently the host polyp exercises a dominance comparable with that of the head of the planaria.

This gradation does not seem to be anatomical in nature, since no definite structural difference is known to exist along the stem of a hydroid, for instance. Consequently, it is thought that the gradation is "physiological," by which we mean that it arises from differences in function rather than from differences in structure. In seeking for the functions that might be the cause of a gradation which affects all regions of the body, we must look to general functions rather than to particular ones. The most general function of the animal body is metabolism—the totality of the chemical processes involved in the building-up and breaking-down of protoplasm. This anteroposterior gradient, according to one theory, consists in a gradation of metabolic rate along the body. The rate is highest in the head region and decreases toward the posterior end.

Various lines of evidence have been accumulated in favor of the **metabolic gradient theory.** Direct measurements of the oxygen consumption of pieces cut from various levels of the body of planarias have shown that anterior pieces carry on respiration at a faster rate than posterior pieces. Such experiments are not as conclusive as they might seem because of the changes which pieces of a co-ordinated many-celled animal conceivably undergo when cut from the body. If planarias are placed in a poisonous solution of sufficient concentration, they die; but death does not attack all parts at the same time. In these low forms death occurs in a regular progression, beginning at the head end and extending gradually backward. This is explained, on the basis of the

Degeneration in a poisonous solution begins at the head end. (Based on Child)

theory, by assuming that the parts with the highest metabolic rate are affected first and most severely by a poison, while less active parts are more slowly affected.

If the formation of a head at the anterior cut end of a piece, and of a tail at the posterior cut end, depend on an anteroposterior gradient in that piece, it should be possible to change this result by **experimentally altering the gradient.** Thus, if pieces are cut so short that there is no appreciable difference between the anterior and posterior cut surfaces, such as very short pieces near the eyes in planari-

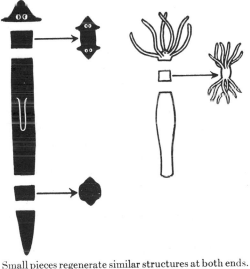

Small pieces regenerate similar structures at both ends. *Left*, a planaria. *Right*, a hydra. (Based on Child)

as, they regenerate a head at both ends of the piece. Pieces can also be obtained from the posterior regions of planarias which will regenerate a tail at both ends. The gradient can readily be changed in coelenterates. Small pieces of a hydra taken from very near the mouth will grow a mouth and a crown of tentacles at both ends. It is possible to obliterate the gradient in pieces of hydroid stems by putting them in anesthetics for a time; when replaced in water, such pieces may grow out several polyps irregularly from whatever part of the piece happens to be uppermost. The polarity of pieces of hydroid stems can be reversed, and a polyp made to grow out at the basal end of the piece by exposing this end to a better oxygen supply and decreasing the oxygen supply at the upper end. Other external factors have been shown to operate in the same way.

One may inquire **how gradients get started** in animals. It seems that they must arise early in development by the action of external factors on protoplasm. The position of the egg in the ovary is one such external condition. The egg is attached by one end to the ovary and is free at the other end. It is known for a good many eggs that this position determines, or at least is correlated with, the polarity of the egg, that is, which end of the egg is to become the anterior (or the mouth end) of the future animal. The polarity is a property of the cytoplasmic background of the egg and is based upon a gradient, a graduated difference of some kind. This gradient is thought to be related to differences in rate of diffusion of oxygen and other substances in the attached and freely exposed ends. Once established, the polarity continues throughout embryonic development.

It has been found that many experiments on regenerating adult flatworms can be duplicated on developing eggs and embryos. By subjecting eggs and embryos to poisonous solutions of proper concentration and at the right time, the development can be greatly modified and all sorts of curious embryos can be obtained. In general, because such solutions act most severely on the parts highest in the gradient, these embryos show suppression of the head region and the sense organs and often have small heads, reduced eyes or eyes fused into one, and so on.

Such results suggest that some kind of gradient is an important factor in embryonic development and furnishes an underlying pattern which controls the orderly development of normal form and proportion. Thus we might think of the several kinds of symmetry as resulting from differences in the number of gradients which act in development. In the truly spherical animal all radii are alike and there is no one main axis of differentiation, though there is some difference between the interior of

Grafting in animals has been studied extensively in fresh-water planarias because these animals have exceptional powers of regeneration. By producing abnormal relationships between the tissues of a single animal or by combining tissues from two different animals we learn a great deal about the development of the normal characteristics of organisms. Studying the results of interspecies grafting is particularly easy with certain planarias not only because grafts between different species take so well but because tissues of host and graft can be distinguished by their different patterns of pigmentation. *Left: Dugesia dorotocephala*, in which the black pigment is distributed over the dorsal surface in small granules against a brownish background. The ventral surface is gray. *Center: Dugesia tigrina*, which has the dorsal surface finely spotted with yellowish-orange pigmented areas against a tan background. The ventral surface is white with gray pigmented areas. In the grafting experiment host and donor were placed on a tray of ice to slow down their movements. Then with a small knife a piece of the head, including the two eyes, was cut from *D. tigrina* and placed in a hole, made just behind the eyes, in the head of *D. dorotocephala*. *Right:* 175 days after the operation the tissues of host and donor are completely fused. The grafted eyes lie behind those of the host. The only evidence of any effect produced by the graft on the host was the growth, in the tissues of the host, of a small projection on either side of the grafted eyes. (Photos of living animals by James A. Miller)

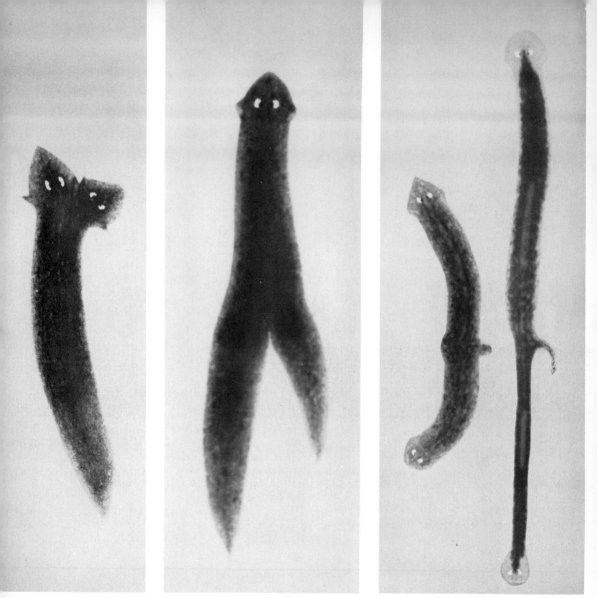

Left: an animal with two well-developed heads produced by a procedure similar to the one on the preceding page. In this case, however, the graft did not unite on all sides with the tissues of the host but has induced the formation of a complete head, having the pointed head shape, pointed sensory lobes, and pigmented pattern of the host (*Dugesia dorotocephala*) rather than that of the donor (*D. tigrina*), which has a more rounded head with blunter lobes and a different pigment pattern. *Second from left:* A two-tailed worm obtained by grafting a head piece from *D. tigrina* into the tail region of *D. dorotocephala*. The graft grew out as a complete head with *D. tigrina* characters. It induced the formation of a pharynx in the host tail. Then the host head and pharyngeal region were removed; and the cut surface, which we might expect to regenerate a new head, produced only a pharynx and tail, supposedly because the tissue had come under the domination of the graft head. After this photograph was made, the two tails fused along their inner borders. When this animal, with the double posterior region containing two pharynxes, later divided asexually, the new animals so produced had four eyes, two pharynxes, and a double digestive tract. *Second from right:* A graft was placed in the tail region and grew out as a head; then the posterior part of the host tore away, leaving only the graft attached to the anterior piece of the host. *Right:* The same worm, here shown greatly extended, has been fed blood to make the pharynxes stand out. The posterior pharynx, which has been induced in host tissue by the graft, takes a direction related to the graft head and opposite that of the original host pharynx. (Photos of living animals by James A. Miller)

the cell and its exposed surface. Such an animal can show only a limited amount of differentiation. In an ameba all points on the surface are alike in that any point is capable of sending out a pseudopod, but the pseudopod proceeds in a definite direction. By means of chemical indicators it has been shown that in an actively moving ameba there is a definite, though constantly changing, physiological gradient—highest at the tip of the leading pseudopod and lowest at the opposite end of the animal. Radial and bilateral animals have more permanent axes of differentiation. In radial types the main axis (also called the "polar axis") is from mouth to base. In bilateral types there are, besides the main anteroposterior axis, two minor axes of differentiation. There is evidence for physiological gradients in

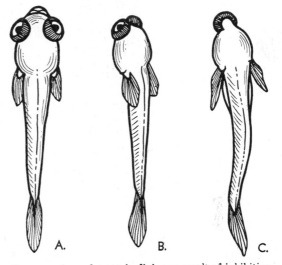

Suppression of eyes in fish as a result of inhibition of the head by a harmful substance (in this case magnesium chloride). **A,** normal individual. **B,** fish with eyes close together as a result of inhibition of growth of median region of head. **C,** fish with single median eye due to inhibition at a very early stage in head development. (After Stockard)

these two axes, one highest in the mid-region and decreasing on the two sides (the mediolateral gradient), and one higher on the ventral surface and decreasing toward the dorsal surface (the ventrodorsal gradient). These two minor gradients are usually masked by the more prominent anteroposterior gradient. And in higher animals even the anteroposterior gradient is obscured by the complexity of adult structure; it can be clearly shown only in the embryo.

THE experiments on regenerating invertebrates have helped us to understand and have been fruitful in suggesting experimental means of approach to the general problems of animal form, growth, and development.

CHAPTER 13

FREE-LIVING AND PARASITIC FLATWORMS

LIVING at the expense of one's neighbor is an old habit among animals. Practically all animals harbor one or more kinds of parasites, and most of these are themselves hosts to still smaller parasites. The total bulk of the parasites residing in one host is necessarily less than that of the free-living animal which provides the food and lodging for so many unwelcome guests. But from the standpoint of actual numbers of organisms, the animal kingdom has many more parasitic than free-living individuals.

Nearly every phylum has its parasitic members, and some phyla have more than their share. Of the three classes of flatworms which compose the phylum Platyhelminthes, two are exclusively parasitic and one consists mostly of free-living animals. The principles of parasitism will be discussed in describing the flukes and tapeworms. But the beginnings of parasitism are to be found in some of the free-living flatworms.

FREE-LIVING FLATWORMS

THE free-living flatworms are much like planarias. Externally they are covered with cilia, the beating of which creates in the water the "turbulence" that suggested the name of the class, **Turbellaria**. The form of the gastrovascular cavity provides a basis for dividing turbellarians into several groups.

The most primitive group consists of tiny worms which have a mouth but no gastrovascular cavity, and hence are called **acoels** ("without a cavity"). Food is swallowed into a solid mass of endoderm cells and there digested. There is no excretory system. The nervous system, which has several radially arranged main strands of nervous tissue and an anterior sense organ (consisting of a cavity containing a hard particle), is similar to that of ctenophores. Acoels are all marine; and because of their small size (usually only about $\frac{1}{10}$ inch in length), they are difficult to see as they swim or creep about among the rocks and seaweed along shores. They interest us chiefly because they represent a stage of complexity somewhere between that of ctenophores and the more highly developed turbellarians like the planaria.

The **rhabdocoels** ("rodlike cavity") have a straight, unbranched gastrovascular cavity. They are advanced in structure over the acoels, having a flame-cell system and a more highly developed nervous system. The worms are tiny, usually from $\frac{1}{10}$ to $\frac{1}{4}$ inch in length. They occur in fresh and salt water. *Microstomum* is a rhabdocoel that undergoes asexual reproduction like the planaria. The parts fail to separate at once, so that there result chains of as many as eight or even sixteen subindividuals, each with its own mouth. This animal is interesting also in that it possesses stinging capsules—not of its own manufacture but stolen from the hydras upon which it feeds. The stinging capsules pass from the gastrovascular cavity of the worm through the mesenchyme to the ectoderm. There they

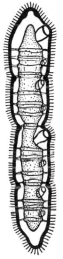

Microstomum, a **rhabdocoel.** Eight subindividuals, each with its own mouth, can be distinguished. (Modified after von Graff)

A land planarian (*Bipalium*) hanging from a branch by a string of mucus as it lowers itself to the ground. Found in greenhouses. Natural size. (After Kew)

Hoploplana, a **polyclad**, has a many-branched gastrovascular cavity.

become lodged among the epithelial cells, ready to be used in defense.

The group of turbellaria to which the planaria belongs is distinguished by a gastrovascular cavity that has three main branches, one anterior to the attachment of the pharynx and two posterior. They are appropriately called **triclads** ("three branched"). Besides the fresh-water forms there are also marine and land triclads.

The **polyclads,** so-named for their many-branched gastrovascular cavity, are exclusively marine. They are very thin, leaflike animals— sometimes almost as broad as they are long—and are, in general, the largest of the turbellaria, some species reaching lengths up to 6 inches. They usually have numerous small eyes and often a pair of sensory tentacles which project from the dorsal surface near the anterior end. Their larvas are free-swimming and have eight ciliated lobes, which are thought by some to indicate a relationship to the eight ciliated rows of ctenophores, though this is very doubtful.

Free-swimming larva of a polyclad. (After Lang)

SUCCESSFUL parasitism requires marked adaptation of the parasite to its host, and this evolves slowly. We can distinguish certain intermediate stages in this evolution among free-living flatworms.

Two animals of different species are often found living together in constant association. One derives benefit from the relationship while the other is apparently not injured. Such an association, called **commensalism,** is illustrated by *Bdelloura*, a flatworm that lives attached to the gills of the horseshoe crab (see chap. 23). Bdelloura is the *commensal* and receives shelter, free transportation, and nourishment in the form of tiny scraps from the food of the horseshoe crab, its *host*.

Sometimes a commensal incidentally benefits the host, which may then become so dependent upon this service that it cannot get along without the commensal. Such an association of mutual benefit is called **mutualism.** An example of this has already been given in the case of the intestinal flagellates of termites (p. 47). A well-known case of mutualism is that of *Convoluta roscoffensis*, a tiny marine acoel. The young convolutas are colorless when first hatched, but shortly they take on a green color as their mesenchyme becomes filled with small, green, plantlike flagellates. After entering the convolutas, the flagellates lose their flagella and undergo other changes, but continue to carry on photosynthesis, producing carbohydrates and oxygen. From the flatworm they receive a sheltered place in the sunlight and a steady supply of carbon dioxide, nitrogen, and phosphorus (from the metabolic wastes of the animal). These last are

Bdelloura, a **triclad** that lives as a **commensal** on the gills of the king crab.

particularly important because nitrogen and phosphorus, in a form available for protein manufacture, are at a premium in the ocean. While the young convolutas feed like other flatworms, the adults do not feed and are completely dependent for their nourishment upon their plant guests. Experiments have shown that the worms live longer in the light (where the contained flagellates can carry on photosynthesis) than in the dark. Convolutas live on sandy shores between tide-levels and regularly migrate with the tide. As a result, the algae are exposed to sunlight for a maximum time. It should be added that the relationship between Convoluta and its algae has two different phases. In the early stages of the association the worms feed, and there is also a passage of fat from the algae to the tissues of the animal. In the later stages, when the worms stop feeding, they digest the algae in their tissues. This finally results in the death of the worms and suggests that the mutualism is not well balanced.

Convoluta, an **acoel** that has a dark spinach-green color from the thousands of green flagellates that live imbedded in the mesenchyme. (After Keeble and Gamble)

Commensalism usually evolves, not in the direction of mutualism, but toward **parasitism**. A commensal that at first takes only shelter, and then scraps of food, finally begins to feed on the tissues of the host body, and the host suffers a certain amount of harm. Should the parasite become so well adjusted that it causes little damage, and, in fact, finally proves to be of some service to the host, the parasitism becomes a mutualism. Thus, the three kinds of relationships are only different stages in the process of living-together, and it is not always possible to draw a sharp line between them.

Mutualism and parasitism can probably arise directly from commensalism, but they may evolve from each other. Since mutualism requires the greater number of adjustments, it is relatively rare as compared with parasitism.

A well-adapted parasite usually lives without causing serious harm to its host, for the success of any parasite depends upon the continued success of the host. A well-adapted parasite, already mentioned, is the trypanosome (p. 44), which lives in the blood of African wild game with no apparent ill effect upon the game. The fact that this same flagellate causes a severe illness, and finally death, when it gets into the blood of man or his domestic animals, is taken as an indication that the parasite has only relatively recently come into contact with these hosts. Neither parasite nor host has had time to make the proper adjustments. Some of the flatworms parasitic in man are fairly well adapted—for example, the tapeworms. The flukes are often less so.

FLUKES

THE flukes (class **Trematoda**) differ from free-swimming flatworms in the loss of external cilia (in the adult) and in the development of a thick outer layer, the **cuticle.** Flukes have one or more **suckers** by which they cling to their host. One usually occurs at the anterior end

External parasitic fluke (*Acanthocotyle*) moving about on the surface of a fish, its host. It holds on by means of the large posterior sucker. (After Monticelli)

surrounding the mouth, another at the posterior end or on the ventral surface near the middle of the body. The mouth leads through a muscular pharynx into a two-forked gastrovascular cavity. There is an excretory system of flame cells and canals. The nervous system is usually well developed, but there are no special sense organs. In all these respects the flukes are relatively simple animals.

However, when we consider the **reproductive system,** which occupies most of the animal's interior, we find a degree of complexity seldom equaled and not surpassed even in the higher animals. In the posterior part of the body is a pair of testes, from each of which a duct leads to the genital pore near the second sucker. In front of the testes is a single ovary,

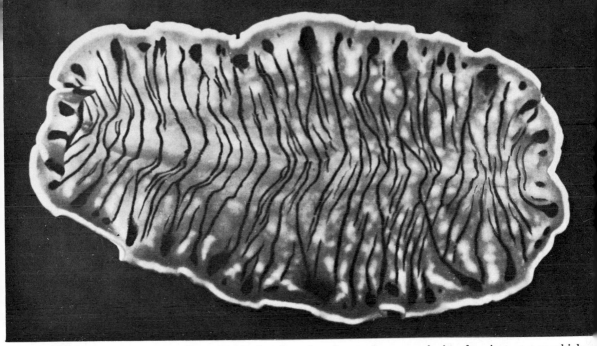

A polyclad (*Pseudoceros*), pale yellow with black stripes, found on rocks or on colonies of tunicates upon which it feeds. About 1 inch long. The head, to the left, bears sensory projections. (Photo of living animal. Bermuda)

A fresh-water triclad (*Dugesia*) fed a dye to make the branches of the gastrovascular cavity stand out. Actual size of worm, ¼ inch. (Photo of stained preparation)

A common polyclad of our West Coast is this extremely flattened leaflike worm. The tiny dots on the head are the numerous eyespots. Actual size of worm, ¹¹⁄₁₆ inch. (Photo of stained preparation)

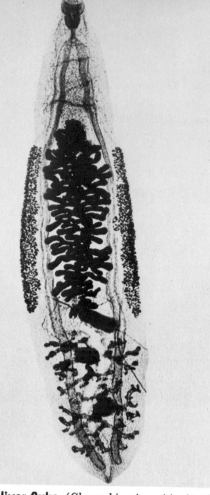

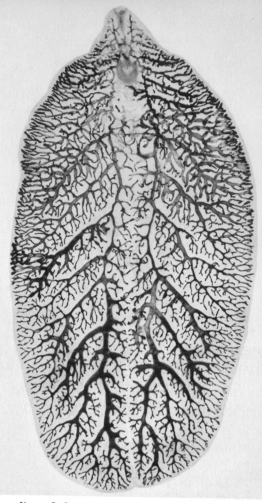

Chinese liver fluke (*Clonorchis sinensis*) showing the two-branched gastrovascular cavity, and sex organs. Actual size, ½ inch. (Photo of stained preparation, courtesy Gen. Biol. Supply House)

Sheep liver fluke (*Fasciola hepatica*) stained to show the highly branched gastrovascular cavity. Actual size, ¾ inch. (Photo, courtesy Gen. Biol. Supply House)

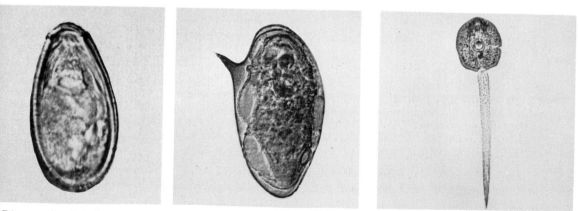

Diagnosis of fluke infection is made by examining the excreta for eggs. *Left:* Egg of a liver fluke. *Right:* Egg of blood fluke (*Schistosoma mansoni*). (Photos, courtesy General Biological Supply House)

Cercaria showing tail and ventral sucker. The species is unknown. (Photo of stained preparation)

from which a long, much-coiled tube, the **uterus,** also leads to the genital pore. In the body margins are numerous yolk glands, whose ducts connect with the uterus. The presence of the uterus is the chief difference from the reproductive system of planarias. In the uterus are stored the immense numbers of eggs found in connection with the parasitic habit.

The *lowest grade of parasitism* is that practiced by the flukes which live as **external parasites** attached to the skin or to the gills of fish, feeding

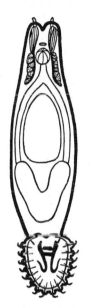

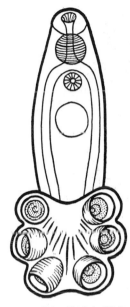

External parasitic fluke (*Gyrodactylus*) holds on to its goldfish host with a large sucker surrounded by hooks. (Modified after Kükenthal)

Parasitic fluke (*Polystoma*) in mouth cavity of a turtle. (Modified after Stunkard)

upon the epithelial tissue or on blood. Hanging on to the outside of a fast-moving object is no easy matter, and these flukes frequently have an enormously developed sucker (or group of suckers) at the posterior end, besides numerous hooks. As they move about on the surface of their host, these flukes might stray occasionally into the cavities which communicate with the exterior: the mouth, the nasal passages, and the urinary bladder. Thus, it is not surprising to find that many aquatic vertebrates regularly harbor parasitic flukes which have become adapted to live in these cavities, where the danger of being swept off is very much less. The flukes living on the skin and gills and those inhabiting cavities which communicate freely with the exterior have simple life-histories, like those of the free-living flatworms.

The flukes which live as **internal parasites,** imbedded in the tissues or clinging to the lining of cavities far from the surface, have little trouble holding on securely; their hooks or suckers are not as elaborate as those of external parasites. But the problem of getting their offspring established in a new host is a much more difficult one. It has been solved by an increase in the number of potential offspring and by the development of complex life-histories. All the flukes which parasitize man are internal.

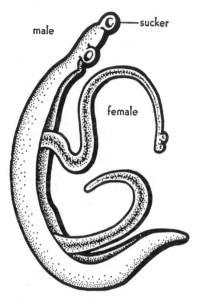

The most important of these are the **blood flukes** or schistosomas, elongated and slender flukes which differ from most in that they are not hermaphroditic but occur as separate males and females. The sides of the male fold over to form a groove in which the longer and more slender female is held. In *Schistosoma japonicum,* the species which we shall consider, the worms live in the blood vessels of the intestine, clinging to the walls of the vessels by means of suckers and feeding on blood.

The female lays her eggs in the small blood vessels of the intestine wall, close to the cavity of the intestine. As the blood

Human blood fluke (*Schistosoma japonicum*). (Modified after Looss)

vessels of the host become congested with eggs, the walls of the vessels rupture, the intestinal epithelium breaks, and the eggs are discharged into the cavity of the intestine. From there the microscopic eggs are carried out in the feces. If the feces were removed by a modern sewage system or deposited in a dry place, that would be the end of the young schistosomas. But in China and Japan, where this parasite flourishes, human feces are used to fertilize the soil, and for that purpose are conserved in reservoirs on the banks of canals or irrigation ditches. This provides the eggs with ready access to water, where they hatch. A ciliated larva, the **miracidium,** emerges and swims about. If the miracidium does not encounter a snail of a certain kind, it perishes after about twenty-four hours. If it comes into contact with the right kind of snail, it burrows into the soft body of the snail and feeds on the tissues. Meanwhile, the cilia are lost and the miracidium is transformed into a sac, called a **sporocyst,** which produces asexual buds internally. These buds, called **cercarias,** resemble the adults in several ways. They have two suckers, a forked digestive tube, and an excretory system with flame cells, but differ in possessing tails. They make their way out of the snail and swim about near the surface of the water, where they come into contact with the skin of a man who is bathing or wading. Millions of Chinese and Japanese are infected during the planting of rice, as they stand barelegged in the shallow water of the rice fields; the diagram which heads the chapter shows the principal stages in the life-history and the method of infection in man. The cercaria attaches itself to the skin and (by means of glands) digests its way through the skin into a blood vessel. It is carried in the blood stream to the blood vessels of the intestine. There the young fluke feeds and grows into an adult worm, finally mating with another that entered at the same time or with one already established from a previous infection.

The presence of schistosomas in man causes a disease (schistosomiasis) characterized by body pains, a rash, and a cough in the early stages, severe dysentery and anemia later on. Victims may live for many years but gradually become weak and emaciated and eventually many die of exhaustion or succumb to other diseases because of their weakened condition.

Control measures for schistosomiasis might reasonably begin with sanitary disposal of human feces. This is not practicable in the orient, however, for the use of human feces as fertilizer is an important part of the economy of the people. It has made possible intensive cultivation of the same soil for thousands of years (whereas in the United States we sometimes deplete the nitrogen content of the soil in two or three generations). In Japan, where infection is restricted to small areas, it has been possible to kill the snails with chemicals. In China, where infection involves vast areas in the Yangtze Valley, it is better to educate the farmers to conserve the feces for a few weeks before using it in the fields, so that the young schistosomas within their protective egg membranes will have died.

Other species of blood flukes infect peoples in the northeastern part of South America, certain places in western Asia, and a large part of Africa. In Egypt the commonest blood fluke lives in the blood vessels of the urinary tract; the eggs are extruded into the bladder and then come out in the urine. The life-history involves a snail. Infection occurs during bathing or from drinking water containing cercarias. The disease is spreading, owing to increasing use of irrigation in farming. It has been estimated that about three-fourths of the population is infected, and one parasitologist has said that Egypt will never be a country of consequence until there is better control of the flukes that are now draining the energy of the people.

WHEN a parasite lives for part of its life-cycle in one kind of animal and spends another part of its life-cycle in another kind of animal, the host that harbors the sexually mature form is said to be the **final host,** and the one that harbors the young stages is called the **intermediate host.** In the case of *Schistosoma japonicum*, man is the final host, and a certain species of snail is the intermediate host.

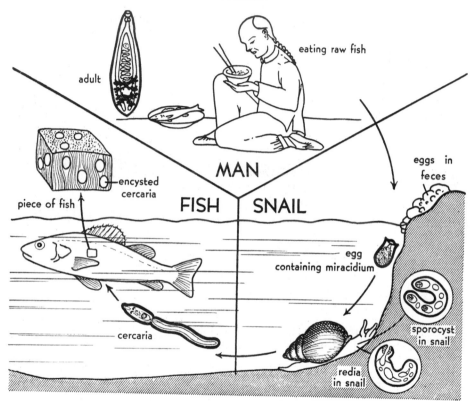

Life-cycle of the Chinese liver fluke (*Clonorchis*). (Based on Faust)

A similar type of life-history is shown by the large **liver fluke,** *Fasciola hepatica*, which inhabits the bile passages of the liver in cattle and sheep and inflicts severe, often fatal damage—with important economic consequences for stock-raisers. Instead of boring directly into the final host, after leaving the snail, the cercarias of this fluke encyst on grasses and other vegetation and are eaten by the final host.

A liver fluke that lives in man and illustrates a life-cycle involving *two intermediate hosts* is the **Chinese liver fluke,** *Clonorchis sinensis*, of China,

Japan, and Korea. The adult is about a half-inch long and has two suckers, one at the anterior end and one a short distance behind this. The cuticle is thick and highly resistant to digestive fluids. The fluke is hermaphroditic, and the fertilized eggs pass from the liver into the intestine and out with the feces. If the feces get into water, as they commonly do, the eggs do not hatch into free-swimming miracidia, as in most

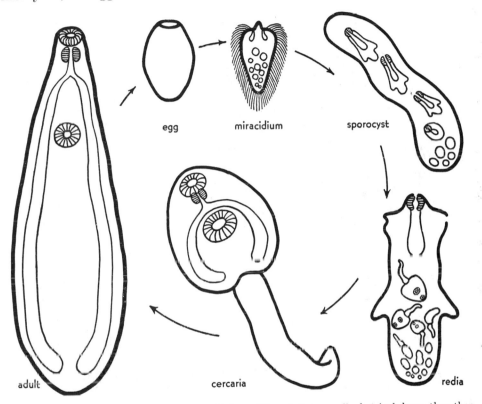

egg   miracidium   sporocyst

adult   cercaria   redia

Typical stages in the **life-cycle of liver flukes.** The adult is usually ½–1 inch long; the other stages are microscopic. (Modified after Faust)

flukes, but are eaten by snails. Within the digestive tract of the snail the egg opens, and the miracidium emerges and makes its way through the wall of the digestive tract into the tissues of the snail. There it becomes transformed into a sporocyst, which produces, instead of cercarias, another asexual form, the **redia.** The development of redias further increases the number of young forms, for each redia subsequently produces many cercarias, which escape from the snail and swim about. The cercarias encyst, not on grass like those of the sheep fluke, but in the muscles of a fish, which thus serves as the second intermediate host. They burrow through the skin of the fish, lose their tails, and secrete about themselves protective capsules. The fish responds by forming an outer capsule around each one produced by a parasite. There they remain until the fish is eaten by the final host, man. In the human stomach the cysts are digested out of the flesh, and in the intestine the capsule is weakened and the young fluke emerges. It makes its way up the bile duct and into the smaller bile passages of the liver, where it attaches by its suckers and feeds on blood. These flukes may persist for many years, causing serious anemia and disease of the liver from blocking of the bile passages.

In this case control should be a relatively simple matter, for it is only necessary to cook fresh-water fish thoroughly to destroy the encysted cercarias. Yet in certain regions in the south of China from 75 to 100 per cent of the natives are infected. Not only do they like to eat raw fish, but, unfortunately, the cost of the fuel necessary to cook the fish is an economic problem.

AS THEY become better adapted, parasites usually lose active means of transportation and depend upon more or less **passive transfer** to new hosts. In the flukes and other parasites passive transfer is frequently achieved through the food habits of the host. Thus the sheep liver fluke is rare in man because man does not usually eat grass, though he sometimes becomes infected by eating water cress upon which cercarias have encysted. He frequently gets the Chinese liver fluke, however, because in some places he habitually eats fresh-water fish containing encysted cercarias. Having once smuggled themselves into a host, the "problem" of an internal parasite (not of the individual animal, but of the species) is how to get the offspring into new hosts. The easiest way of leaving a host is with the outgoing feces, just as the easiest way of entering is by way of the mouth. But passive modes of transfer are very hazardous. The chance of having the eggs or the larvas eaten by the right kind of host at the proper stage in the life cycle is very small. Only parasites which produce enormous numbers of potential young can survive this kind of life-cycle. It is not surprising, therefore, that most parasites seem to live only to reproduce, the reproductive organs occupying most of the animal's body.

The chance that an egg will reach a suitable spot for hatching is remote, and the probability that the miracidium will find a snail within a short time (the time being limited by the small amount of energy available in the food stored within the egg) is even more remote. But if even a single miracidium manages to enter a snail, it can multiply within its intermediate host by asexual means and so compensate for the enormous loss of potential individuals by the random distribution of the eggs. It has been estimated that a single miracidium will give rise, through several generations of sporocysts and redias, to as many as ten thousand cercarias. Of all these, only a few will reach new final hosts.

PARASITISM often results in weakness or disease of the host, but the **effects of parasitism on the parasite** are even more marked. The parasite retains the general plan of organization of its phylum, but it becomes so completely adapted to its peculiar environment that it usually loses many of the structures characteristic of its free-living relatives. In adult flukes we saw a loss of external cilia and of sense organs, structures related to locomotion. In even more highly adapted parasites, like the tapeworms, there is a loss of still more structures.

TAPEWORMS

THE tapeworms (class **Cestoda**) are usually long, flat, ribbon-like animals, some species of which live as adults in the intestine of probably every species of vertebrate.

The most common tapeworm of man is the so-called "beef tapeworm" (*Taenia saginata*). It maintains its place in the intestine, despite the constant flow of materials, by means of four suckers on the minute knoblike **head** (or scolex). Behind the head is a short neck or growing region, from which a series of body **sections** (proglottids) are constantly budded off. The sections closest to the neck are the youngest ones; those farthest away, the most mature. Thus the body widens gradually along its length, and the sections are in all stages of development.

The body is covered externally by a protective cuticle, as in flukes; and there is no ectodermal epithelium. (After secreting the cuticle the ectoderm cells sink into the mesenchyme.) Unlike the flukes, which feed actively on the tissues of their host and do their own digesting, tapeworms have no mouth and *no trace of a digestive system*. They live in the intestine of their host, where digested food is readily available; there they simply "soak up" their nourishment—truly the laziest way of living.

Beneath the cuticle are longitudinal and circular muscles. The **nervous system** is like that of turbellarians and flukes, but less well developed. From a small concentration of nervous tissue in the head two longitudinal nerve cords run backward through the body. Between the nerve cords, and parallel with them, run two longitudinal **excretory canals,** connected with each other by a crosswise canal near the posterior border of each body section. The smaller branches of the excretory system end in flame cells.

The **reproductive system** lies imbedded in the mesenchyme and is so highly developed in mature sections that the tapeworm is sometimes de-

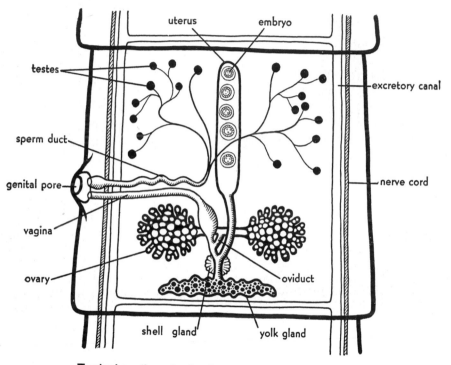

Typical section of a beef tapeworm (*Taenia saginata*)

scribed as nothing but a bag of reproductive organs, a complete set of which, both male and female, develop at some time in every section. The male sex organs start to grow first. They consist of numerous small **testes,** scattered throughout the mesenchyme and connected by many fine tubes with a single large convoluted **sperm duct,** the end of which is modified as a muscular organ, the penis, for the transfer of sperms. The sperm duct opens into the **genital chamber,** which connects with the outside through a **genital pore.** Running parallel with the sperm duct and also opening into the genital chamber is the **vagina,** a female duct which

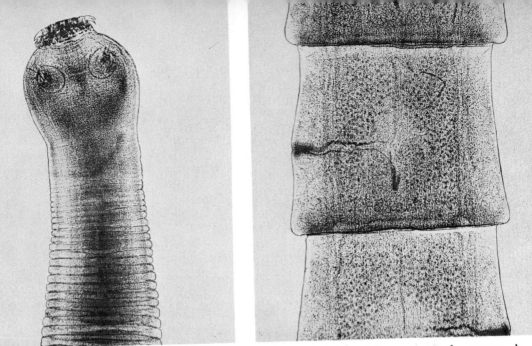

1. Head of tapeworm (*Taenia serrata*) from a dog, showing suckers, hooks, and young sections which develop in region behind head. Actual dia. of head 1/20 inch.

2. An immature section with male organs developed. The female organs are only beginning to appear.

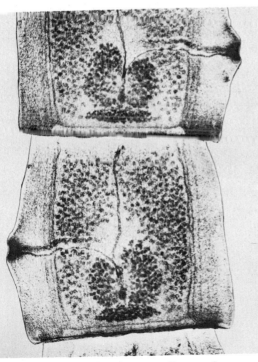

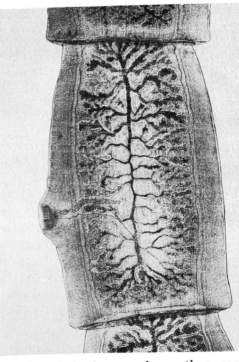

3. A mature section showing both sets of sex organs well developed. To identify the organs see the diagram of a typical section in the text.

4. A ripe section is not much more than a sac containing the enormously enlarged uterus filled with eggs. Actual width 1/5 inch. (Photos of stained preparations, courtesy Gen. Biol. Supply House)

144-1

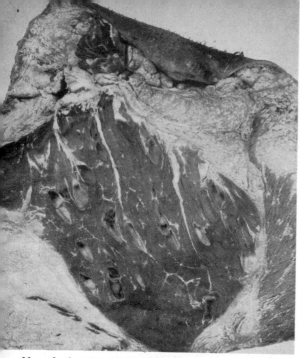

Measly beef, or beef containing the bladders of the beef tapeworm, is now rarely found on the market, thanks to government meat inspection. Meat which is cooked until it has lost its red color is safe. (Photo, courtesy Army Medical Museum)

Echinococcus cysts in section of the liver of a person who accidentally swallowed the mature sections or eggs. The adult parasites live in the intestine of dogs. Actual size of cysts ½–2 inches in diameter. (Photo of specimen in Pathology Museum, University of Chicago)

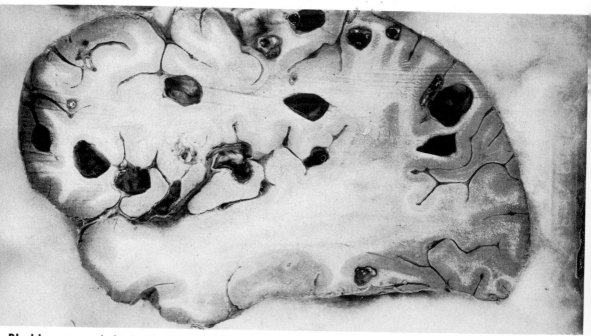

Bladder worms in brain of a woman, thirty-four years old, a resident of the Chicago region, who was brought to the hospital with a history of epileptic fits. These became more frequent until three days before her death, when convulsions set in every half hour. The brain (shown here in longitudinal section) contains 100-150 of these bladders. (Photo of specimen in Pathology Museum, University of Chicago)

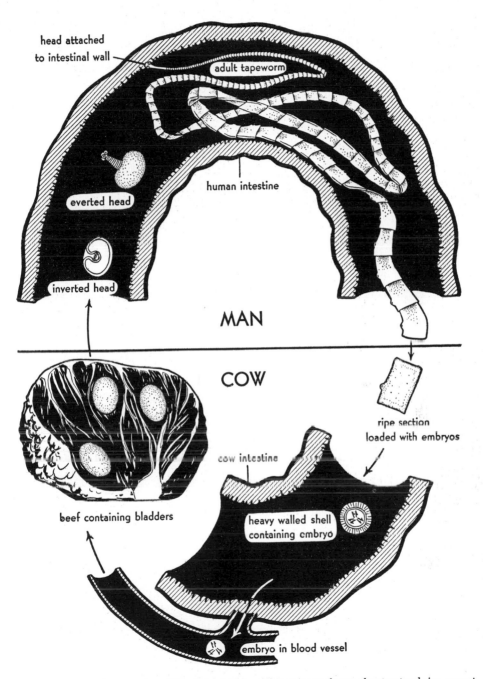

Life-cycle of beef tapeworm (*Taenia saginata*). All stages are shown about natural size, except the six-hooked embryo which is microscopic.

receives sperms. Self-fertilization can occur within the same segment, or cross-fertilization can take place between the segments of different worms when two or more are present in the same host. But the most common method is transfer of sperms from one section of the worm to a more mature section farther down the length of the same worm. This is possible when the animal is folded back on itself for part of its length. As the section matures, the male sex organs degenerate while the female sex organs develop. Eggs are produced in a pair of **ovaries** and pass into the **oviduct.** There they are fertilized by sperms that have entered through the vagina and have been stored in an enlarged portion of its inner end.

The fertilized eggs are combined with yolk cells from the **yolk gland** and are then covered with a shell secreted by the **shell gland.** The completed eggs pass forward into the **uterus,** which at first is a single sac (as shown in the diagram) but later develops numerous branches. Eventually all the female organs degenerate except the uterus, which becomes enormously distended with eggs that are already undergoing development into embryos. At this stage the "ripe" sections, each containing many thousands of young embryos, detach from the worm and pass out with the feces.

Transfer to a new final host is entirely passive and involves the eating habits of man and cow. The cow eats vegetation on which human feces have been deposited. In the intestine of the cow the eggshell is digested off. The embryo, which is armed with six sharp hooks, bores its way through the wall of the intestine and into a blood vessel. In the blood stream the **six-hooked embryo** is carried to a muscle. There it remains and grows into a sac or **bladder,** from the inner wall of which is developed the inverted head of the future tapeworm. When man eats raw or "rare" beef, the inclosed bladder is digested off; the head everts and attaches to the intestinal wall by means of its suckers. Nourished by an abundant food supply, it soon grows a long body and produces eggs.

The bladders of the beef tapeworm occur most frequently in the jaw muscles and in the muscles of the heart; these are the parts of the cow usually examined by meat inspectors. The bladders are almost half an inch long but can readily be overlooked. Meat inspection in Western countries and in the United States has greatly reduced the occurrence of this once common parasite. But there are still many cases, and it is best to avoid eating beef that is not cooked thoroughly. In parts of Africa where sanitation is poor, and in Tibet, where beef is prepared by broiling large pieces over an open fire, a large proportion of the population is infected. Among the Hindus of India, who consider the cow sacred and have religious restrictions against eating beef, to be caught with a beef tapeworm would undoubtedly prove embarrassing.

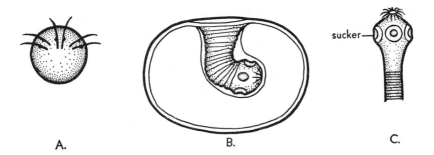

Pork tapeworm (*Taenia solium*). **A**, six-hooked embryo. **B**, bladderworm. **C**, head of adult show
ing suckers and hooks. (After various sources)

Likewise, the pork tapeworm (*Taenia solium*) is rare among Jews and Mohammedans,
who avoid the meat of the hog, and very common in parts of Europe, where pork is eaten
without thorough cooking. Meat inspection has made it uncommon in the United States.
The pork tapeworm resembles the beef tapeworm closely and has a similar life-history.
except that the bladders develop in pigs. It is especially dangerous, however, because
self-infection with the embryos can occur, and the bladderworms then develop in man
If these settle in the muscles, no great harm results. Sometimes, however, they lodge and
grow in the eyeball, interfering with vision. Certain cases of insanity or epilepsy are really
due to bladderworms in the brain.

A very thin person is frequently accused by his friends of harboring a tapeworm, and
it is true that infected individuals are sometimes emaciated. The anemia and the nervous
disorders that sometimes occur are not due as much to loss of food as to the poisonous
substances given off by the parasite. Also, the mere bulk of the worm, especially when
folded back on itself many times, may block the intestine and produce serious temporary
illness.

The presence of a tapeworm can be detected by the appearance in the feces of the
white, ripe sections loaded with embryos. The only way to get rid of the parasite is to
take by mouth some drug which kills the head and causes it to detach from the intestinal
wall, whereupon the whole worm is evacuated with the feces.

Many tapeworms have more than one intermediate host. The "broad
fish tapeworm" (*Diphyllobothrium latum*), which may be $\frac{3}{4}$ inch wide and
60 feet long, is the largest and the most injurious tapeworm that lives in
man. The life-history requires that the eggs reach water, that the larvas
which hatch are eaten by copepods (small crustaceans), and that the
copepods are eaten by fish. Man gets the parasite when he eats raw or
imperfectly cooked fish.

The fish tapeworm occurs in many places all over the world and has been known for
centuries in the Baltic region of Europe, where in some localities nearly all of the people
are infected. In relatively recent years Baltic immigrants to our Great Lakes region have
brought these tapeworms with them and have established them by infecting the fish in

the lakes of Minnesota, northern Michigan, and Canada. Since these lakes supply millions of pounds of fresh fish to other parts of the country, and since visitors to the region carry tapeworms home to their own localities, this parasite is spreading in the United States. Pikes and pickerels are the fish most commonly infected with the bladderworms, but others may be as well. Those who eat fish from these regions should never taste the raw fish during the preparation and should be careful to cook the fish very thoroughly; smoked fish may not be safe.

Sometimes man is the intermediate host for a tapeworm that lives its adult life in some other mammal. *Echinococcus granulosus* is a minute tapeworm (with only three or four sections) that lives as an adult in the

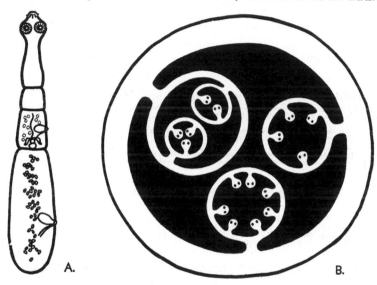

A.   **B.**

Man is the intermediate host for *Echinococcus*. **A**, adult, $\frac{1}{8}$ to $\frac{1}{4}$ inch long, lives in the dog. **B**, cyst from liver of man. (Modified after Leuckart)

intestine of the dog and occurs only as a larva in man. In spite of the small size of the adult, the larva is enormous. Human infection results from drinking contaminated water or from allowing dogs to lick the face and hands. Because of the unclean habits of dogs their tongues are likely to carry tapeworm eggs. The young larva develops into a hollow bladder. From the inner walls of this grow smaller bladders, and within each of these are produced numerous heads. The whole structure is known as a "hydatid cyst" and may grow to the size of an orange or even larger. When such a cyst develops in the brain, the results are extremely serious. Some cases of "epilepsy" are due to hydatid cysts. From the "point of view" of the parasite, development of the cysts in man is unfortunate because man is rarely eaten by dogs, and the cysts cannot reach their final host. The most common intermediate hosts are sheep and cattle. The parasite is common in the great cattle- and sheep-raising regions of the world, of which the United States is one.

N O ONE has ever seen a free-living animal evolve into a parasite. But there are so many animals that lead lives which are transitional between these two extremes that we feel fairly safe in hazarding the guess that parasitism starts as a harmless association or commensalism in which one animal takes shelter in the home of, or on the body of, another. Next, the commensal takes small scraps of the food of its host. When it begins to feed on the tissues of the host, the commensal becomes a parasite. External parasites are little changed from their free-living relatives except for the development of hooks or suckers for holding on. Internal parasites usually show marked structural adaptations to their special environment. The nervous and muscular systems, so important for an active free-living life, may become degenerate. Many highly adapted parasites lose the free-swimming young stages and depend entirely upon passive transfer from host to host. In some intestinal parasites digestive organs are reduced and in others are lost altogether. On the other hand, the reproductive system of parasites is so highly developed that most of the energy of these animals is directed toward one main activity—the production of tremendous numbers of eggs to offset the losses incurred in the hazardous transfer from one host to another. Asexual reproduction within the body of an intermediate host is another method for increasing the number of young forms, and therefore the chance that a reasonable number will reach the final host.

The young stages of many parasites live imbedded in such tissues as liver, muscle, brain, etc. But the adult, which produces the eggs, must live in or near some cavity which has direct access to the outside. The digestive cavity is the one most frequently occupied, since the digestive tract is the easiest and consequently the most popular highway for the entrance and exit of parasites, particularly for those that depend upon passive transfer.

The relation of a parasite to its host requires not only marked adaptation on the part of the parasite but often also an adjustment on the part of the host. The host may secrete a capsule about the larva imbedded in its muscles; this helps to confine the activities of the larva. Or the host may develop a resistance to the toxic substances given off by the adult.

The *parasite-host relationship is usually specific.* Some parasites can live in a variety of closely related hosts, but most can develop in only one particular species. Some can grow in species other than their normal hosts; but when they do so, there is a lack of mutual adjustment and the host or parasite may suffer unduly.

There is a tendency among most people to look upon parasitism as an aberrant way of life and upon parasites as being somehow "immoral" or at least less "respectable" than their free-living relatives. But since there are more parasites than free-living individuals, a parasitic existence must be considered a "normal" way of life. Who can say that the parasite, the very existence of which depends upon doing as little harm as possible to its host, is a less "considerate" creature than the voracious carnivore, which kills its victim outright?

GNATHOSTOMULIDS

The phylum **GNATHOSTOMULIDA** is not the most recently created phylum, but it is the only one whose members seem to have escaped detection until the 20th century. And no wonder. They live on marine shores in fine grained sand, but are most abundant, probably more so than any other metazoans, as one goes deeper into the sandy mud to the black odoriferous layer which even zoologists tend to avoid because its odor of hydrogen sulfide suggests that it will harbor little besides bacteria of decay.

Gnathostomulids are microscopic, ranging from $\frac{1}{2}$ mm to several mm long. They move so rapidly among the particles of mud and sand grains in a microscopic field that they were probably passed over as minute turbellarians. And they can only be recognized when alive, so that preserved collections did not reveal them. At first they were classed with the turbellarians because they are ciliated, flattened and elongated little hermaphroditic worms without an anus. But it is now evident that they have many distinguishing features, especially the halo of stiff sensory cilia on the head, a skin consisting of a single layer of cells each with a single cilium, and a unique chewing apparatus with little toothed jaws.

CHAPTER 14

ONE-WAY TRAFFIC—PROBOSCIS WORMS

THE proboscis worms are common along seashores under stones and among seaweeds; a few live in fresh water or damp soil. Their elongated flattened bodies range in length from less than an inch to many feet, and they are often colored a vivid red, orange, or green, with contrasting patterns of stripes and bars. Their most distinctive character is the **proboscis,** a long, muscular tube which can be thrown out to grasp prey. The aim of the proboscis is said to be "unerring," and this is the meaning of **NEMERTEA,** the name of the phylum to which these worms belong.

Like planarias, they are bilaterally symmetrical. The anterior end is more or less marked off as a head and bears numerous simple eyes and specialized sensory cells. The proboscis worms are not a very large group; they are not usually seen by visitors to the seacoast, nor do they have special economic or medical importance. They are described here because they are the lowest animals to possess two important advances in construction over the flatworms. These structural improvements are present in all higher animals and are, apparently, essential to any increase in complexity over the flatworm plan.

First is an increased efficiency of the **digestive system.** The mouth is near the anterior end and opens into a long, straight **intestine** having short side branches. This intestine extends the length of the body and at the posterior end communicates with the exterior through an opening called the **anus.** This is the first phylum of animals encountered in which all of the members have a digestive tract with two separate openings, one at the anterior end exclusively for taking in food, and the other at the posterior end for the exit of indigestible materials. This "one-way traffic" has great advantages over the general "traffic jam" in which the food finds itself in the gastrovascular cavity in coelenterates and flatworms, where the newly ingested food becomes mixed with partly digested food and indigestible residues. In a gastrovascular cavity the lining cells must be both digestive and absorptive. The whole system is inefficient and an obstacle to differentiation of the digestive epithelium. In a one-way system the food passes along a digestive tract which can be differentiated into various parts or organs with specialized functions: one region for food intake, another for digestion, still another for absorption, and so on

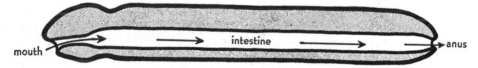

Diagram of the **digestive system** of a nemertean.

until the end, the anus, which is for elimination. All animals above flatworms have an anus, which makes possible a continuous digestive canal through which food passes in only one general direction.

The proboscis of nemerteans, which is often as long as the body, lies inside a muscular sheath just above the digestive tract. When a likely victim approaches, such as a nereis (an annelid worm), the proboscis is extended quickly, wraps around the prey, and entangles it with the aid of a sticky mucous secretion. In some nemerteans the proboscis bears a sharp stylet which pierces the body of the prey and makes a wound into which is poured a poison from glands in the proboscis. The proboscis is fastened to its sheath near the anterior end and is turned inside out when protruded. It is drawn back by a retractor muscle attached to its posterior end.

The second important structural advance pioneered by the nemerteans is a new system, the **circulatory system,** which takes over the circulatory function of the old gastrovascular cavity. This makes it possible for the intestine to become more efficient in digestion, and leaves the distribution of food and oxygen and other substances to a system more fit for the job. The circulatory system of nemerteans consists of three lengthwise muscular tubes, the **blood vessels.** These lie in the mesenchyme, one on each side of, and one just above the intestine, and connect with each other

by transverse vessels. They contain a fluid, the **blood,** which is generally
colorless and contains cells. In some species the cells are red from the
same substance (hemoglobin) that colors human blood red. Hemoglobin

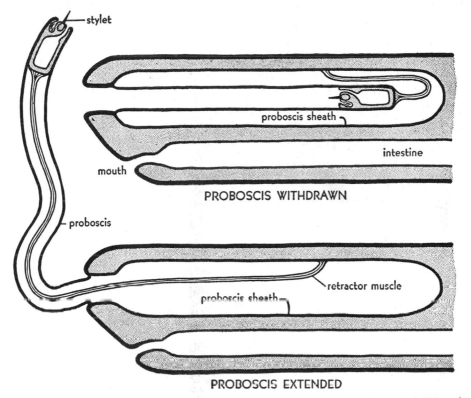

PROBOSCIS WITHDRAWN

PROBOSCIS EXTENDED

Diagram of the anterior end of a nemertean showing **proboscis** withdrawn and extended. When the
stylet is lost, it is replaced from a sac of extra stylets in the tip of the proboscis. (Based on Coe.)

Diagram of the **circulatory system** of a nemertean.

combines readily with oxygen, making the blood more efficient as an oxy-
gen-carrier. It is most characteristic of vertebrates but occurs in many
groups of invertebrates, of which the lowest are the nemerteans. The circulation of nemerteans is primitive in several important respects. There

is no special pumping organ or heart to circulate the blood. The two outer
lengthwise vessels are contractile, but blood flow depends mostly on
pressure from muscular waves that pass along the body wall. Also, the
blood vessels are not finely branched; therefore materials must move
longer distances to and from cells by the slow process of diffusion. Like
most structures when they first arise, the circulatory system of nemerteans
lacks specialization and does not do as good a job as the more complex
circulatory systems of higher groups.

scribed above, is very similar to that of a planaria and includes all of the same organ-
systems. The animal is covered completely with a ciliated epithelium which contains

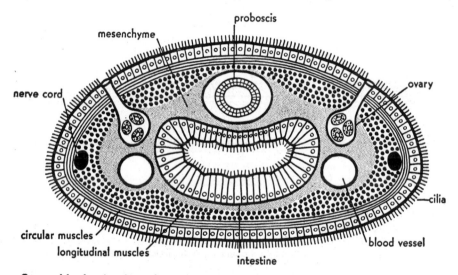

General body structure of a nemertean shown in cross-section. (Based on W. R. Coe)

many gland cells. Under the epithelium are thick muscular layers, circular and longitudinal, by which the highly contractile animal can execute agile movements of all sorts.
The digestive tract consists of an epithelium. Some muscle cells are usually associated
with it, but food is moved along chiefly by means of muscular contractions in the body
wall which pass down the animal from front to rear and force the food along as they press
on the intestine. These muscular waves also assist the flow of blood. Between the gut and
the body wall there is a thick layer of mesenchyme cells. Imbedded in the mesenchyme
is the circulatory system, already described, and also the excretory system, consisting of
a pair of lateral canals with side branches ending in flame cells. Wastes are removed from
the mesenchyme and from the blood and pass into the canals, which open on the surface
by pores. The nervous system is similar to that of flatworms, but the brain is more massive
and forms a ring around the digestive tract; longitudinal nerve cords run the length of
the worm on each side.

The eggs and sperms are produced in little sacs which lie in the mesenchyme between the branches of the intestine. Each opens directly to the outside through its own pore on the surface. There is none of the complicated sexual apparatus seen in flatworms, the sex cells being simply shed to the outside, and in this respect the nemerteans are less spe-

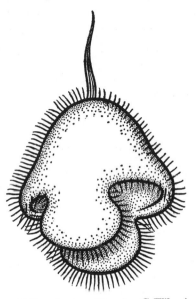

Pilidium larva. (Based on C. Wilson)

cialized than the flatworms. In marine nemerteans there is a ciliated larva shaped like a helmet with earlaps and known as the **pilidium larva.** It has a ventral mouth but no anus, and at the end opposite the mouth is a sense organ topped by a tuft of long, stiff flagella. These features suggest that the nemerteans probably arose from an ancestral stock related to the flatworms and ctenophores. The pilidium is similar to the trochophore larva of several of the higher phyla. By a series of rather complicated changes the pilidium larva develops into the adult worm.

CHAPTER 15

ROUNDWORMS

MOST people are host at some time or other to the cylindrical white worms called "roundworms." Of the fifty different species of roundworms that have been found in man, only about a dozen are common parasites. Of these, some are harmless and do not even make their presence known, while others cause mild or very serious diseases.

Many roundworms look like animated bits of fine sewing-thread; and from the Greek word for thread, "nema," has been derived the technical name of the group, phylum **NEMATODA.**

This is by no means a small or obscure phylum. Nematodes are so abundant that a spadeful of garden soil teems with millions of them. When we see a sick dog, our first guess is that he has more roundworms than he can stand. And even in the best-regulated cities roundworms occur in the drinking-water. The widespread occurrence of this group inspired a leading student of the nematodes to write:

If all the matter in the universe except the nematodes were swept away, our world would still be dimly recognizable, and if, as disembodied spirits, we could then investigate it, we should find its mountains, hills, vales, rivers, lakes, and oceans represented by a film of nematodes. The location of towns would be decipherable, since for every massing of human beings there would be a corresponding massing of certain nematodes. Trees would still stand in ghostly rows representing our streets and highways. The location of the various plants and animals would still be decipherable, and, had we sufficient knowledge, in many cases even their species could be determined by an examination of their erstwhile nematode parasites.

Some roundworms occupy very specific niches. One species occurs practically only in the appendix of man. Another has never been found anywhere except on the felt mats on which Germans set their mugs of beer. However, the group as a whole lives anywhere that other animals can live. Any collection of earth or of aquatic debris from an ocean, a lake, a pond, or a stream, if examined with a lens, will reveal the tiny white worms thrashing about in a

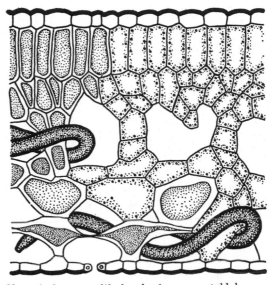

Nematodes parasitic in plants cause untold damage. As small larvas, they enter through the breathing pores of the leaves. They suck up the cell sap, causing a wilting and withering of the leaves, stunting the plant, and sometimes inducing the growth of galls. The worms shown here (*Aphelechus*) are in a section of the leaf of a dahlia. They have already injured the cells to the left. (After H. Weber)

way so characteristic of roundworms that it immediately identifies them.

Nematodes are so remarkably alike that a description of an **ascaris** roughly fits almost any other roundworm. The elongated cylindrical body is pointed at both ends. Curiously enough, it is entirely devoid of cilia, outside and in. The body is covered with a thick, tough **cuticle** secreted by the underlying ectoderm, which has many nuclei but is lacking in cell walls and is called, therefore, a **syncytium.** Under this is a **longitudinal muscle layer,** divided up into four lengthwise bands by four projections of the syncytium. These bands can be seen on the outside as light lines—a dorsal, a ventral, and the slightly more prominent right and left **lateral lines.** The muscles are primitive and consist of large lengthwise cells with bulblike cytoplasmic expansions which project into the interior. The stiff cuticle and the lack of circular muscle fibers permit bendings of the body in only the dorsoventral plane; and as a result, nematodes move in a very erratic and apparently inefficient manner by simply thrashing about. When the worms are free in the water, the whip-like contortions of the body do not result in locomotion; but when they

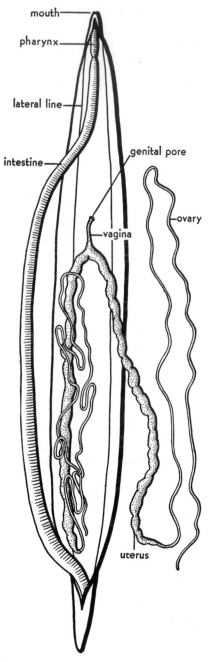

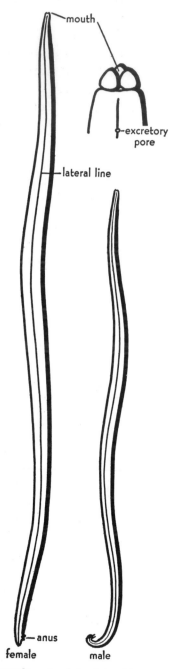

Dissection of a female Ascaris. The animal has been slit open along the mid-dorsal line, revealing the internal organs.

Ascaris, natural size, taken from the intestine of a pig. The upper figure on the right shows the anterior end, enlarged.

are in the soil, in the contents of the intestine, or in the tissues of the body, the solid particles afford friction and the worms manage to move along fairly well.

The **mouth** is at the anterior tip, encircled by **sense organs** in the form of protuberances, and leads into a short muscular **pharynx** by means of which the worm sucks in food. The **intestine** is made up of only one layer of cells and opens through the **anus** near the posterior tip. In parasitic forms which live constantly bathed by the digested food of the host the intestine has no digestive gland cells.

Between the intestine and the muscular layer there is a fluid-filled space formed by the coalescing of vacuoles in a syncytial mesenchyme. The fluid in this space, propelled by the movements of the body, aids in distributing food and oxygen. In higher animals this space develops in a different way, with the result that it is completely lined by mesodermal epithelium and becomes a new cavity called the "coelom."

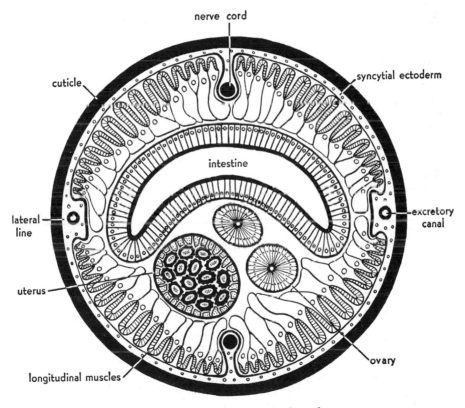

Cross-section of a female Ascaris.

The **nervous system** consists of a ring of nervous tissue, around the pharynx, from which longitudinal trunks run backward. The principal trunks are dorsal and ventral, and run in the dorsal and ventral thickenings of the ectoderm. The **lateral line,** a thickening on each side, contains a canal which is probably excretory, although there are no flame cells, as in flatworms. The two canals unite near the anterior end to form a single tube, which opens ventrally by an **excretory pore.**

The **reproductive system** lies in the space between the intestine and the muscle layer. The sexes are separate, and the males are usually

smaller than the females. The reproductive system of each sex consists of a long tube—single in the male, paired in the female—which coils back and forth in the body space. The two **ovaries** are long slender tubes which gradually widen into **oviducts** and finally into large tubes, the **uteruses,** where the eggs accumulate. The two uteruses unite into one short duct, the **vagina,** which leads to the female **genital pore,** situated in the anterior part of the worm on the ventral side. The **testis** consists of a long coiled tube in which the sperms are formed. It gradually enlarges into a **sperm duct,** which opens near the posterior end. The sperms are transferred to the female with the aid of a pair of horny bristles.

The **sperm of Ascaris** is ameboid rather than flagellated, as in most animals.

O F THE parasitic nematodes, one of the largest is *Ascaris lumbricoides*, which inhabits the human intestine. The adult is generally several inches to more than a foot in length. The males are smaller than the females and can be distinguished readily by their curved posterior ends. The oval eggs are easily recognized, in microscopic examination, by their warty shells. Two hundred thousand may be laid daily by each female. They pass to the exterior in the feces, are deposited on the ground, and there develop inside their shells into little worms which are infective to man when the eggs are swallowed. With so many eggs it would seem that everybody would be infected; but hazards occur which are fatal to the developing worm, such as drying, temperatures that are too low (below 60° F.) or too high (above body temperature), or sanitary laws and habits of men. On hatching, the little worms do not remain in the intestine but take a sort of "tour" through

The **egg of Ascaris** has a shell so resistant that the embryo may continue to develop for a time when placed in concentrated solutions of poisonous chemicals.

the body. They burrow through the intestine into the blood vessels and are then carried about to various organs, but take hold only in the lungs. They bore through the lung tissue into the bronchial (air) tubes, ascend into the mouth, and are then swallowed back into the stomach, whence they pass again into the intestine. There they remain, growing rapidly to adult size and feeding upon the digested food of their host. They resist digestion themselves by secreting a substance which counteracts the action of the host's enzymes, but if the worms die they are digested by their host.

The greatest damage to the host is done during the migrations of the young; the adult worms in the intestine seem to be relatively harmless unless they occur in large numbers. Up to five thousand worms have been found in one host, but even a hundred worms may block the intestine completely and cause the death of the host. Sometimes they wander about the body into the liver, the appendix, the stomach, and even up the esophagus and out through the nose, to the horror of the surprised host.

Infection occurs chiefly among people with bad sanitary habits, though it may be obtained in the best establishments from eating inadequately washed fresh salad vegetables grown in soil contaminated with human feces. The worms are common in our southeastern states, where the children, particularly, are likely to deposit their feces in the yard adjoining the house. The eggs are spread around by pigs, by the family dog, or by the children themselves, who finally get the eggs on their hands and carry them into the kitchen.

An ascaris which cannot be distinguished by its structure from the one that lives in man is very common in pigs. However, the ascaris eggs from pigs do not ordinarily develop to maturity in man, nor do those from man's feces infect the pig.

MUCH more serious than the big ascaris is the tiny **hookworm** (*Necator americanus*, a name which means "the American killer"). The mouth cavity of the worm contains plates by which the worm grasps a bit of the intestinal lining of the host and holds on while it sucks in blood and tissue fluids. The eggs pass out in the feces and fall on the ground, where they hatch into larval worms. These live in the soil for some time, feeding and growing. After they have attained a certain size and have stored up food, they cease to feed and are capable of infecting man. They in-

Hookworms, natural size.

vade their human host by burrowing through his skin, and infection most often occurs from the habit of going barefoot in localities where the soil is likely to contain human feces. After entering the skin, the worms pursue the same course as described for the ascaris, eventually reaching the intestine.

Hookworm disease in the United States is largely confined to rural parts of our south-eastern states. Adequate moisture and temperatures between 68° and 86° F. are necessary for the development of the worms in the soil; hence, hookworm disease of man (and also of other mammals) is restricted to regions such as the South, where these conditions prevail. The symptoms of the disease are widely known: anemia, diarrhea, and general lack of physical and mental energy, leading to a retardation of physical and mental development, so that an infected child of fifteen years of age may appear to be only ten years old. The "poor white trash" of our South have suffered such inefficiency for generation after generation. The resulting poverty, ignorance, and deterioration of culture only accentuate the condition. The fact that these poor people have been held in contempt by their more fortunate neighbors, who attribute the condition to "natural-born shiftlessness," has not been helpful. Negroes harbor the worms but are not so susceptible to their harmful effects. During the last two decades health agencies have done much to stop this drain on the national economy and culture.

Simple treatment with drugs eliminates most of the worms, but in addition to treatment it is necessary to prevent new worms from entering. Wearing shoes and avoiding contact of the skin with infected soil is one factor. A second is the sanitary disposal of feces which contain eggs. Though common in the United States and Europe, it is rare or absent in Australia and in most of the Orient.

THE **trichina worm** (*Trichinella spiralis*) is a much dreaded parasite of man. It is usually obtained by eating insufficiently cooked pork but occasionally is contracted from other kinds of meat—bear meat, for example. The worms become sexually mature in the human intestine. The eggs hatch in the uterus of the female worm, which thus gives birth directly to larval worms. The larvas gain access to the blood and lymph vessels of the human intestine and are carried about through the body. They leave the vessels and burrow into the muscles—usually those of the diaphragm, ribs, tongue, and eyes. In the muscles the larvas increase about ten times in size, to a length of 1 mm, and then encyst, curling up and becoming enclosed in a thick wall formed by the host tissue. They develop no further and eventually die, unless the flesh (containing the cysts) is eaten by a suitable host. Pigs obtain the worms by eating the flesh of other animals, usually fragments of slaughtered pigs or rats, containing **encysted larvas. In the body of the pig the worms go through the same history as recorded above for man. The muscles of an infected pig contain numerous encysted trichina worms; and if man eats infected pork that has not been thoroughly cooked, a trichina infection usually results.**

The adult worms do no harm, and after a few months disappear from the intestine. The greatest injury occurs during the migration of the larvas, when half a billion or more of them may simultaneously bore through the body. At this time there are excruciating muscular pains, muscular disturbance and weakness, fever, anemia, and swellings of various parts of the body. It is during this stage of the disease that death may occur, and about a third of the sufferers die. If the victim survives this period, the larvas become encysted and the symptoms subside, though there may be permanent damage to the muscles. In less heavily infected cases the symptoms may be mild and are likely to be diagnosed as "intestinal trouble." Serious cases are often diagnosed as typhoid fever. Thus, the actual occurrence of this disease is much higher than is generally supposed. Autopsies have shown that about 20 per cent of the population have suffered from trichinosis at some time or other.

It is possible that the religious laws of the Jews prohibiting the eating of pork resulted from experience with trichina infections, although at that time nothing was known about the worms themselves. At the present time the United States government does not inspect pork for the occurrence of encysted trichina worms, since such inspection requires microscopic examination, and light infections could be readily overlooked anyway. Inadequate inspection is worse than none, because it gives a false sense of security to the consumer. The absolute prevention of trichinosis lies with the consumer, who has only to cook all pork thoroughly. It is important not to roast pork in pieces so large that heat does not penetrate to the center. Large public barbecue picnics are often a source of epidemic trichinosis. Many of the worst cases have been due to "homemade" pork sausage which was improperly prepared. An ounce of infected pork sausage may contain one hundred thousand encysted worms. All market animals are parasitized in some way or other, and it is understood that the consumer will prepare his food properly to safeguard himself against infection.

THE **filaria** (*Wuchereria bancrofti*) is a roundworm of great importance as a human parasite in tropical and subtropical countries. This worm differs from the preceding ones in that an intermediate invertebrate host is involved in the life-cycle. The adult worms look like coiled strings as they lie in the lymph glands or ducts of an infected person. The female is 3 or 4 inches long, and the male about half this length. The female gives birth to small larvas, known as **microfilarias,** which get into the blood vessels and develop no further unless sucked up by a mosquito of the right species. Within the mosquito the larvas continue development and migrate to the biting apparatus. When the mosquito bites another person, the worms creep out onto the skin of the victim and penetrate near the bite. The chief consequence of filaria infection is the blocking of the lymph channels. This results in immense swelling and growth of the affected parts—a condition known as *elephantiasis.*

Microfilaria in blood. The worm is incased in a transparent sheath, really the inner lining of the egg. Three red blood corpuscles are shown to give scale. (Modified after Faust)

THE guinea worm (*Dracunculus medinensis*) is one of the more serious discomforts of life in India, Arabia, Egypt, and Central Africa. The male worm is not well known but is probably only about an inch long. The female is from 2 to 4 feet long and 1/25 inch in diameter. It usually

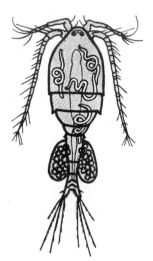

lives under the skin but sometimes lies near the surface, where it appears like a coiled varicose vein. When mature, the female approaches the surface of the skin, usually that of the arms or legs, and secretes a toxic substance which causes a blister to form. The blister breaks, exposing a shallow depression or ulcer with a hole in its center. When this ulcer is suddenly plunged into cold water (as by women when they wash clothes in the river), a milky fluid containing large numbers of tiny larvas is ejected from the hole in the ulcer. The larvas swim about in the water until they find a cyclops, a small crustacean, into which they enter and in which they undergo development. When man drinks unfiltered water containing a cyclops, the larvas are introduced into their final host.

A cyclops containing four larval guinea worms (*Dracunculus medinensis*). (After Martini)

In some places one-fourth of the population is incapacitated during part of the year by the guinea worm. Some of the symptoms, which appear at the time the blisters are formed, are vomiting, diarrhea, and dizziness. Native medicine men usually extract the worm by slowly and painfully winding it out on a stick. This often results in infection, followed by loss of the limb or by death. This method is quite successful, however, if done by a doctor, who uses the proper precautions against infection. Control of the disease would be very easy if infected natives could be taught to stay out of the water and if communities could be induced to filter their drinking water. In India this is difficult because of the religious traditions that surround the ways in which the people obtain and use water.

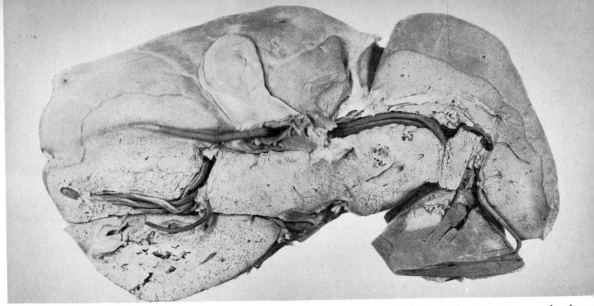

Ascaris usually lives in the cavity of the intestine, where it does relatively little harm; but it may cause death if it wanders into the body tissues. This is a human liver cut away to show ascaris worms which have entered through the bile duct. (Photo, courtesy Army Medical Museum)

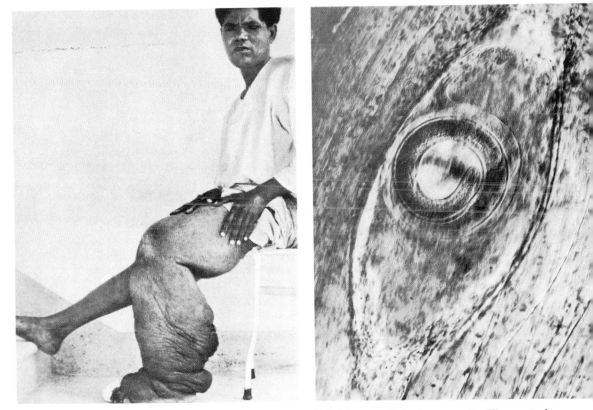

Elephantiasis is caused by certain filaria worms which live in the lymph glands and block the lymph passages. This results in diversion of lymph into the tissues and in the enormous growth of connective tissue. (Photo made in Puerto Rico by O'Connor and Hulse.)

Trichina cyst in hog muscle. The cyst does no harm, and the worm eventually dies; the damage is done by the boring of millions of these larvas before they encyst. Actual size of cyst, $\frac{1}{50}$ inch long (Photo of stained preparation by P. S. Tice)

164-1

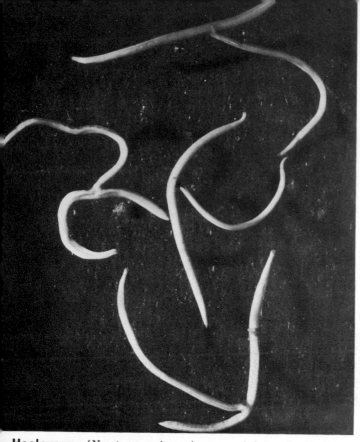

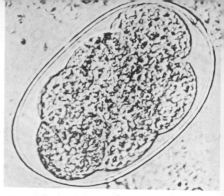

Developing hookworm embryo at time it leaves the body of the host is in the 4- or 8-cell stage. (Army Med. Museum)

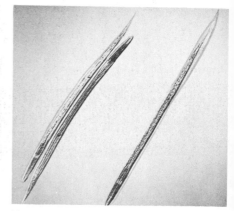

Hookworms (*Necator americanus*) removed from the intestine. The worms are shown copulating; the shorter male (actual length 5/16 inch) has an expansion at the posterior end by which it holds the female (7/16 inch long). (Photo, courtesy Army Med. Museum)

Hookworm larvas from soil. Length 1/50 inch. This is the infective stage. (General Biological Supply House)

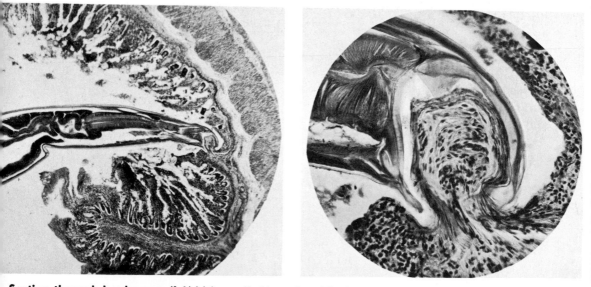

Section through hookworm (*left*) biting wall of intestine. The American hookworm holds on by sharp cutting ridges and feeds on blood and tissue juices. *Right,* closeup of portion of same section to show the head of the worm with a bit of intestinal lining in its mouth. (Photo, courtesy Army Medical Museum)

HAIRWORMS

IT IS an old belief that the so-called "horsehair snakes" arise from horsehairs that have fallen in the water. The horsehair snakes are neither horsehairs nor snakes, but are members of the small phylum **NEMATOMORPHA** ("form of a thread"). It is not difficult to understand how the erroneous notion of their origin got its start when we consider that these worms, which live in almost any body of fresh water, are often found in drinking-troughs, and that many of them are about 6 inches in length, black or brown in color, and, though somewhat thicker, look not unlike the hairs of a horse. Too, it seemed necessary to explain why one should see no trace of them on one day and then find numbers of these worms in the same place on the next day. We now know that this sudden appearance of the worms is due to the fact that the larvas develop as parasites in insects and the adults emerge full-grown from their insect hosts. They probably drop into the water when their hosts approach a pond or stream, or perhaps are swept into the water by a heavy rain.

The hairworms resemble the roundworms in structure and sometimes are included as one of the classes of the phylum Nematoda. Their life-history differs from that of parasitic nematodes in that the adults are all free-living. The difference is not important, however, because the adults in many cases lack a mouth; and even those with a mouth probably do not feed. Thus, the free-living adult may be looked upon as only a repro-

ductive stage, though it may sometimes last for months. The female lays
the eggs in long egg-strings which she winds about water plants. The
larvas that hatch have a spiny proboscis by means of which they bore
their way into the body of an aquatic insect larva. The next stages are
not well known, but we find the mature hairworms in the bodies of land
beetles, crickets, and grasshoppers. Perhaps the transfer to the land host
is effected when the aquatic larvas mature, go on land, and are eaten by
a beetle; or the first insect host may die and its parasitic hairworm
larvas escape and bore their way into the second host. In the body of the
second insect host they develop into adults, which finally return to the
water to mate and lay eggs.

Gordius (shown in the illustration at the beginning of this section) is a
genus of hairworms which wriggle about in ponds and ditches all over
the world. The name comes from the fact that the adults are often found
together in masses so tangled as to suggest a "Gordian knot."

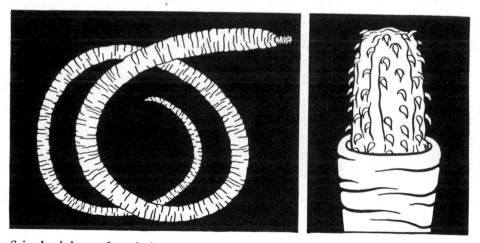

Spiny-headed worm from the intestine of a pig. Natural size.   Head, enlarged. (Modified after
Van Cleave)

SPINY-HEADED WORMS

THESE elongate, cylindrical worms live as parasites in the intestine of
vertebrates and used to be considered as a class of nematodes. But
as it is difficult to reconcile their unique body plan with that of roundworms or of any other group, they are now set aside by themselves as the
phylum **ACANTHOCEPHALA,** a name that means "spiny-headed" and
refers to their most characteristic structure, an anterior retractile pro-

boscis armed with rows of stout recurved hooks. Behind the proboscis is
a short neck region and then the body proper, which is roughly cylindrical.
By means of the burrlike proboscis the worm clings to the intestinal lining
of its host, absorbing nourishment through the delicate cuticle. There is
no trace of a digestive tract.

Acanthocephalids shed their eggs in the feces of the host. If the host is
an aquatic vertebrate, the eggs are probably eaten by a crustacean or an
aquatic insect, and in these animals the larvas develop. They get back
into a vertebrate when the intermediate host is eaten by the vertebrate
final host. The life-history is similar for land vertebrates, but it involves
land insects. A species common in rats and another which lives in pigs
are both occasionally found in man. The rat parasite is from 2 to 10
inches long. The eggs are shed in the feces; and when rat feces are eaten
by cockroaches, the larvas develop. Rats become infected by eating cockroaches, which sometimes form their chief article of diet. Man probably
becomes infected when he unwittingly eats an infected cockroach. The
acanthocephalid of pigs is a huge worm over a foot long, with a pinkish
wrinkled body. Pigs become infected by eating grubs (larvas of the June
beetle) which they find as they root about in the soil.

CHAPTER 16

LESSER LIGHTS

THE animal kingdom is divided into about twenty phyla. The exact number depends upon how many different plans of organization the classifier thinks there are. Some of the phyla are more important than others—at least to man—and among those usually considered of less importance, six will be discussed in this chapter.

These owe their inclusion among the "lesser lights" to one or all of the following reasons: they have a small number of species or of individuals; the members are of small size; they constitute no important source of food or of disease for man; and they illustrate no principle of theoretical interest that is not as well shown by other phyla.

ROTIFERS

WHENEVER a body of fresh water is examined for free-living protozoa, one is almost certain to find, in addition, microscopic animals about the size of protozoa but consisting of the equivalent of many very small cells and with a grade of structure a little more complicated than that of flatworms in some respects, less so in others. These are the rotifers. Because they are microscopic and play no important role in man's economy, these abundant animals are little known except to zoölogists and amateur microscopists, who seldom fail to be fascinated by their great variety of shapes (many of them truly fantastic) and by their rapid and often seemingly incessant motions.

Rotifers can be recognized at once by the presence at the anterior end of a **crown of cilia,** which serves as the chief organ of locomotion and also as the means of bringing food to the mouth. In some forms the beating of the cilia, which are arranged around the edge of one or more disk-shaped lobes, gives the appearance of a revolving wheel—hence the name of the phylum, **ROTIFERA,** which means "wheel-bearers."

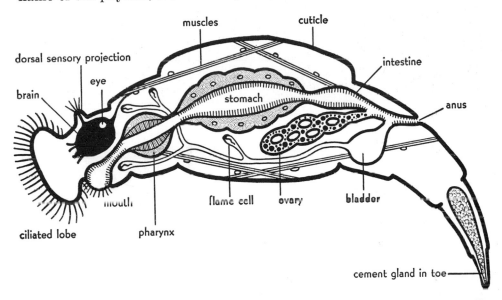

Rotifer, showing structure. Only a few of the nuclei are shown (e.g., muscles and stomach wall). (Combined from several sources)

Rotifers vary in shape from wormlike bottom-dwellers, or flower-like attached types, to rotund forms that float near the surface; but all are bilaterally symmetrical. In many species (as in *Philodina*, shown in the illustration above the chapter heading) the body is elongated and is roughly distinguishable into three regions: a **head** which bears the mouth and cilia; a main central portion, called the **trunk;** and a tapering portion, known as the **foot.** At the end of the foot are the "toes," pointed projections from which open cement glands that secrete a sticky material used to anchor the rotifer during feeding. The toes aid in a second method of locomotion in which the animal proceeds "inchworm fashion." It stretches

The **teeth of the pharynx** are the most distinctive structures of rotifers and are used as a basis upon which to distinguish one species from another. (After Harring and Myers)

out, takes hold at the front end, releases the toes, and contracts the body; then it fastens the toes again, and extends, etc. The whole body is inclosed in a transparent, flexible cuticle, which is folded into sections that can be "telescoped" one into the other when the animal contracts.

When **feeding,** a rotifer remains attached to a bit of debris, and the rapid beating of the cilia draws a current of water toward the mouth. Protozoans and microscopic algae are swept through the mouth into a muscular pharynx (or mastax), which contains a chewing apparatus consisting of little, hard "teeth," operated by muscles. In some rotifers the long pincer-like teeth can be extended through the mouth and used, like a forceps, for catching prey. The pharynx leads into a straight digestive tract which opens by an anus at the junction of trunk and foot.

In **general structure** a rotifer is similar to flatworms, nemerteans, and nematodes. The cuticle serves as a place of attachment for the muscles by which the animal moves. Muscular activity is co-ordinated by a simple nervous system which centers about a brain in the anterior end. Many rotifers have simple eyes and sensory projections called "antennas." The excretory system is like that of the flame-cell systems of flatworms and nemerteans; but in addition the terminal portion of the excretory tube is enlarged into a bladder which pulsates, ejecting its contents into the most posterior part of the intestine. There is no circulatory system in rotifers, and this is what one would expect of such minute organisms. Substances simply diffuse the microscopic distance from the gut to muscles and other tissues.

Rotifers are peculiar in that their bodies are not divided up into distinct cells but, like some of the tissues of sponges, flatworms, and nematodes, consist of **syncytia,** that is, protoplasmic masses containing a number of nuclei. Cell walls are present in embryonic stages but later disappear. It is also a striking fact that the number of cells of the late embryo, or the number of nuclei of the adult, is constant (about one thousand) for each individual of a species; and, further, each nucleus occupies a definite position, so that all the nuclei of a rotifer can be numbered and mapped. Such cell constancy also appears to a limited extent in some other phyla.

Rotifers **reproduce** sexually. The sexes are separate; but the males are generally small and degenerate, sometimes entirely lacking the digestive and excretory systems. Such individuals can live for only a few days. During most of the year, in the typical life-history, females give rise to other females by way of eggs that are not fertilized. This development of eggs without fertilization is called **parthenogenesis** and occurs also in other phyla. As the sexual season approaches, certain of the females lay eggs which are smaller and differ in other ways from the usual female-producing eggs. If not fertilized, these smaller eggs hatch into males. The males then fertilize the females, after which fer-

Male rotifers are usually small and degenerate. (After Hudson and Gosse)

tilized eggs are laid. These are distinguished from the parthenogenetic ones by a hard thick shell, often ornamented. They can withstand drying, freezing, and other unfavorable conditions, and after a resting period hatch into females. In one group of rotifers males have never been seen, and perhaps they do not occur. In this group the eggs develop without being fertilized and always become females.

Some rotifers can withstand **drying** even more than can many protozoans and nematodes. In this almost completely dried state they may

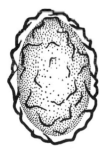

The fertilized female-producing **egg of a rotifer** has a hard, thick shell which protects the egg during the resting period. (After H. Miller)

live for years. As soon as moisture appears, they swim about and feed actively. Because of this capacity to resist drought, rotifers can live in places that are only temporarily wet, such as roof gutters, cemetery urns, rock crevices, among moss, and in similar places. When the water evaporates, the animal contracts to a minimum volume and loses most of its water content. Sometimes the animal itself dies but its contained eggs survive until moisture returns. There are some marine rotifers, but the group is much more abundant in fresh water. Because of their small size and their capacity for withstanding temporary drying, rotifers have been distributed the world over, chiefly by wind and by birds. If environmental conditions are similar, a lake in Africa will contain the same species of rotifers as a lake in North America.

GASTROTRICHS

ALMOST any aquatic debris that contains rotifers will also contain a few members of the small phylum **GASTROTRICHA.** These minute many-celled animals are about the size of rotifers and resemble them in many details of structure. They have no crown of cilia but swim by means of tracts of cilia on the ventral surface. The cuticle is often clothed with scales or bristles, and gastrotrichs are likely to be confused with ciliated protozoans.

The digestive system is a straight tube with a muscular sucking pharynx more like that of nematodes than of rotifers. In the species commonly seen in fresh water the tail end of the body is forked; and at the tip of each fork is the opening of a cement gland, which serves the same function as in rotifers.

Nearly one-half of the known gastrotrichs live in the ocean; these are hermaphroditic. The rest live in fresh water; and with possible exceptions, these are females which apparently reproduce parthenogenetically; no males have ever been seen.

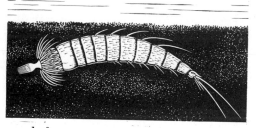

KINORHYNCHS

IT TAKES patience, and luck, to find kinorhynchs—unless one has a bicycle pump. Bubbling air through a bucket of slimy mud, from an estuary or a shallow marine shore, will bring kinorhynchs to the surface. From there they can be lifted off with a sheet of paper and rinsed into a dish. All members of the phylum **KINORHYNCHA** are marine and microscopic. They live mostly in black mud, taking in organic particles with a sucking pharynx like that of gastrotrichs and nematodes, which they resemble also in many other internal characteristics. The thick cuticular covering of the body is yellow or brown, and spiny, but lacks cilia. The animals cannot swim but inch their way along by extending and retracting the spiny head, which acts as an anchor against which the muscles can shorten, drawing the body forward.

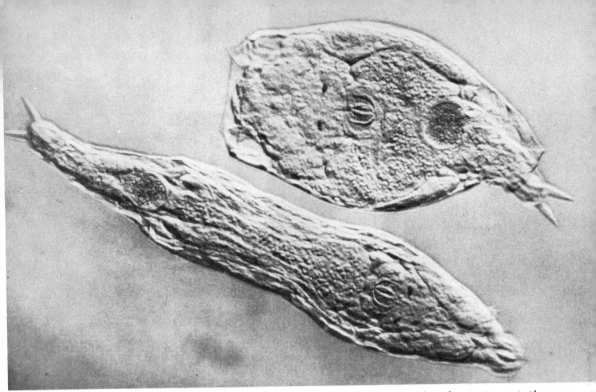

Wormlike rotifers can be seen in almost any drop of pond water, either creeping about on vegetation or remaining temporarily attached to the substratum by means of a sticky substance secreted through the tips of the two pointed "toes." When fixed, the beating cilia on the lobes at the head end sweep small animals and plants into the mouth. When moving, they either swim by beating the cilia or crawl about like a leech, alternately extending and contracting the body and taking hold by the toes and then letting go. The lower animal shown here is fully extended, with the head end on the right and the toes on the left. The upper animal is contracted, with the head and foot (except the toes) telescoped into the trunk region. (Photo of living animals by P. S. Tice)

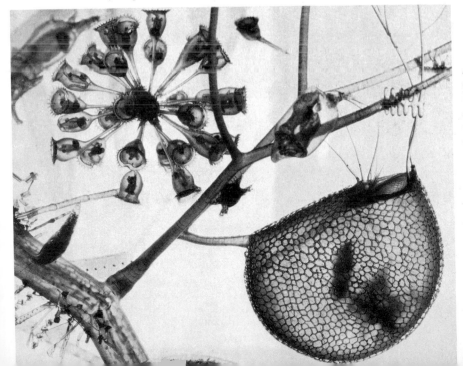

A colony of rotifers, *Conochilus*, appears in the upper left-hand corner of the picture. The colony swims through the water as a revolving sphere, with the members radiating from a common center at which their stalks are attached to each other. They all feed independently. (Photo of glass model, courtesy American Museum of Natural History)

Tube-dwelling rotifer, *Floscularia,* is sessile. It builds a protective tube by cementing together minute balls of debris. Protruding from the tube is the four-lobed ciliated crown edged with cilia that create the food-bearing currents. (Model)

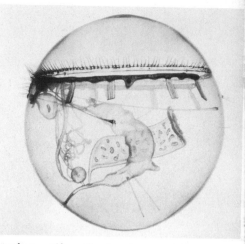

Floating rotifer, *Trochosphaera,* which lives at the surface of ponds and streams. The spherical body ($\frac{1}{50}$ inch in diameter) is propelled about by the band of cilia above its equator. (Glass model)

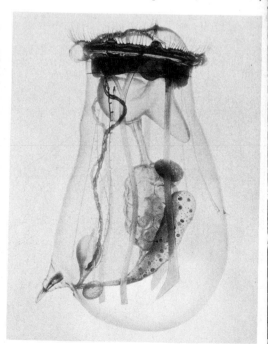

Floating rotifer, *Asplanchnopus,* through whose transparent body wall one can often see unborn daughters, and within them developing grandchildren! (Glass model.) (All photos on this page courtesy American Museum of Natural History)

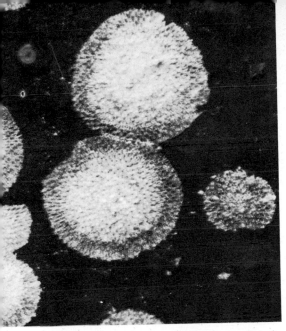

ncrusting bryozoan colonies which grow as flat circular patches on rocks and seaweeds are common on both our seacoasts. (Photo of colonies on seaweed. Mount Desert Island, Maine)

Erect branching bryozoan colonies are often mistaken for small delicate seaweeds. *Bugula*. (Photomicrograph. Beaufort, N.C.)

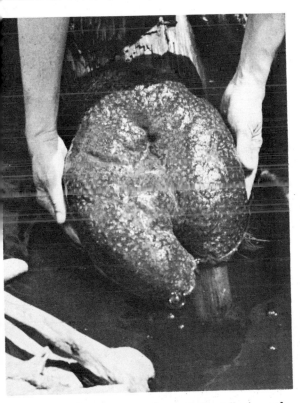

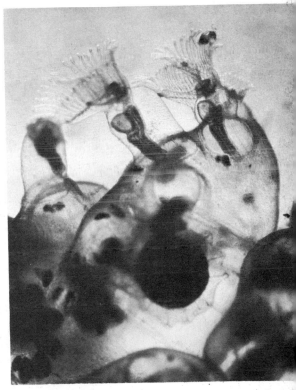

Fresh-water bryozoan, colony *Pectinatella*, is a gelatinous mass encrusting sticks and stones in lakes and streams. (Photo of living colony. Pennsylvania)

Fresh-water bryozoan, (from colony, *left*) showing several individuals each with a statoblast at its base.

Brachiopods are exclusively marine animals, most of which live attached to rocks by a stalk that passes out through a hole in one valve. *Terebratulina*, shown here about $2\times$ natural size, was brought up in a dredge from about 30 feet of water. *Left*, dorsal view; *right*, ventral. (Photos of living animal. Mount Desert Island, Maine)

Brachiopod laid open to show the internal structure, more like that of bryozoans than like clams, which most brachiopods superficially resemble. *Laqueus californicus*, shown here about twice natural size, lives in deep water off the West Coast. In its natural position it lies with the ventral shell (*above*) uppermost and the valves slightly agape. The two coiled ridges, supported on calcareous extensions of the dorsal valve, are bordered by delicate tentacles whose cilia maintain a circulation of water and sweep food organisms into the mouth. (Photo of preserved animal)

BRYOZOANS

S OME of the small and more delicate "seaweeds" admired by visitors to the seacoast are not seaweeds at all but are the branching colo-
nies of members of the phylum **BRYOZOA** a name that means "moss animals" and refers to the plantlike appearance of many bryozoans. Because of the colonial habit of its members, some prefer to call the phylum, POLY-ZOA ("many animals"). Some colonies are shrublike and hang from blades of kelp or grow out from crevices of rocks; others form flat incrusting growths on seaweeds and rocks; and some fresh-water bryozoans grow as gelatinous masses around stems and twigs that have fallen into the water.

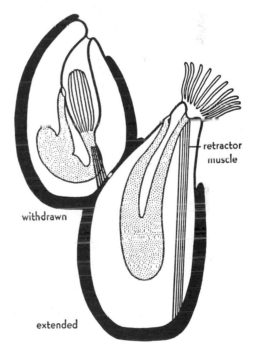

Two members of a bryozoan colony, one extended and one withdrawn. (Modified after Delage and Herouard)

Each individual of a colony lives in a protective case of hard material, calcareous or horny, into which it can withdraw completely. (The fresh-water bryozoan colony shown in the heading is *Plumatella*, which has a delicate and transparent horny covering). At first glance the animals resemble hydroids, for at the anterior end they have a set of tentacles borne on a circular or horseshoe-shaped ridge, called the "lophophore." However, they are considerably advanced over the hydroids and have a grade of structure more like that of rotifers. Although,

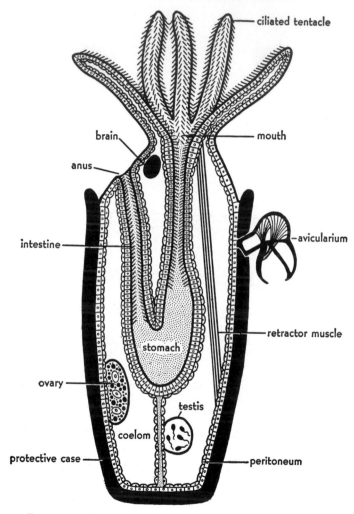

Bryozoan, showing structure. (Combined from several sources)

as in hydroids, the members of a bryozoan colony arise from each other by budding, they conduct their activities independently. When undisturbed, the animals emerge "cautiously" and spread their tentacles in the water; but at the slightest vibration they retreat into their cases.

The tentacles are ciliated and, when spread in the water, create currents

which drive microscopic organisms into a
mouth situated within the ring of tentacles.
The food is moved through the U-shaped digestive tract by means of cilia. The anus
opens near the mouth, but just outside the
circle of tentacles. The proximity of mouth
and anus does not seem to us particularly desirable, but apparently it is a satisfactory
adjustment for an animal that lives in a
case with only one main opening.

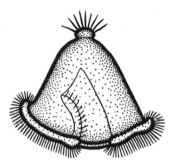

Bryozoan larva. (Modified after
Prouho)

The hard case that incloses a bryozoan is
secreted by the underlying ectoderm. This ectoderm and a layer of mesodermal cells constitute the thin body wall. Between the body wall and
the digestive tract there is a large fluid-filled space, completely lined with
mesoderm. Such a body cavity lined with mesoderm is called a **coelom**
("hollow"), and the coelomic lining is called the **peritoneum.**

The bryozoans show a remarkable reorganization at various times,
when the tentacles, gut, and other internal organs degenerate, forming a
compact mass known as the "brown body." New organs are regenerated
from the body wall; and the brown body, which comes to lie in the
stomach of the regenerated individual, is eliminated through the anus.
Since these bryozoans lack an excretory system, it is possible that the
formation of the brown body is related to excretion.

Some bryozoans illustrate **polymorphism.** In these we find, attached to
the normal feeding individuals, highly specialized individuals which
resemble a bird's head and so are called "avicularia." Each avicularium
has a pair of jaws, operated by muscles, which can
snap shut upon any small animal that wanders over,
or chances to alight on, the colony. Presumably, the
function of these individuals is not to aid in feeding but to prevent larvas (and other small animals)
from settling upon, and interfering with, the feeding activities of the colony.

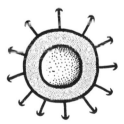

The **statoblast** of *Pectinatella*, a common fresh-
water bryozoan, is about
$\frac{1}{25}$ inch in diameter and
has a row of anchor-
shaped hooks. (After
Kraepelin)

Growth of the colony is by asexual budding. New
colonies are provided for by sexual reproduction.
The ovaries arise from the peritoneum of the body
wall. The testes usually form on the peritoneum covering the strand of mesenchyme-like cells that fastens the intestine to the body wall. Eggs and sperms

are shed into the coelom, where fertilization takes place. The ciliated larva of marine bryozoans is free-swimming and resembles a **trochophore,** a kind of larva found in many invertebrates (see chaps. 17 and 19).

Fresh-water animals usually do not have free-swimming larvas. Fresh-water bryozoans have, instead, buds known as **statoblasts.** These, like the gemmules of fresh-water sponges, consist of a mass of cells surrounded by a protective covering. They survive the winter and develop into new bryozoan individuals.

ENDOPROCTS

FORMERLY included in the same phylum with the bryozoans are the members of the small phylum **ENDOPROCTA** ("inside anus") in which the anus opens within the circle of tentacles—in contrast with the Bryozoans, in which the anus opens outside the circle of tentacles. The name BRYOZOA is often replaced by ECTOPROCTA ("outside anus"). Unlike the ectoprocts, which have no special excretory system, the endoprocts have a flame-cell system. In the endoprocts mesenchyme fills the space between the gut and the body wall, and there is no coelom. This has been taken as an indication that the endoprocts are more primitive than the ectoprocts, for a coelom, unless secondarily reduced, is present in all higher phyla.

BRACHIOPODS

ONE of the early investigators, who pried open the shells of a brachiopod and looked inside, thought that the two spirally coiled ridges within the shell were "arms" by which the animal moved and that they corresponded to the foot of a clam. From this mistaken notion came the name of the phylum, **BRACHIOPODA,** which means "arm-footed." The shells of a clam are right and left, while those of a brachiopod represent dorsal and ventral surfaces. The gape of the brachiopod shell is at the anterior end, and the hinge is at the posterior end; the shells can be opened and closed by means of muscles. The posterior end of the body is extended into a stout muscular stalk (peduncle) by which the animal is attached, usually to a rock. The shells are secreted by two folds of skin which inclose the main part of the body.

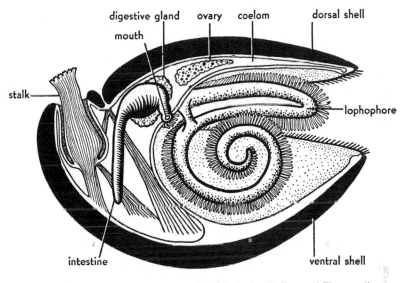

Brachiopod, showing structure. (Modified after Delage and Herouard)

Within the shells the most conspicuous structures are the two spirally coiled tentacular arms, or **lophophore,** which are thought to correspond to the lophophore of a bryozoan. Running along each arm is a ciliated groove, and on one side of this a row of ciliated tentacles. The beat of the cilia sweeps minute organisms into a mouth situated between the bases of the arms, from there into a stomach supplied with digestive glands, and finally into an intestine. The water currents also maintain a steady supply of oxygen.

The circulatory system is simple and contains a contractile muscular enlargement called the "heart." Just as in bryozoans, there is a true coelom lined with mesoderm. The excretory tubes lead from the coelom to the large cavity in which the arms lie and which connects directly with the outside.

The sexes are usually separate, and the sex organs lie near the intestine. The eggs or sperms are discharged into the coelom and reach the outside through the excretory tubes. The ciliated larva resembles a trochophore.

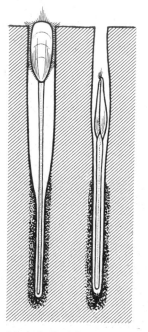

There are two main groups of brachiopods. In the more primitive group, of which *Lingula* is an example, the two shells are of a horny texture, somewhat rectangular in shape, and of equal size. They are held together only by muscles; there is no hinge. The stalk is usually very long and passes out between the shells. In the more advanced and larger group of brachiopods, represented both in the illustration at the beginning of the section and in the diagram, the shells are calcareous and are hinged together. The ventral one is larger than the dorsal, and at its posterior end has a kind of upturned "beak" through which the short stalk passes. The dorsal shell bears two calcareous coiled projections which serve as an internal support for the tentacular arms. In this group there is no anus.

Lingula lives in vertical burrows in the sand, attached to the bottom by the long stalk. (Modified after Francois)

Brachiopods are all marine and are not very widespread. Only about two hundred living species are known. But there was a time in past geological ages when there were at least three thousand species, of which we now have good fossil records. At that time brachiopods played a very important role in the invertebrate world, comparable to that of the clams and oysters of the present.

Modern species of *Lingula* are almost identical with species which we estimate, from the fossil record, to have lived almost 500,000,000 years ago. This is a record for conservatism among animals, and *Lingula* has the "honor" of being the oldest-known animal genus. Fossil brachiopods are of great value to geologists, as they constitute one of the most important criteria for dating rock strata. (See also chap. 27.)

PHORONIDS

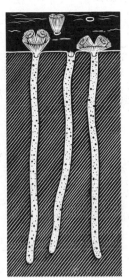

THE phylum **PHORONIDEA** is a small one, consisting of only about a dozen species. These wormlike animals are all marine, sedentary, and tubedwelling. A common species of our West Coast lives in straight cylindrical tubes imbedded vertically just below the surface of the substratum in mud and sand flats. The animals have a horseshoeshaped lophophore, spirally coiled at the ends, which bears ciliated tentacles that catch food. In this food-catching organ, in the U-shaped intestine, and in other respects phoronids resemble bryozoans. The larva is of the trochophore type.

ARROW WORMS

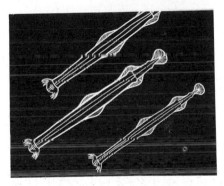

IN THE open ocean we find transparent, slender animals, usually 1–3 inches long, that look like cellophane arrows as they dart after their prey. Though at certain seasons they occur in incredible numbers, and at such times form a large part of the food of fish, the arrow worms are members of a phylum, the **CHAETOGNATHA,** which has relatively few species. The name means "bristle-jawed" and refers to the curved bristles, on either side of the mouth, that aid in catching prey. The body is divided into head, trunk, and tail and has finlike projections, which probably serve as balancers. The brain is well developed, and there is a set of eyes. The anus is situated at the junction of trunk and tail, about a third of the way from the posterior end. The three body regions are separated internally by transverse partitions, and there is also a longitudinal partition which separates the coelom into right and left halves. The animals are hermaphroditic: both male and female sex cells arise from the lining of the coelom. The body plan is so different from that of other groups that it is difficult to say what relationships they have to other invertebrates. In certain details of development the chetognaths resemble some of the members of the phylum to which man belongs.

PRIAPULANS

CYLINDRICAL, fleshy, wormlike animals up to nearly 6 inches in length, buff or brownish in color, and covered with spines and wart-like bumps can be dug up in cold or cool marine water, from shallow shores down to 1500 feet. They lie buried, with the mouth close to the surface of the mud, awaiting slowly passing prey. There are only two genera, with only a few known species, yet they are distinctive enough to be set aside in the phylum **PRIAPULA**. *Priapulus caudatus*, the best known species, has a heavily spined, eversible proboscis used in feeding and burrowing. The trunk has 30–40 superficial rings around the body and many small spines, and from the posterior end there protrude two appendages that look like bunches of grapes.

POGONOPHORES

POGONOPHORES ("beard bearers") are marine worms of cylindrical shape, and exceedingly long and slender dimensions, that live in closely fitting, delicate, secreted tubes imbedded in bottom sediments. They are most abundant on deep continental slopes, where temperatures are low and organic content high, but may be found at shallower depths in cold latitudes. The common name, beard worms, refers to the closely packed "beard" of tentacles at the front end, but some members of the phylum **POGONOPHORA** have only a single tentacle, of a diameter half that of the body. The longer worms may reach more than 12 inches, and the smallest may be less than 0.5 mm in diameter. The tubes are longer than the worms they contain and composed of repeated rings or funnel-shaped units. They look so much like straw or like the fibers of rope, that they were, until recent decades, discarded by zoologists dredging for invertebrates. The internal anatomy is comparable with that of moderately complex invertebrates, and the blood contains dissolved hemoglobin. But they are the only *free-living* metazoans without a mouth or internal digestive epithelium. It is presumed that the worms absorb all their nourishment through the epidermis.

The most typical mollusks
are the chitons, which creep about on the rocks of the seashore, rasping off fragments of algae. Most are active at night and, as daylight appears, they return to a particular "home" position in a sheltered place on the rocks. Chitons can be seen at low tide, each clinging tightly to its spot on the rock with its broad muscular foot. (Photo of living animals. Panama)

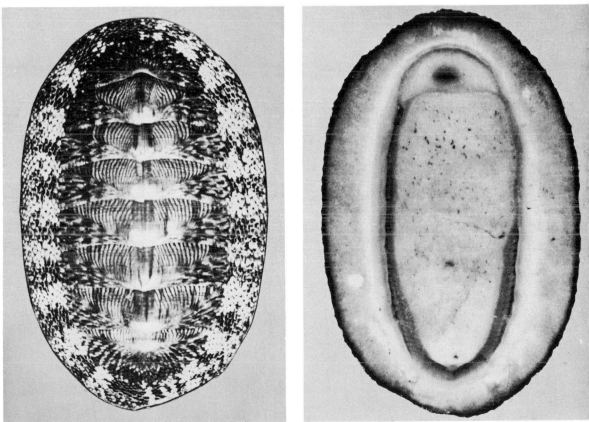

The chiton gives us some idea of the kind of animal from which snails, clams, and squids evolved. The upper surface (*left*) is protected by eight shells which overlap like the tiles on a roof. The undersurface (*right*) is occupied mostly by the large, oval, fleshy foot, in front of which is the degenerate head bearing the mouth. Surrounding the foot is the mantle, a fleshy fold which roofs over the body and secretes the shell. (Photos of living *Chiton*. Bermuda)

A chiton rolls up when detached from a rock—about the only defensive trick this sluggish animal has when removed from the rocky surface that normally shields the soft parts underneath. In the picture at the left the ventral edges of the mantle are pulled aside to show the mantle cavity, in which lie two narrow rows of gills. This specimen, *Cryptochiton*, about 10 inches long, is a member of the largest species of chiton known. The eight shells are not visible because they are imbedded beneath the surface. (Photo of living animal. Pacific Grove, California)

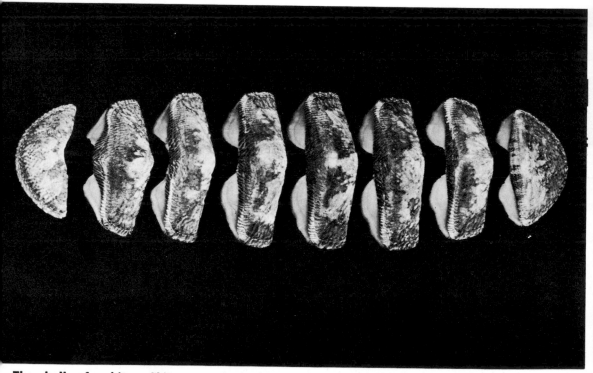

The shells of a chiton, *Chiton*, were obtained by throwing a dead animal on the lawn and allowing ants to clean off the flesh. If the chiton had died in the water, hermit crabs would be likely to do the same job of scavenging. The possession of eight shells is a stable character that goes back to antiquity; there are fossil chitons with eight shells that are at least four hundred million years old. The shells are arranged in order, the most anterior one at the left. (Actual width of center shell 1 3/16 inches. Bermuda)

CHAPTER 17

SOFT-BODIED ANIMALS

THE second largest and second most familiar invertebrate group is the phylum **MOLLUSCA.** The name means "soft-bodied"; and because of their soft, fleshy bodies, which are of relatively large size, the mollusks, more than any other invertebrates, are widely used as food by man. Some of the better-known mollusks are snails and slugs, clams and oysters, octopuses and squids.

Despite the lack of similarity in the external appearance of a snail, a clam, and a squid, their body plan is fundamentally the same and differs radically from those of all of the other invertebrate groups. The typical features of a mollusk are much modified, and some are even lost, in a highly specialized animal like a clam. They are less changed—from what we think was the condition of the primitive molluscan ancestor—in the chitons.

THE **chitons** are sluggish animals which browse on the algae that grow on rocks near the seashore. When disturbed, they clamp down upon the rock so tenaciously with their powerful muscles that it takes much persistence—and often a chisel—to pry them loose.

The body is bilaterally symmetrical. At the anterior end is a reduced and inconspicuous head, which is probably a secondary adaptation to a sedentary life; it is the chiton's main disqualifying character as a typical mollusk. The primitive molluscan ancestor of the chiton probably had a prominent head with sense organs—more like a snail's head than that of a chiton. The ventral surface is largely taken up by a broad, flat muscular creeping **foot,** abundantly supplied with a slimy secretion. The **visceral mass** (containing most of the organs) lies dorsal to the foot and is completely covered by a heavy fold of tissue which extends around on each side of the foot, much as a roof covers a barn. This fold is called the **mantle,** and the part under the "eaves" is the mantle cavity. On its

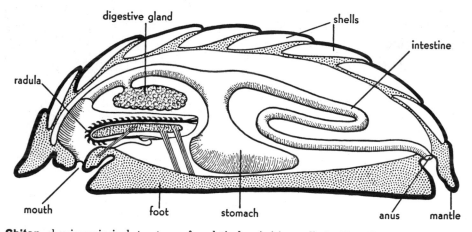

Chiton, showing principal structures of a relatively primitive mollusk. The gills are not shown (see the photographs at the end of chapter 18)

upper surface the mantle secretes a **shell,** which in most of the chitons consists of eight separate plates, overlapping from front to rear like the shingles on a roof. Between the mantle and the foot in the mantle cavity on both sides is a row of **gills,** thin-walled structures used in breathing.

The **digestive system** is a tube extending from the mouth, in the head, to the anus, at the posterior end of the animal. The mouth leads into a muscular chamber, the pharynx, in which is found the **radula.** This is a horny ribbon covered with many rows of hard recurved teeth. A complicated array of muscles pulls the radula back and forth over a cartilaginous projection, much as a cloth is pulled over a shoe in polishing it. When feeding, the chiton protrudes the radular apparatus through the mouth; and as the teeth of the radula move over the surface of plants, they rasp off small fragments. Behind the radula the esophagus opens into the stomach, from which a long intestine runs to the anus, at the posterior end.

The **circulatory system** is better developed than that of nemerteans. There is a specialized pumping organ, the **heart,** and extensively branched blood vessels which carry blood from the heart to all parts of the body and then back again. The heart lies in a cavity, the **pericardial cavity,** which is a part of the body cavity, or coelom.

There are two **excretory organs** (the kidneys), consisting of a glandular epithelium which extracts nitrogenous wastes from the blood passing through them. The waste material is discharged to the outside by way of pores near the anus.

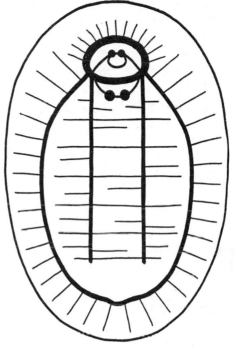

Nervous system of the chiton. (Based on several sources)

The **nervous system** is a ring of nervous tissue around the esophagus, connected with two pairs of longitudinal nerve cords which go to the muscles of the foot and mantle. It is a "ladder type" of nervous system, not very different from that of nemerteans.

Eggs and sperms are shed into the sea water. The fertilized egg develops into a **trochophore larva,** similar to that of several phyla already

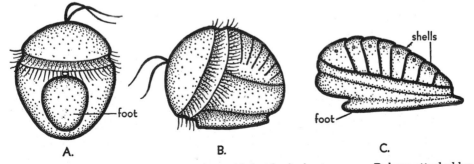

Development of a chiton. A, trochophore with foot beginning to appear. **B,** larva attached by foot; shells starting to develop. **C,** shells clearly marked; adult appearance indicated. (After Heath)

mentioned and also to that of the marine annelids (discussed in chapters 19 and 20). The typical trochophore is a spherical larva with a prominent band of cilia about the equator that serves as a locomotor organ and also to bring food to the mouth. Internally, the trochophore has a stomach and an intestine which connects with the exterior through an anus. There also develops a larval excretory organ with flame cells; it disappears when the larva changes into the adult. At the top pole of the larva is a group of sensory cells connected to a tuft of long cilia.

THOUGH less common than their more conspicuous and more economically valuable relatives, the chitons are described here because they display the molluscan body plan in its most typical form. The body consists of three main regions: a ventral muscular foot, a dorsal mass containing the viscera, and covering this a fleshy mantle which secretes the protective shell. A unique molluscan character is the radula. Not all mollusks have a radula, but nothing like it is found anywhere else in the animal kingdom. A shell is not peculiar to mollusks, for hard protective coverings have been developed by many groups. In most of these the outer covering seriously limits the activities of the animal. In mollusks there is a compromise, and the shell does not always completely incase the animal. In a snail, for instance, the head and foot can be extended when the animal is moving, withdrawn when danger threatens.

The phylum **MOLLUSCA** is divided into seven classes and that comprised of the chitons is the class **Polyplacophora** ("many-shell bearers").

Two smaller classes, which live on deep marine bottoms, are not likely to be seen by readers of this book and are mentioned only briefly.

The **solenogasters** are small and wormlike. A primitive kind of radula is their most obvious claim to inclusion in the mollusks. But they lack any shell, and the class name is therefore **Aplacophora.** They also lack head, mantle, and foot, but resemble chitons in that the external cuticle is studded with calcareous spicules.

The genus *Neopilina,* with only a few known species, comprises the class **Monoplacophora** ("single-shell bearers"). The shell is a cone with apex very far forward. Until the 1950's the monoplacophorans were known only from fossils and thought to be extinct. Then dredging brought up live specimens from deep bottom trenches off Costa Rica and Peru. A neopilina has a head (without eyes or tentacles), a mantle, a broad foot, etc. It differs from chitons and other mollusks in the repetition of unvarying numbers of gills, shell muscles, excretory organs, and nerve connectives, a condition called **segmentation** and seen in more highly developed form in the segmented worms, or annelids, described in a later chapter.

CHAPTER 18

TWO WAYS OF LIFE—CLAM AND SQUID

THE success of the molluscan plan is attested by the fact that there are over 100,000 species of mollusks, a phylum number second only to that of the species of arthropods, which include the ubiquitous insects. How various kinds of mollusks have adapted the same body plan to their specialized and very different ways of life is the subject of this chapter.

GASTROPODS

THE class **Gastropoda** includes such common animals as snails and slugs, limpets and whelks, and is by far the largest of the classes of mollusks.

Most gastropods show all of the chief molluscan features: a protective shell, a mantle, a large fleshy foot, a dorsally placed visceral mass, a radula, and usually one or more gills. In the possession of a well-developed head with eyes and sensory tentacles they are more like our idea of the primitive mollusk than is the chiton. But they are highly modified in the possession of a coiled shell and an asymmetrical organization of the visceral mass. The asymmetry results from atrophy of most of the visceral organs of one side, leading to the coiling of visceral mass and shell. The spirally coiled shell is a very compact arrangement for the disproportionately long visceral mass of gastropods. If the viscera and shell were in

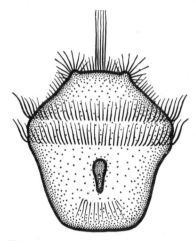

the form of a long straight cone, they would be almost unmanageable and a serious impediment to locomotion.

In the chiton the mouth and anus occur at opposite ends of the body. In most gastropods the anus opens anteriorly and lies above the head. The advantages of this arrangement are clear enough in an animal that lives in a shell with only one opening. Of the exact manner in which this has been brought about we are less certain. The explanations that have been given are based upon the development of gastropod larvas.

Trochophore of a gastropod. (After Patten)

The molluscan larva starts out as a **trochophore** with an equatorial girdle of cilia. As growth continues, the band of cilia becomes expanded, often into very large ciliated lobes, which serve to propel the larva about and bring food to the mouth. This second free-swimming larva with the expanded ciliated zone is called the **veliger** and is peculiar to mollusks. It occurs in all classes except the one to which belong the squids and octopuses. While the trochophore has the larval organs seen in marine larvas of this type, the veliger is characterized by the development of adult organs, such as the shell and foot.

The veliger is at first bilateral and has an anterior mouth, a posterior anus, a dorsal shell, and a ventral foot, as in the chitons. As development proceeds, the digestive tube is bent downward and forward until it lies near the mouth. This approximation of mouth and anus occurs in some members of all classes of mollusks except the Amphineura. But what follows is peculiar to gastropods. While the head and foot remain stationary, the visceral mass is rotated through an angle of 180°, so that the anus and the mantle cavity that surrounds it are carried upward and finally come to lie dorsal to the head. In addition, the organs on one side of the body fail to develop; and as a result of this unequal growth, the visceral mass and mantle (and the shell secreted by the mantle) become spirally coiled.

In most gastropods it is the organs of the original left side that degenerate. But since the visceral mass is rotated through 180°, the adult appears to lack the kidney, the gill, and one of the chambers of the heart (atrium) on the right side. The nervous system of these gastropods remains bilateral and uncoiled but becomes twisted into a figure-of-eight when the viscera rotate. In some gastropods there is a reversal of the rotation of the viscera (see below) and the nerve cords are untwisted.

Only the more primitive gastropods have a free-swimming trochophore. Most marine forms pass through the trochophore stage while still confined within the protective capsule in which the eggs are laid, and emerge as veligers. Some gastropods pass through even the veliger stage within the capsule and emerge as young adults. In land and fresh-water

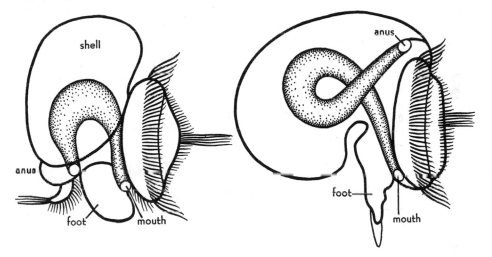

Left, a **veliger** before the rotation of the viscera. *Right,* a veliger after the rotation of the viscera; the anus now lies above the mouth. (Based on Patten and Robert)

gastropods, development of the eggs is modified, and usually there is no recognizable veliger stage. In some cases the eggs are not laid at all but develop within the body of the parent.

For all its advantages, a shell is a handicap to active locomotion. And there is a tendency among many groups of gastropods toward reduction or even complete loss of the shell, accompanied by an uncoiling and an untwisting of the visceral mass. That these forms have descended from typical gastropods is shown in their larval development. The larvas have a coiled shell and undergo twisting, followed by an untwisting and by loss of the shell and uncoiling of the viscera. As is generally true in animal evolution, organs once lost are not regained, although some substitute may be formed eventually from some other region of the body. Thus, gastropods descended from forms which have undergone atrophy of one side of the body, have only one gill and one kidney (instead of a pair of each), even though, as adults, they are uncoiled, and have reverted to an external bilateral symmetry.

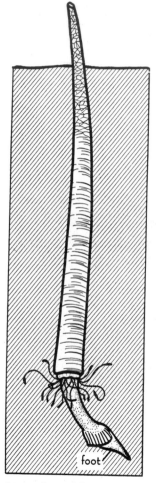

Gastropods are most successful in the water; but many, like the snails and slugs, have invaded the land. These are called the *pulmonate* gastropods because they have a modified mantle and mantle cavity which acts as a "lung" for air-breathing. Many pulmonates have gone into fresh water; but, as has been pointed out before, organs once lost do not reappear again, and, though aquatic, these snails have no gills. They must come to the surface periodically to take air into the lung. Some freshwater snails have gills, but these have descended directly from marine snails instead of land types.

TOOTH SHELLS

A SMALL class of mollusks are the "tooth shells" or **Scaphopoda,** in which the shell is shaped like a miniature elephant's tusk, tubular and open at both ends. There is a mantle, a poorly developed head bearing a number of extensible filaments that serve as sense organs and aid in capturing prey, a radula, and a muscular burrowing foot. The gills are lost, and the mantle serves as the respiratory organ. A current of water is maintained in and out of the upper end of the shell, as the animal lies buried almost completely in the sand. The larva goes through free-swimming trochophore and veliger stages.

A **scaphopod** lies buried in the sand with only the upper end of the shell protruding into the water. (Modified after Sars)

BIVALVES

THE clam, oyster, scallop, and others of the mollusks with two shells are often called "bivalves" (two valves) and comprise the class **Pelecypoda.** The name of the group means "hatchet foot" and refers to the shape of the foot in many pelecypods. Most members of the group are marine, but some clams are very abundant in fresh waters. The description that follows applies in general to almost any of our common fresh-water clams.

The **clam** is flattened from side to side. The two shells or valves, which represent right and left sides, are fastened to each other dorsally by an elastic horny **ligament.** The gape of the shells is ventral. Near one end of the ligament is an elevated knob, the **umbo.** The end of the animal nearer the umbo is the anterior end. At the opposite or posterior end are the openings through which currents of water enter and leave the clam.

horny outer layer

prismatic layer

pearly layer

The **shell** consists of three layers.

The umbo represents the oldest part of the **shell.** As the animal grows, the mantle secretes successive layers of shell, each projecting beyond the last one laid down. This results in a series of concentric lines of growth which mark the external surface of the shell and represent the successive outlines of its ventral margin.

The shell consists of three layers. The dark, horny, *outer layer* (periostracum) forms the ligament and protects the calcareous shell from being dissolved by carbonic acid in the water. It is thin and is usually eroded from the older parts of the shell, such as the umbo. The *middle layer* (prismatic layer) consists largely of crystals of calcium carbonate arranged perpendicularly to the surface of the shell. The innermost, or *pearly layer* (nacreous layer) consists mostly of thin sheets of calcium carbonate laid down parallel to the surface of the shell. The first two layers are secreted only by the edge of the mantle, and

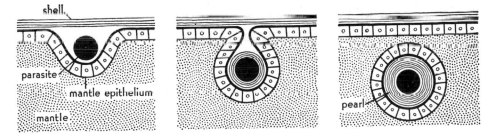

Formation of a pearl. *Left,* a parasite lodges between the shell and the mantle epithelium. *Middle,* it is almost completely inclosed in a sac formed by the epithelium, which secretes thin, concentric layers of pearly substance. *Right,* a pearl of good size has surrounded the parasite and prevented it from harming the clam. (Based on Haas)

hence show the concentric markings of discontinuous growth. The inner, pearly layer is laid down by the whole surface of the mantle and has a smooth, lustrous surface.

A *pearl* may be secreted by the mantle as a protection against some foreign body, usually a parasite such as the larval stage of a fluke. The larva enters the mantle and becomes inclosed in a sac formed by the growth of the mantle epithelium, which secretes thin, concentric layers of pearly substance around the foreign body.

When undisturbed, the clam lies partly buried in the sand or mud with the ligament up and the shells slightly agape ventrally. In this position the animal protrudes its fleshy foot and burrows through the mud like an animated plowshare. First, the pointed foot is extended forward into the mud and anchored by a turning or by a swelling of the free end (due to an influx of blood into a cavity within the foot). Then, as the muscles of the foot contract, they draw the body of the clam forward. Such a slowly moving animal, with a shell that is heavy and cumbersome to carry about, could hardly run down its prey. Instead, like so many other sedentary animals, the clam has evolved a method of drawing water through its body and straining out the microscopic organisms and other nourishing organic particles contained in the water. For protection it relies on its heavy shell and retiring habits.

The shells are held agape by the elasticity of the ligament. They are closed by the contraction of two large **muscles.** Near these are smaller muscles which extend and retract the foot; and attached to the shell along a line close to, and parallel with, its ventral margin is a row of small muscles which retract the edge of the mantle. When the shell is lifted back (after cutting the muscles), its inner surface will show "scars," which represent the former attachments of all these muscles. Also, it will be seen that the dorsal margin of the shell has long ridges and irregular toothlike projections, the **hinge teeth,** that fit into grooves or pits in the opposite shell. This interlocking arrangement fits the two valves together.

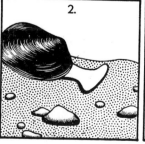

A **moving clam** extends its foot into the sand . . . .

the tip of the foot swells and acts as an anchor . . . .

the muscles of the foot contract, drawing the body of the clam forward . . . .

The visceral mass lies dorsally, most of it between the two large muscles that close the shell. The **mantle** covers the visceral mass and extends ventrally as two mantle lobes, one just underneath each shell. The space between the mantle lobes is the **mantle cavity.** At the posterior end the lobes are thickened locally and approximated at certain points to form the openings for entrance and exit of water. These extend out just beyond the margins of the shells, when the shells are agape. A current of water passes into the mantle cavity through the ventral or *incurrent opening* and out through the dorsal or *excurrent opening.*

On removing one mantle lobe, the mantle cavity and its organs are exposed. From the mid-ventral region of the visceral mass the foot extends into the mantle cavity and out between the shells. Between the foot and the mantle lobe, on each side, a pair of sheetlike double folds, the **gills,** hang freely into the mantle cavity. The gills have a sievelike structure, being perforated by microscopic **pores,** and are covered with cilia. The beating of these cilia draws water through the incurrent opening and into the mantle cavity. The water passes through the microscopic pores, leaving suspended particles on the surface of the gills. Within the gills the water flows up the **water tubes,** vertical channels formed by partitions that subdivide the cavity between the inner and outer walls of a gill. The water tubes open dorsally into the **dorsal gill passages,** which run one above each gill and open posteriorly near the excurrent opening, through which the water leaves the clam.

The rate at which water passes through the mantle cavity has been measured in certain marine clams and oysters. For an animal of average size, the minimum rate averages about 2.5 liters (almost 3 quarts) an hour.

Food particles left on the surfaces of the gills by this steady stream of water are distinguished, mostly by their small size, from silt and other

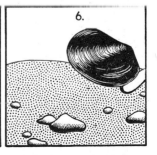

the foot is again extended . . . .   the tip is anchored . . . .   and the body is again drawn forward

undesirable materials during their passage to the mouth. Heavy particles of sand or mud simply drop from the surface of the gill to the edge of the mantle, are carried backward by cilia on the mantle, and are expelled posteriorly. The lighter particles become entangled in mucus secreted by the gills and are carried, always by beating cilia, to the ventral edge of the gill and then forward until they meet the ciliary tracts on the **palps,** a pair of folds on each side of the mouth. Further sorting occurs

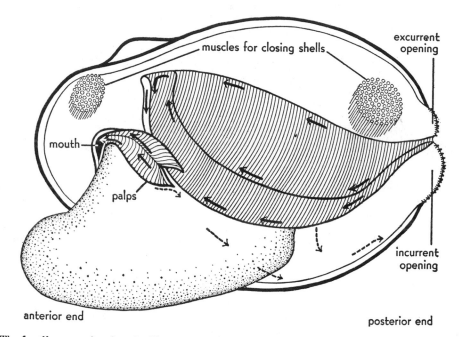

The **feeding mechanism** is ciliary; food particles, caught on the surface of the gills, are carried to the mouth, as shown by the *solid arrows.* Rejected particles are removed from gills and palps, as shown by *dotted arrows.*

here, and the larger particles are carried to the tips of the palps and then dropped off into the mantle cavity, from which they are removed.

Selected materials are carried to the deep groove between the two palps. This groove leads directly into the mouth, which lies between the two "lips," or ridges, that connect the palps of one side to the palps of the opposite side. There is no radula, nor could it be of any use to the animal that feeds only on microscopic particles. The food, entangled in strings of mucus, goes into the mouth and through a narrow tube, the esophagus,

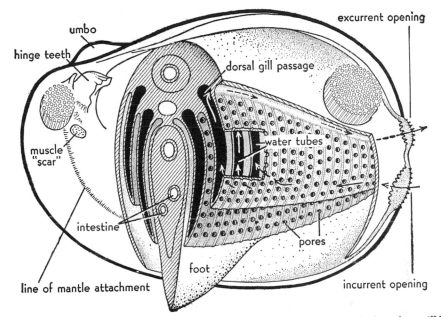

The direction of the **respiratory currents** of the clam is shown by the arrows. A piece of one gill has been cut away to show that the space between the inner and outer walls of the gill is divided into vertical water tubes. The pores of the gill are microscopic and are shown here greatly enlarged. The *anterior part of the clam has been removed.* In the *sectioned surface* the relations of the water tubes to the dorsal gill passages can be seen. The intestine appears more than once in the section because it coils back and forth in the foot.

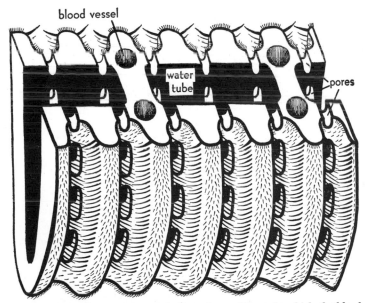

A **small portion of a gill** enlarged to show how the partitions, in which the blood vessels run, divide the space between the inner and outer walls of the gill into vertical water tubes which communicate with the mantle cavity through microscopic pores.

to a saclike stomach, which is connected by ducts to a large **digestive gland.** This gland surrounds the stomach and is the main organ of digestion. From the stomach the intestine runs ventrally, makes several coils through part of the foot, and then runs dorsally again, passing through the cavity that surrounds the heart and appearing to pass through the heart itself (actually, the heart is wrapped around the intestine). The anus opens near the excurrent opening, and the feces are carried away in the outgoing current.

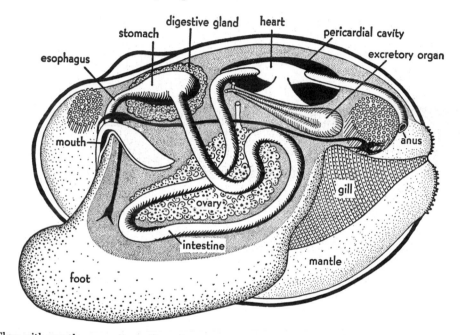

Clam with mantle, one pair of gills, and part of foot cut away to show the **digestive system** and other organs.

Digestion in fresh-water clams is not well understood; most of what we know about the physiology of digestion in pelecypods is based on studies of certain marine clams. As we might expect in animals that eat only finely divided food, the digestion is mostly intracellular. Food from the stomach enters the digestive gland, the cells of which readily ingest and break down solid particles. Protein and fat digestion are exclusively intracellular, and the cells of the gland also absorb carbohydrates. The only extracellular enzyme is the carbohydrate-digesting one, set free in the stomach by the dissolution of the *crystalline style,* a gelatinous rod that lies in a pouch off the intestine and projects into the stomach. The style-pouch is lined with cilia, the beating of which rotate the style and move it forward so that its free end is constantly rubbed against a special portion of the stomach wall. In this way the head of the style is worn away and its material mixed with the stomach contents.

The **circulatory system** consists of a heart and blood vessels. The heart has three chambers: right and left thin-walled *atria* (singular atrium), which receive the blood, and a single muscular *ventricle* which pumps the

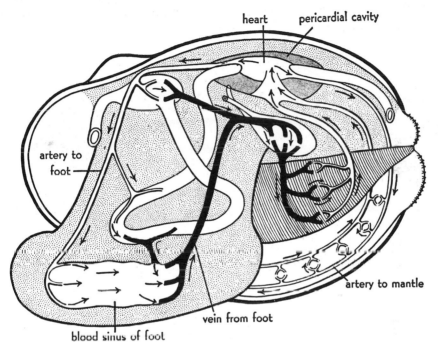

The **circulatory system** of the clam is an *open* system with blood vessels that supply and drain irregular channels and sinuses in the tissues. The filling of the large sinus in the foot helps to produce the swelling of the tip of the foot in locomotion.

blood. The heart lies in the pericardial cavity, which is lined with an epithelium and filled with fluid. It represents a remnant of the body cavity or coelom, which is so extensive in annelids (see chap. 19) and other groups. The heart pumps blood both forward and backward through arteries. Anteriorly, it supplies the viscera and foot. Posteriorly, it supplies the mantle. Many of the arteries lead, not into fine capillaries, but into irregular channels in the tissues. These channels, called *blood sinuses*, lack the epithelial lining of true blood vessels. From the sinuses the blood flows into veins and then back to the heart. The blood is colorless and contains ameboid cells.

The **excretory organs** lie, one on each side, just below the pericardial cavity. Each is like a tube bent back on itself, with the two parts lying parallel and one above the other. The lower part, or *kidney* proper, has glandular walls. At its anterior end it connects with the fluid-filled pericardial cavity. At its posterior end it is continuous with the thin-walled

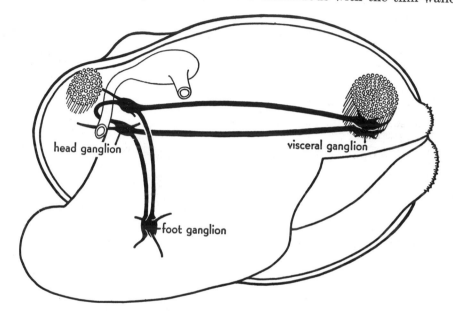

The **nervous system** of the clam has a pair of ganglia for each main region of the body. The head ganglia correspond to the double but fused "brain" of other animals. They lie on each side of the mouth, joined by a commissure that runs around the esophagus. They send nerves to palps, anterior shell muscle, and mantle. From each head ganglion a connective runs ventrally to the ganglia which supply the muscles of the foot. Two connectives also run from the head ganglia to the visceral ganglia, which send nerves to the digestive tract, heart, gills, posterior shell muscle, and mantle.

bladder which lies above the kidney and opens anteriorly into a dorsal gill passage. The kidney extracts waste products of metabolism from the blood and from the pericardial fluid. The wall of the bladder is ciliated and maintains an outgoing current.

The body is muscular and performs co-ordinated movements; but the **nervous system** is reduced, as would be expected in so sluggish an animal. There is no head, nor would it be of much use to an animal that lives with its anterior end buried in the mud. Each of the three main regions of the body—"head," foot, and viscera—has a pair of ganglia; and from the head ganglion two long nerves (connectives) run to each of the other two pairs of ganglia. The **sense organs** are poorly developed. Near the foot ganglia is a pair of hollow vesicles, lined with sensory cells and containing a limestone con-cretion, which are thought to be balancing organs. A patch of yel-low epithelial cells (the osphra-dium) lies on the visceral ganglia and is thought to be sensitive to chemicals in the water that enters the incurrent opening. The man-tle has scattered sensory cells, which are most abundant on the small projections along the edges of the mantle at the openings for

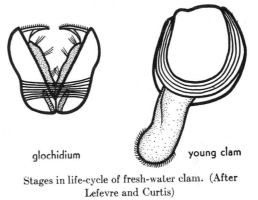

glochidium   young clam

Stages in life-cycle of fresh-water clam. (After Lefevre and Curtis)

the water currents. They probably respond to touch and to light. When a clam is irritated, the foot and mantle edges are withdrawn, and the two valves close very tightly—or, as we say, "shut up like a clam."

The **reproductive system** consists of a glandular mass, which surrounds the coils of the intestine that lie in the foot, and opens near the external pore of the bladder into a dorsal gill passage. The sexes are separate. The male sheds sperms from the testis into the outgoing water. They enter the female through the incurrent opening, pass through the pores on the surface of the gills, and reach the interior of the gills, where the eggs are held and fertilization effected. The zygotes develop within the gills to bivalved larvas called **glochidia** (singular, glochidium, "point of an ar-row"). Tremendous numbers of glochidia are produced and expelled into the water, where they slowly sink to the bottom; but most of them die. To develop further, they must, within a few days, become attached to the fins or gills of a fish and live as parasites until they have developed into young clams. Then they drop from the fish and take up the independent life of the adult clam.

In some fresh-water clams there is no parasitic stage; the young de-velop within the body of the mother. Marine pelecypods shed eggs or sperms into the water, where fertilization takes place. There is first a

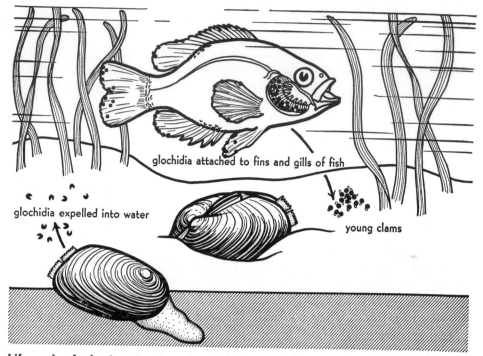

glochidia attached to fins and gills of fish

glochidia expelled into water

young clams

Life-cycle of a fresh-water clam. The glochidium clamps its valves tightly into the tissues of its host and in some way stimulates them to grow around it, thus forming the so-called "blackheads" of fish. After 3–12 weeks of parasitic life, the young clam falls off and becomes independent. (Based on Lefevre and Curtis)

trochophore and then a veliger larva. (The glochidia of fresh-water clams correspond to the veligers of marine types.)

CEPHALOPODS

THE most highly organized mollusks are the nautiluses, squids, and octopuses—all marine and members of the class **Cephalopoda.** The name means "head-footed," for in these animals the foot, which is divided up into a number of "arms," is wrapped around the head.

As in gastropods, all degrees of reduction of the shell can be found. While the nautiluses have a large, calcareous, external coiled shell, the squids have only a thin horny vestige of a shell imbedded in the mantle, and the octopuses have no shell at all.

The **squid** is one of the most highly developed invertebrates. Certain of its structures will be described here to illustrate the ways in which the squid has adapted the molluscan body plan to an active, predatory life.

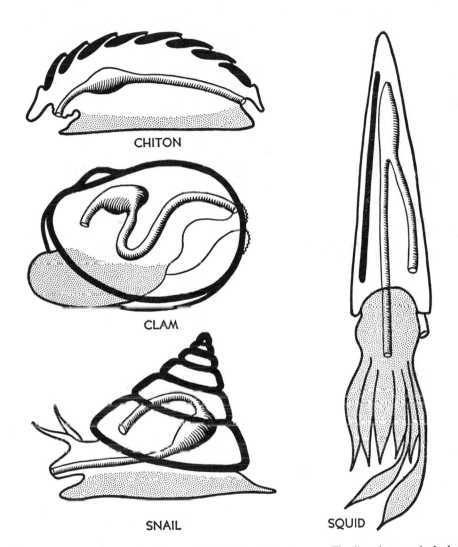

CHITON

CLAM

SNAIL

SQUID

The **molluscan body plan** has been modified in the various groups. The digestive tract is shaded, the foot is stippled, and the shell is indicated by the heavy black line.

Unlike most bilateral animals, which are elongated in an anteroposterior direction, the long axis of the squid is dorsoventral. To compare
the body with that of a clam, one would have to place
the squid so that the foot was down and the pointed
end up. The functional upper surface of a swimming
squid is structurally the anterior surface. The functional under surface is structurally posterior. Thus, a
squid usually swims with the ventral surface forward,
the dorsal surface hindmost, the anterior surface up,
and the posterior surface down.

The squid relies for protection not on a heavy shell
but chiefly on its ability to leave the scene of danger
in a hurry. The **shell** is vestigial and is represented
by a feather-shaped horny plate buried under the
mantle of the anterior surface. The **mantle** is thick
and muscular and has taken on the protective function which in other mollusks is served by the shell.
The mantle is also the chief swimming organ. At the
dorsal end, its anterior surface is extended into a pair
of triangular folds or "fins," which can be undulated
to move the animal slowly and to change its direction
of movement. Ventrally the mantle ends in a free edge,
the **collar,** which surrounds the "neck" between the
head and visceral mass. The collar articulates by three
interlocking surfaces (ridges which fit into grooves)
with the visceral mass and with the **funnel,** a conical muscular tube that projects beyond the collar on
the underside of the head. When the mantle is relaxed,
water enters the mantle cavity around the edge of the
collar; and when the mantle contracts, the edge is
tightly sealed and water is forced out through an
opening in the funnel. When the squid is excited, the
mantle is contracted strongly, forcibly expelling a jet
of water from the funnel. This pushes the animal in
the direction opposite to that in which the jet is expelled. When the tip of the funnel is bent backward,
the squid darts quickly forward to seize its prey. When
the tip of the funnel is directed forward, the animal
shoots backward like a torpedo; and this is its usual

The **shell** of the squid
is a thin, horny plate
which lies buried under
the mantle of the anteri-
or surface.

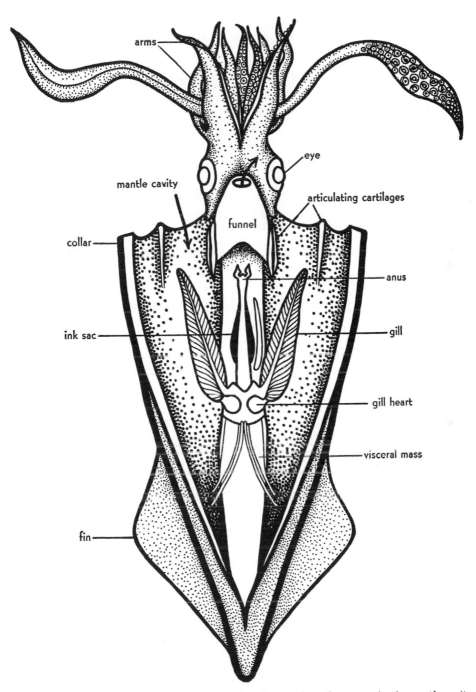

Squid with mantle slit open along the posterior surface to show the organs in the mantle cavity. Water enters the mantle cavity around the edge of the collar and leaves through the funnel, as shown by the arrows.

behavior in escape. When attacked, it may emit, from a special **ink sac** which opens into the funnel near the anus, a cloud of inky material. The "ink" is thought to serve as a "smoke screen"; but it has also been suggested that it forms a dark object that distracts the enemy while the squid goes off in another direction.

The "foot" of the squid is subdivided into the funnel and ten sucker-bearing **arms** which surround the mouth. When the animal is swimming, the arms are pressed together and aid in steering. Two of the arms are different from the rest and can be extended forward to seize the prey with their **suckers** and to draw it toward the mouth. There it is held firmly by the other arms, while two strong horny **jaws** in the mouth kill the prey, biting out large pieces, which are then swal-

Left, an arm of the squid, showing numerous suckers. *Right, above*, a sucker showing the muscular stalk by which it is attached. *Right, below*, sucker showing the toothed, horny ring with which it is lined.

lowed so rapidly that the **radula** (which is quite small in the squid) is probably seldom used.

Three rows of **teeth from the radula** of the squid.

The active life of the squid would not be possible with the slow type of respiration that serves the clam. The contraction and expansion of the mantle provide a steady and effective circulation of water through the mantle cavity, in which lie the two **gills.**

The **circulatory system** is also much improved and provides for the rapid distribution of oxygen through the tissues. The blood flows within vessels, which are lined throughout with an epithelium—not into irregular unlined spaces among the tissues, as in the clam. The tissues are permeated with networks of very small vessels, the **capillaries,** through the thin walls of which gaseous exchanges take place rapidly. There

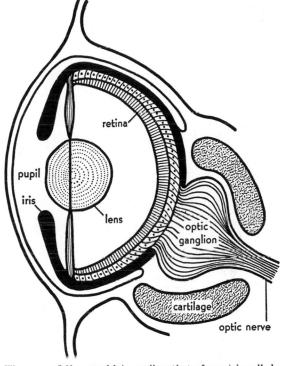

The **eye of the squid** (as well as that of man) is called a "camera eye" because it is built on the same principle as a camera, which consists of a dark chamber to which light is admitted only through an opening (pupil) in the diaphragm (iris). Behind this opening is a lens which focuses the light on a light-sensitive film (retina).

are separate pumping mechanisms for blood going through the gills and that going out to the tissues. The deoxygenated blood returning from the tissues enters two *gill hearts,* each of which pumps the blood through one gill. This gives the blood a fresh impetus, so that it passes through the gills at higher pressure. Freshly oxygenated blood from the gills enters a single *systemic heart,* from which it is pumped out again to the tissues.

The **nervous system** of the squid is very highly developed—in sharp contrast with that of its slow-moving relative, the clam. A large brain encircles the esophagus and lies between the eyes. The brain is unusual in that it consists of several pairs of ganglia all fused together, and therefore has several centers of nervous control, which in lower invertebrates are spread out over the animal. Besides olfactory organs and a pair of structures that probably aid in balancing, the squid has two large **image-**

perceiving eyes. They are remarkably like the human eye in construction but are developed in quite a different way. When two similar structures having a similar function appear in two distantly related groups, so that there is no possibility of a common ancestor which could have possessed such a structure, then the structure must have evolved independently. Thus the eye of the squid and the eye of man are said to have arisen by **convergent evolution.**

Luminescent squids are mostly deep-sea forms which live in perpetual darkness. It seems reasonable that the *light-producing organs* serve as lanterns, but this would be difficult to prove. (Based on Chun)

The two groups of animals which show the best development of eyes, the vertebrates and the cephalopods, also have the most highly developed light-producing organs. In some fishes and squids these organs are amazingly complex, having, besides the photogenic cells which produce the light, lens tissue, reflector cells, and a layer of pigment to screen the animal's tissues from the light.

The occurrence of light-producing organs in two distantly related groups like squids and fish is no more strange than the convergent evolution of the two groups toward the same general type of eye. For **bioluminescence,** or the production of light by living organisms, is a widespread phenomenon found in bacteria and fungi and in almost every phylum of animals. Its distribution among animals seems quite hit-or-miss, following no special evolutionary lines but accompanying certain ways of life. No luminous fresh-water animals are known, but animal light is extremely common in marine forms, particularly coelenterates and comb jellies. The "burning of the sea" at night is caused mostly by

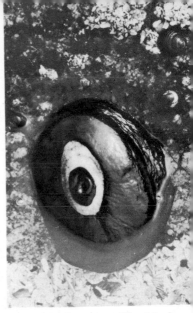

The limpet is a gastropod, a member of the class of mollusks which typically have coiled, asymmetrical shells. But limpets have reverted to a simple cap-like shell. The underside shows the bilateral head with two sensory tentacles, large muscular foot with which it creeps sluggishly about on rocks, the mantle, and to one side of the head, the gill. (Photo of living animal. Mount Desert Island, Maine)

Key-hole limpet. The black mantle almost covers the shell, which has a central opening for escape of respiratory current. (Living *Megathura*, 5 inches long. Pacific Grove, Calif.)

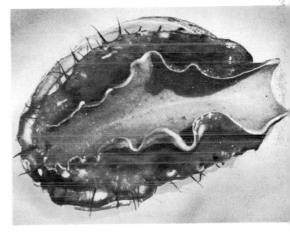

Abalone, *Haliotis* (8 inches long), is a gastropod with a flattened spiral shell. The row of holes in the large whorl is for the outgoing respiratory current. The underside of the abalone (*above, right*) shows the large muscular foot and numerous sensory projections. Huge abalone shell heaps (*left*) indicate the extent to which these large and delicious mollusks are eaten. The colorful pearly layer of the shell is made into jewelry. (Photo of living animal. Pacific Grove, California)

The whelk, *Buccinum.* Many gastropods feed like chitons, rasping off fragments of vegetation by means of a horny toothed ribbon, the radula; but the whelk is a carnivore, grasping its prey with the large muscular foot and then attacking it with a long extensible proboscis which has the radula at its tip. (The organ seen protruding from the edge of the shell is not the proboscis but the siphon, a tubular prolongation of the mantle for directing water to the gill.) It uses its proboscis to bore a hole through the hard armor of a recently dead crab or lobster. But, in attacking a scallop, it simply waits until the valves are agape, and sticks the edge of its own shell between the open valves to prevent them from closing. Then it inserts the proboscis and rasps away the soft parts. (Photo of living animal about natural size, by D. P. Wilson, Plymouth, England)

Periwinkles (*Littorina*) live at the water's edge and spend much of their time in air. They resist drying by retiring into the shell, closing the opening with a horny plate on the foot, and secreting a mucous seal around the shell opening. (Photo of living animals by Beaufort B. Fisher. Pacific Grove, California)

Nudibranchs are sluglike marine gastropods that have small coiled shells as embryos but later lose their shells, uncoil, and develop symmetrically on the two sides—at least externally. They have lost the true molluscan gills and breathe through the finger-like projections along the sides of the body. Most of them live among sea-weeds, on which they feed; but some are carnivorous like the grey sea slug, shown here, which feeds on living anemones, one of which is shown in the picture. Some nudibranchs eat sponges. Their delicate colors, soft textures, and often bizarre shapes place the nudibranchs among the most beautiful of all invertebrates. (Photo of living *Aeolidia papillosa* 2× natural size by D. P. Wilson. Plymouth, England)

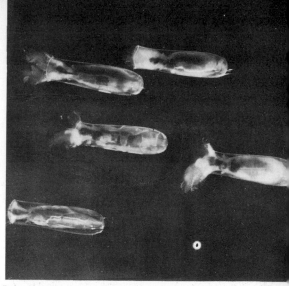

Fresh-water snail (*Ampullarius* from S. America) viewed from below as it glides on a piece of glass by means of its broad, muscular foot. The eyes are on short stalks at the bases of the tentacles. This snail breathes by taking air into its lung and also by means of a gill. (Photo of living animal by P. S. Tice)

Pteropods are marine gastropods that swim in the open ocean by flapping finlike extensions of the foot. The uncoiled, vaselike shell is thin and transparent. Pteropods usually swim together in enormous numbers and in northern waters some species are so abundant as to furnish food for whales. (Photo by William Beebe)

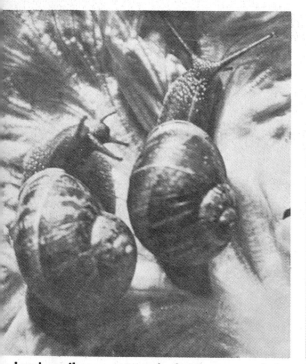

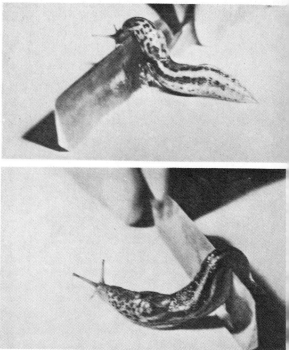

Land snails are gastropods that have part of the mantle cavity modified as a lung for air-breathing. They use their radulas to rasp off fragments of plants. This pair soon reduced to a few shreds the piece of lettuce on which they are shown. (Photo of living animals. Pacific Grove, California)

Slugs are land gastropods that have lost the external shell, having only a thin plate imbedded in the mantle. The slime they secrete and upon which they glide is lubricating and protective, as is demonstrated by these pictures of a slug passing unharmed over the sharp edge of a razor. (Photos by W. La Varre)

Tree snail, *Liguus fasciatus roseatus*, lives in trees, feeding on bark and leaves. It deposits its eggs in the leaf-mold. After about 8 months, minute snails emerge. During the dry, cool winter, the snail attaches itself firmly to a tree by means of a secretion that hardens into a parchment-like substance which prevents drying. With the return of the rainy season, the cement softens and the animal begins an active life again. Land life is possible for these soft-bodied animals because of the abundant mucous secretion and the shell which conserves moisture. (Photo of living animal, about 2× natural size. Royal Palm State Park, Florida)

Fresh-water clams can plow slowly through the sand or mud with the muscular foot; but they spend most of their time in one place, with the ligament up, the shells slightly agape, and the openings for water currents protruding (as in the clam at the left). Water is drawn in, strained of its load of minute food organisms, and then expelled. (Photo of living animals, courtesy Shedd Aquarium, Chicago)

Horse-hoof clam (*Hippopus*) lies with ligament down. It has a degenerate foot and never moves. A foot long, and one of the largest of clams, it is dwarfed by the giant clam, *Tridacna*, which may be 5 feet long and weigh 500 pounds. (Photo of living animal by Otho Webb. Australia)

Marine clam with its two long separate siphons extended. About natural size. (Photo of living animal. Pacific Grove, Calif.)

Scallops (*Pecten*) swim about erratically by clapping the two shells. Both mantle edges have a row of steely blue eyes, which show here as bright spots. The large muscle that closes the shells is the only part of a scallop that is eaten. (Photo of living animals. Woods Hole, Massachusetts)

Gaper clam (*Schizothaerus*) is one of the largest bivalves on the West Coast, the shell alone reaching a length of 8 inches. It lives deeply buried with the long siphons extending to the surface of the mud. People concerned chiefly with their own digestion find "gapers" good eating. Biologists interested in the digestive processes of bivalves, find these large clams good subjects for experimentation. This specimen (prepared by T. L. Patterson) is anesthetized with ether; a rubber tube, with a balloon tied to its end, has been pushed into the incurrent siphon, across the mantle cavity, and through the mouth into the stomach. The tube is connected with a device for recording changes in air pressure exerted on the balloon by contractions of the clam's stomach.

Rock-boring bivalves hold on with the foot and bore through solid rock by movements of their roughened shells. When imbedded, only the two siphons protrude. (Photo of living animals. Pacific Grove, California)

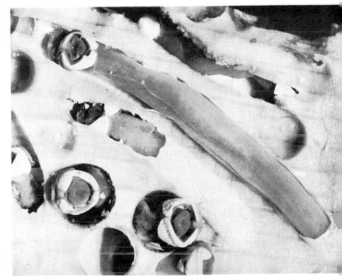

Wood-boring bivalves (*Teredo*) exposed in their burrows in a piece of infested wood that has been split open. Popularly known as "shipworms," they are not worms at all, but greatly elongated clams. The two shells, which inclose only a very small part of the anterior end of the body, have a ridged and roughened surface and are used for the boring. The animals feed on wood particles, as well as on minute organisms brought in by the respiratory current. Every year shipworms do millions of dollars' worth of damage to wooden wharf pilings and ships. (Photo of preserved specimens, 3 × nat. size)

Clams, oysters, and other mollusks are our largest invertebrate source of food. In the United States, squids are used by the ton for fish bait, and to a lesser extent as food for man. In oriental and Mediterranean countries, squids, cuttlefishes, and octopuses are popular articles of the human diet.

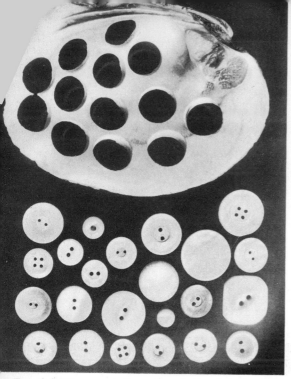

Pearl buttons cut from the shells of fresh-water clams. Thousands of tons of shells were used annually for this purpose, and in the U.S. came mostly from the Mississippi Valley. (Photo by Cornelia Clarke)

Pearls from fresh-water clams are irregular, but occasional valuable ones are found. Pearls are protective secretions made of the same substance that lines the shell of the bivalve. (Photo by Cornelia Clarke)

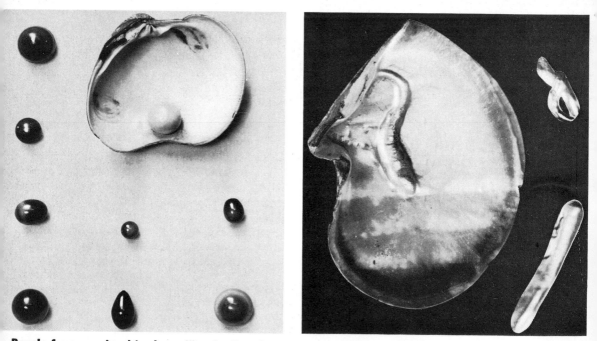

Pearls from marine bivalves, like the "pearl oyster," are the most valuable. About one thousand "pearl oysters" are opened to find one pearl. *Left:* Highly valued black pearls from the Gulf of Mexico. *Right:* Mother-of-pearl covers a fish which became lodged between shell and mantle. (Photo, courtesy Am. Mus. Nat. Hist.)

The cuttlefish, *Sepia,* resembles the squid in structure and habits. The shell, a calcareous plate imbedded in the fleshy mantle, is the cuttlebone given to cage birds as a source of lime salts. The contents of the ink sac provide a rich brown pigment, sepia, used by artists. (Photo of living animals by Raoul Barba. Monte Carlo)

Squid (*Loligo*) of East Coast. (Living animal. Woods Hole, Mass.)

Sucker Marks, $\times$ ½, on the skin of a whale tell of an encounter between this largest of vertebrates, and the largest of invertebrates, a giant squid, which may be over 50 feet long. (Photo by L. L. Robbins)

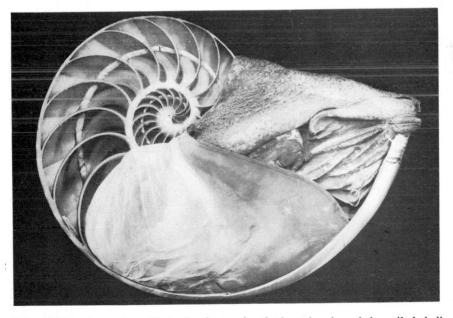

The nautilus, in section. The animal occupies the last chamber of the coiled shell and protrudes its arms to catch crabs and other animals. It lives in deep water in the South Pacific. (⅔ natural size.) (Photo, courtesy Amer. Mus. Nat. Hist.)

Large squid. (Pacific Grove, California)

204-9

The octopus is a cephalopod with no trace of a shell. It moves by pulling itself over the rocks with its arms or by forcibly expelling water from the funnel. Its sinister reputation may be deserved by certain of the giant octopuses; but most of them, like this species (*Octopus vulgaris*, ⅓ natural size), make for the nearest rocky crevice at the approach of a large animal like man. The sucker-bearing arms seize crabs, whose shells are then broken open by a pair of horny jaws and the radula. (Photo of living animal by D. P. Wilson. Plymouth, England)

Eggs of the octopus are each incased in a capsule, and in this species are laid in a cluster. One end of each capsule is attached to a stone or other object. The female octopus broods over the eggs. The development of the octopus (and of other cephalopods) is highly modified by the large amount of yolk in the egg and is different from that of all other mollusks. There is no free-swimming larval stage, the young octopus hatching directly from the egg capsule. In the young octopuses shown here on either side of the egg cluster, notice the prominent pigment bodies, whose contraction or expansion, under nervous control, effect rapid changes in the color of the animal. (Photo of preserved specimens, courtesy Gen. Biol. Supply House)

luminous protozoa, chiefly flagellates (*Noctiluca, Ceratium, Gonyaulax*). Among luminescent coelenterates we find many jellyfishes (*Liriope*), many hydroids (*Obelia*), siphonophores, scyphozoa (*Aurelia, Cyanea*), and some gorgonians. Practically all the common ctenophores of our coasts produce light (*Pleurobrachia, Mnemiopsis,* and *Cestum*). Other luminescent invertebrates are certain bryozoans, numerous annelids (polychetes), echinoderms (brittle stars), arthropods (crustacea, fireflies, glow worms), besides many mollusks (mostly cephalopods). Luminous bacteria and fungi emit a continuous light and are responsible for the luminosity of decaying flesh and rotting wood, but most invertebrates shine only when stimulated. In flagellates, medusas, and comb jellies any disturbance of the water, as by a passing boat, will cause the animals to flash; in the case of the firefly (which is really a beetle) there is a definite rhythm of flashing determined by internal stimuli and controlled by the nervous system.

The value of luminescence to living organisms is not clear in most cases. It is thought that light from such forms as luminous bacteria is a by-product of metabolism and has no significance for the life of the organism. It has been suggested that the light organs of squids or fish serve as lanterns, to attract prey or repel enemies; but it is difficult to explain why such annelids as *Chaetopterus*, which passes its whole life in an opaque tube, should be luminous. In the case of the firefly and of certain marine annelids the light does seem to serve as a signal to bring the sexes together for mating.

Some important steps have been made in our understanding of the physicochemical nature of animal light. Long ago it was shown that luminous wood stops glowing when placed in a container from which the oxygen is removed. Later it was shown that luminescence in the pelecypod, *Pholas dactylus*, is the result of the interaction of two substances which were extracted from the luminous tissues of the animal. If a hot-water extract and also a cold-water extract are prepared and allowed to stand until the light disappears from the cold-water extract, when the two are mixed together light will be produced. The hot-water extract is supposed to contain a substance, *luciferin*, which is not destroyed by heating; the cold-water extract contains an enzyme *luciferase*, which is destroyed by heating. When the extracts are mixed, the luciferin, in the presence of the enzyme, luciferase, becomes oxidized, with the production of light. (Luciferin is at first also present in the cold-water extract; but it soon becomes oxidized, only luciferase remains, and the light disappears. Then, when luciferin is again added, the light reappears until all of the luciferin is oxidized.) Similar substances have been demonstrated in fireflies, crustacea, and some other invertebrates. In still others luminescence is quite a different process.

Another example of convergence between squids and vertebrates is the development of **internal cartilaginous supports**. The squid has a number of internal cartilages which support muscles and form interlocking surfaces, but most interesting in this connection is the large cartilage which incloses and protects the brain, reminding us of the vertebrate brain case. The squid, perhaps more than any other invertebrate, has evolved along the same lines followed by the fast-moving predatory aquatic vertebrates: large size, streamline shape, rapid locomotion, internal skele-

The **giant squids** are the largest of all invertebrates. (After an old engraving from Figuier)

tal supports, very efficient respiratory and circulatory systems, large brain, and highly developed sense organs.

The giant squids are responsible for many of the "sea-monster" stories. One of the authentic cases was a squid encountered by the French battleship, "Alecton," in the Atlantic in 1860. The monster was 50 feet long, exclusive of the arms, and was 20 feet in circumference at its largest part. Its weight was estimated at 2 tons. The resistance of such an animal, even though sick, as this one probably was, can be judged from this old account: "The commandant, wishing in the interests of science to secure the monster, actually engaged it in battle. Numerous shots were aimed at it, but the balls traversed its flaccid and glutinous mass without causing it any vital injury. . . . . They succeeded at last in getting a harpoon to bite, and in passing a bowling hitch round the posterior part of the animal. But when they attempted to hoist it out of the water the rope penetrated deeply into the flesh, and separated it into two parts, the head with the arms and tentacles dropping into the sea and making off, while the fins and posterior parts were brought on board: they weighed about forty pounds."

BY COMPARING the clam and the squid, we see that the fundamental body plan of an animal may become so modified in adaptation to a special way of life that many of its structures reflect the kind of life it leads rather than its relationship to its more typical relatives.

CHAPTER 19

SEGMENTED WORMS—NEREIS

WE FREQUENTLY refer to the "average man" or the "average student." Biologists who have given some thought to the selection of an "average animal" have found little difficulty in deciding upon some kind of segmented worm as the animal which would occupy the middle position on a scale of increasing complexity from protozoa to insects or vertebrates. In spite of minor differences of opinion in choosing a particular form, the final vote would go neither to the earthworm nor to the leech, the segmented worms most familiar to everyone, but to some less specialized worm, like the **nereis.**

The Nereids of Greek mythology were sea nymphs, usually represented in the female human form. Their invertebrate namesakes are marine worms, probably beautiful only to zoölogists, but certainly graceful as they swim through the water by gentle undulations of the body. The common nereis of our New England coast grows to a foot or more in length and lives under stones or in temporary burrows in the mud or sand between tidemarks.

The most noticeable feature of the nereis, as of the earthworm, is the ringing of the body, which is not merely external but involves nearly all of the internal structures. The name of the phylum to which the earthworm, the leech, and the nereis belong is **ANNELIDA** ("ringed"). The ringed condition is more often known as **segmentation,** and each ring is called a **segment.**

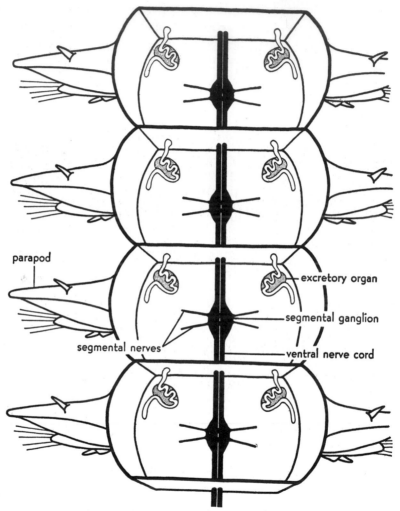

parapod

excretory organ

segmental ganglion

segmental nerves

ventral nerve cord

Diagram to emphasize the repetition of parts in the annelid body plan.

Except for the head and the last segment, all the segments of the nereis are externally alike. They have on each side a projecting appendage or **parapod** ("side foot"), consisting of flattened fleshy lobes from which protrude bundles of horny **bristles.** In addition to the undulations of the whole body, a swimming nereis uses its rows of parapods as a series of locomotory paddles. The bristles are sharp and probably serve a protective function, as well as enabling the animal to obtain a hold on the smooth walls of its burrow.

Although well equipped for swimming, the nereis spends most of the time in its burrow in the sand, with only the head occasionally protruding above the surface. Gentle undulatory movements of the body create a current through the burrow, bringing the worm chemical stimuli from

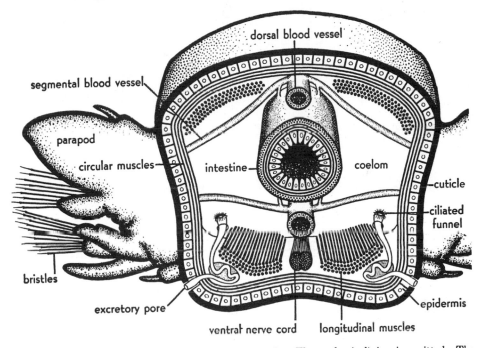

A three-dimensional **cross-section through a nereis.** The coelomic lining is omitted. The muscles of the intestinal wall are shown.

food organisms located near by and also constantly renewing the water for respiration. In feeding, the worm has been seen to extend the anterior part of the body from the burrow, seize its prey with two strong, horny jaws (borne on the end of an eversible pharynx) and drag it into the burrow.

The outer covering of the nereis is a horny but flexible **cuticle,** which is secreted by the underlying ectodermal epithelium, or **epidermis.** Beneath the epidermis is a layer of circular muscles, then a layer of longitudinal muscles, and finally a thin lining layer of mesoderm cells (the coelomic lining discussed below). Together these various layers constitute a definite **body wall.** They run the length of the worm and are divided up by the segmental partitions. More clearly segmental are the bundles of oblique muscles which run in each segment from the mid-ventral line to the parapods, which they move.

To describe one segment of a nereis is to describe nearly the whole worm. Only the **digestive system** shows much differentiation from anterior to posterior ends. The mouth leads into a pharynx, on the inner walls of which are the two large jaws already mentioned. When the pharynx is turned inside out and extended through the mouth, the jaws grasp the food, which is then swallowed by withdrawing the pharynx.

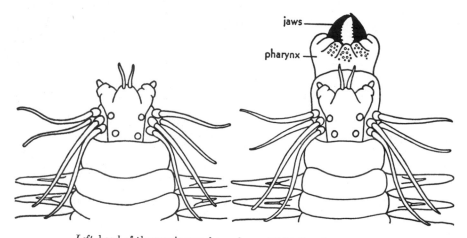

Left, head of the nereis seen from above. *Right,* the pharynx everted.

Behind the pharynx the digestive tube narrows to an esophagus, which runs through several segments and into which opens a pair of glandular pouches, the digestive glands. From the esophagus a long, straight intestine runs the length of the body to the anus in the most posterior segment. The structure of the digestive system presents a definite advance over the condition in the nemerteans in that outside of the digestive epithelium, but in the wall of the digestive tube itself, are thick muscle layers. The contractions of these muscles produce a succession of rhythmic waves of constriction, a type of muscular activity called **peristalsis,** which push the food along, independently of movements of the whole body.

Between the digestive tube and the body wall of the nereis is a definite space, called the *body cavity* or **coelom,** which is lined completely by a sheet of mesoderm cells, the *coelomic lining*. One important advantage of the coelom is that it separates the intestine from the body wall and thus permits a freer play of the body-wall muscles. As already mentioned, this also allows the muscles of the digestive tube to push food along independently of the movements of the body. The coelom is filled with a fluid which contains ameboid cells and many dissolved substances. The coelomic fluid bathes all of the internal organs and thus serves a role similar to that of the circulatory system, even though it has no direct connection with that system. The coelom also plays a role in excretion and in reproduction, as we shall see later.

Behind the esophagus the coelom is not a continuous space but is divided up, by the partitions of coelomic lining, into a series of chambers that correspond to the external segmentation.

The coelom arises by the formation of a pair of spaces in the embryonic mesoderm of each segment of the body. These spaces enlarge, and

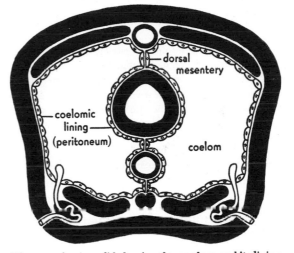

Diagram of an annelid showing the **coelom** and its lining.

are lined by a thin layer of mesoderm, giving rise to a series of coelomic sacs. The inner walls of these sacs envelop the digestive tube; and where they meet in the mid-line, they form a double layer of coelomic lining, the *mesentery*, which supports the gut above and below. In the nereis, and in many other annelids, the ventral mesentery (the part below the digestive tube) is present only in the embryo. It disappears in the adult, and right and left coelomic spaces are confluent below the digestive tube.

The presence of a coelom is considered of such importance that we often divide animals into two large groups, those with a coelom and those without it, categories which correspond roughly with what we mean when we talk of the "higher" and "lower" invertebrates. A space between the digestive tube and body wall occurs in many of the phyla we have studied already. But in such groups as the roundworms and rotifers it has no definite mesodermal lining and therefore cannot be considered a true coelom. In bryozoans, brachiopods, arrow worms, phoronids, and mollusks the body cavity is a coelom. In mollusks it is reduced, for the most part, to the cavity surrounding the heart.

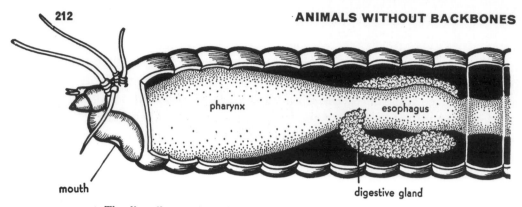

mouth   digestive gland

The **digestive system** of the nereis is specialized only at the anterior end.

The general type of body structure seen in the nereis—with a muscular body wall separated from a muscular digestive tract by a space lined with mesoderm—occurs in all vertebrates. In man the coelom is divided into an abdominal cavity, a cavity surrounding the heart, and two cavities which contain the lungs. The coelomic lining of animals is also called the "peritoneum"; and when it becomes infected in man, as from a ruptured appendix, the serious condition that results is known as "peritonitis."

The **circulatory system** of the nereis does a much better job than the crude apparatus which we saw in nemerteans. The main vessels are a median dorsal vessel, which runs above the digestive system, and a median ventral vessel, which runs just beneath the digestive system. These two longitudinal vessels are connected with each other through the transverse segmental vessels which they give off in each segment. Dorsal and ventral branches of the segmental vessels go to the intestine, parapods, and body wall and there branch and rebranch repeatedly, finally joining each other by way of an intricate network of very fine vessels, the **capillaries.** The walls of the capillaries are composed of only a single layer of flattened epithelial cells and are similar in structure to the capillaries of man. Their thin walls permit a rapid exchange of dissolved food substances, nitrogenous wastes, and respiratory gases. Their extensive ramification insures that substances are delivered almost "at the door" of every cell and do not have to move long distances by the slow process of diffusion.

Extensive branching does not of itself make a good circulatory system, for, as the name of the system implies, the blood must be in constant circulation. In the nemerteans the weakly contractile lateral vessels are inadequate to the task, and movement of blood depends mostly on muscular waves in the body wall. In the nereis the median dorsal vessel and the lateral branches have well muscularized contractile walls. Rhythmic

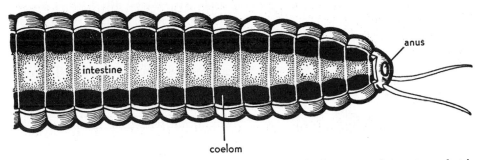

The intestine of the nereis is not imbedded in mesenchyme, as in flatworms and nemerteans, but is surrounded by the coelomic cavity. Most of the middle portion of the worm has been omitted.

waves of muscular contraction, of a peristaltic nature like those of the intestine, run along the dorsal vessel from behind forward, driving the blood anteriorly. The blood flows posteriorly in the noncontractile ventral vessel.

Besides a good circulatory system to distribute the oxygen after it has entered the blood, large and active animals require an extensive **respiratory surface** which is freely exposed to oxygen, either of the air or dissolved in water. In the nereis the amount of body surface exposed to the environment is enormously increased by the thin, flattened parapods, within each of which is an extensive network of capillaries. The capillary beds of the parapods and of the dorsal and ventral body walls lie very

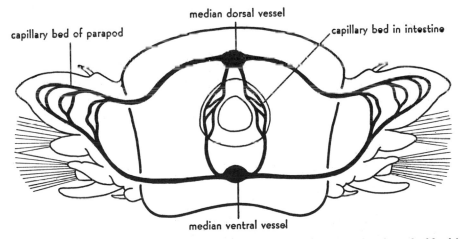

The main **segmental blood vessels** of the nereis are those to the parapods, where the blood is aerated, and those to the intestinal wall, where food enters the blood to be distributed to other tissues.

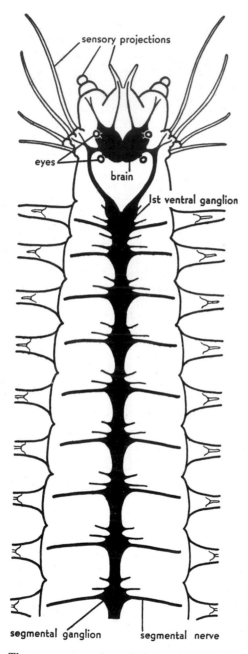

The **nervous system** of the nereis is clearly segmental. All segments are alike, except at the anterior end.

close to the surface and, as the blood passes through them, it receives oxygen from the surrounding water and gives up carbon dioxide collected from the tissues. The oxygen-carrying capacity of the blood is increased by the presence of hemoglobin, which is dissolved in the fluid instead of being contained within cells, as in some nemerteans and in vertebrates.

The **excretory system** is segmentally arranged. A pair of excretory organs lies on the floor of nearly every segment. Each organ consists essentially of a tube which opens at one end by a ciliated funnel into the coelom and at the other end by a pore to the exterior. Wastes extracted from the blood which passes through the excretory organ and also from the coelomic fluid, with which the organ communicates through its internal open funnel, are swept to the exterior by means of cilia lining the excretory tube.

The excretory organs of annelids are called **nephridia** (singular, nephridium). In the nereis the excretory tube is coiled through most of its length, and the coils are compacted within a granular oval mass. The external pore lies at the base of the parapod. The internal end runs forward, passes through the coelomic partition, and opens into the coelomic chamber just anterior to that in which lie the main body of the organ and the external pore. Microscopic particles in the coelomic fluid are wafted into the opening of the funnel by beating of the cilia around its edge.

The concentration of nerve cells into a **central nervous system** with a brain and nerve cords marked the advance of the flatworm nervous system over that of coelenterates. In the nereis this tendency toward centralization is carried still farther. The head bears two pairs of eyes and several pairs of projections, which are sensitive to touch and to food and other chemicals in the water. With so many sense organs at the head end, there is an increase in the size of the brain. But the brain is only the first and the largest of a series of compact masses of nerve cells, or **ganglia,** which occur in each segment of the body. The

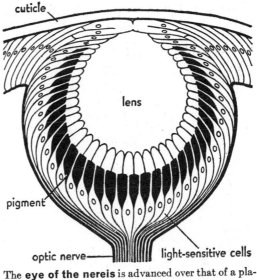

The **eye of the nereis** is advanced over that of a planaria. In addition to the layer of pigmented light-sensitive cells (retina) there is a gelatinous lens which concentrates the light upon the specialized rodlike ends of the sensory cells. The other ends of the cells are continued as nerve fibers which run in the optic nerve to the brain. (Based partly on Kukenthal)

head ganglion or brain lies above the pharynx and connects, by a ring of nervous tissue, with another large ganglion lying below the pharynx. From this first ventral ganglion the nerve cord runs posteriorly, and in each segment enlarges into a segmental ganglion, which gives off nerves to the muscles of the body wall and parapods. Each segmental ganglion is like the governor of a state; while the brain, which receives and co-ordinates the various impulses coming from certain of the sense organs on the head, is like the chief executive of a nation.

The primitive brain, as we saw it in the planaria, served chiefly as a sensory relay—a center for receiving stimuli from the sense organs and then sending impulses down the nerve cord. This is also true of the nereis, for, if the brain is removed, the animal can still move in a co-ordinated way—and, in fact, it moves about more than usual. If it meets some obstacle, it does not withdraw and go off in a new direction but persists in its unsuccessful forward movements. This very unadaptive kind of behavior shows that in the normal nereis the brain has an important function which it did not have in flatworms— that of *inhibition* of movement in response to certain stimuli.

The **reproductive system** is very simple. Sex cells are budded off from the coelomic lining in most of the segments. Sexual maturity generally

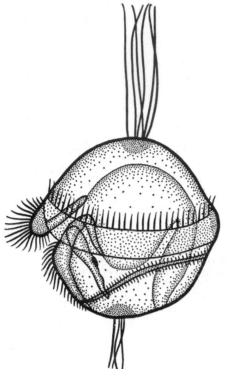

occurs at some definite season of the year, and at this time the worms leave their burrows and swim near the surface. The males are attracted to the females; and as the females burst and shed their eggs in the sea water, the males discharge their sperms. After the sex cells are shed, the worms die.

The fertilized egg of the nereis develops into a ciliated larva called a **trochophore.** This larva is of considerable theoretical importance because the same type occurs in several phyla. Few animals seem farther apart in adult structure than a segmented worm and a snail. Yet their early stages of development are al-

An annelid **trochophore.** (After Woltereck)

most identical, cell for cell; and the trochophores that result are similar in many respects. Beyond the trochophore stage, however, marked differences begin to appear and the adults are very unlike. The close relationship thought to exist between these two phyla would never have been suspected except for the similarities of their trochophores.

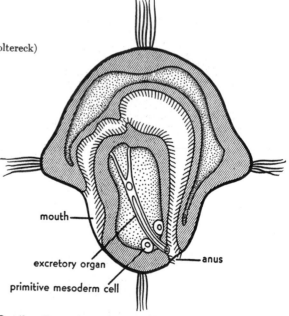

Section through a typical trochophore. (Modified after Shearer)

The most remarkable of the embryological resemblances that link the annelids and mollusks is the origin of the mesoderm in the two groups. The early stages of development of certain embryos have been followed so closely that each cell has been numbered and mapped. As a result of this extremely painstaking kind of work it is possible to trace the "cell lineage" of any portion of the early embryo. The adult mesoderm comes from a single cell (the "4d" cell), which arises in the same way in both annelids and mollusks. This cell divides into a pair, the *primitive mesoderm cells* (shown lying against the wall of the intestine and near the opening of the larval kidney in the diagram of the trochophore). These give rise to two bands of mesoderm, which finally become hollowed out to form the coelom of the adult.

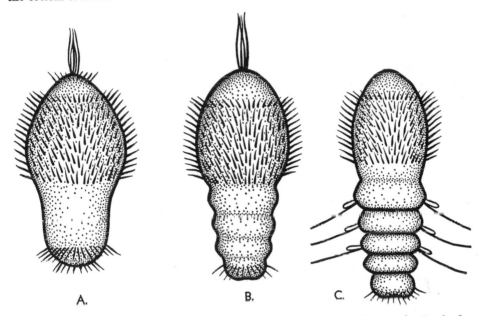

A.   B.   C.

Development of an annelid. A, the trochophore is elongating at its lower pole. **B,** the first segments are indicated, which in **C** are clearly constricted and already bear the larval bristles. (Modified after Mead)

When the young nereis hatches from the egg membrane, it has already passed the trochophore stage and has three segments with bristles. The "typical trochophore" usually described is present in only a few annelids, but it is thought to represent the more primitive condition. Its most characteristic structure is a ciliated band about the equator, which serves as the chief organ of locomotion and also directs a food-bearing current toward the mouth, which lies just below. At the upper pole is a group of sensory cells from which arises a tuft of cilia. Some trochophores have a tuft of cilia at the lower pole also. Internally the larva has a complete digestive tract with esophagus, large bulbous stomach, and a short intestine that opens through an anus at the lower pole. The larval excretory organ contains a flame cell and an excretory tube which opens near the anus.

The development of the annelid trochophore into the adult worm begins with the elongation of the lower region of the trochophore. The elongated region becomes constricted into segments which soon develop bristles. The ciliated bands disappear, and the upper part of the trochophore becomes the head. The young worm then settles to the bottom, takes up a burrowing life, and continues to grow throughout life by the addition of new segments in a region just in front of the last segment.

Various hypotheses have been devised to explain the origin of segmentation, but there is not enough evidence to be able to decide among them. One states that, primitively, all of the organ systems had repeated parts which were spread out over the entire body. The development of crosswise partitions, however they arose, divided the body into segments, each segment receiving representatives of each system. One basis for this hypothesis is the fact that in the planaria and many other animals, structures such as the testes, yolk glands, and the cross-connections between the two nerve cords are repeated along the body, and all that is needed is the development of partitions to produce a segmented condition. Further, in the embryological development of the nereis and other segmented forms, the partitions appear after the basic segmentation is already laid down.

Another hypothesis assumes that each segment represents a subindividual which was produced by asexual budding, as in the planaria, and which failed to detach. In the development of such a flatworm as *Microstomum* (see p. 133), chains of as many as sixteen subindividuals form before any break away. A segmented animal, according to this view, is a chain of completely co-ordinated subindividuals. This hypothesis suffers the disadvantage that in a developing annelid the segments do not arise in this way.

Segmentation in one phylum of animals is not necessarily the same as in another. Hence, we cannot expect to devise one explanation for all the events of segmentation, and we cannot point to the common ancestor of all segmented animals, because segmentation probably arose independently in more than one line of evolution.

SEGMENTATION seems to have the same general advantages as the dividing-up of the animal body into cells, namely, there is the possibility for the different segments to specialize in different functions. In the nereis the segments are practically all alike, and this is the primitive condition. In other segmented animals there are varying degrees of specialization, some of which are very extreme.

CHAPTER 20

EARTHWORMS AND OTHER ANNELIDS

THE earthworm caught by the early bird is no early worm but one that stayed out too late, for earthworms are nocturnal animals, emerging only at night and retreating underground in the morning. Even at night they usually do not leave their burrows but protrude the anterior part of the body in search of the seeds, leaves, and other parts of plants on which they feed while the posterior end maintains a firm hold on the burrow. These retiring habits have probably contributed to the marked success of an animal that is quite helpless above ground. Since earthworms are adapted to living on so abundant and widely distributed a food as the decaying organic matter of the soil, it is not surprising that they occur in countless numbers in moist soils all over the world. Ever since Darwin made their activities the object of a careful study and concluded that "it may be doubted if there are any other animals which have played such an important part in the history of the world as these lowly organized creatures," it has been recognized that the work of earthworms is of tremendous agricultural importance.

Earthworms spend most of their time swallowing earth below the surface and depositing it on the surface around the mouths of their burrows in the form of the "castings," familiar to everyone. In loose soil the burrow may be excavated simply by pushing the earth away on all sides, but in compact ground the soil must actually be swallowed. In moist and rainy weather the worms live near the surface, often doubled up on themselves so that either mouth or anus can be protruded. But in cold weather they plug the opening of the burrow and retreat into its deepest part, which usually ends in a chamber where one or several worms, rolled up together into a ball, pass the winter. In very hot weather, also, they live far from the surface, thus avoiding drying.

The swallowing of earth is not alone a means of digging burrows. The soil passed through the digestive tract contains organic materials of various kinds: seeds, decaying plants, the eggs or larvas of animals, and the live or dead bodies of small animals. These are digested, while the main bulk of the soil passes through. When leaves are abundant on the surface, the worms drag them into the burrows, and few castings are thrown up. When few leaves are taken in as food, the amount of castings increases.

The *effects of the worms on the soil* are many. The earth of the castings is exposed to the air, and the burrows themselves permit the penetration of air into the soil, improve drainage, and make easier the downward growth of roots. The thorough grinding of the soil in the gizzard of the worm and the sifting out of all stones bigger than those that can be swallowed is the most effective kind of soil "cultivation." The leaves pulled into the ground by earthworms are only partially digested, and their remains are thoroughly mixed with the castings, adding organic matter. The excretory wastes and other secretions of the worms also add organic material, enriching the soil for future plant growth. In this way earthworms have helped to produce the fertile humus that covers the land everywhere except in dry and certain other unfavorable regions.

The quantity of earth brought up from below and deposited on the surface has been estimated to be as high as 18 tons per acre per year, or, if spread out uniformly, about 2 inches in 10 years. Seeds are covered and so enabled to germinate, and stones and other objects on the surface become buried. In this way ancient buildings have been covered and so preserved, much to the advantage of archaeologists.

Externally the earthworm differs from the nereis in its *adaptations to a subterranean life*. As in other burrowing animals, the body is streamlined and has no prominent sense organs on the head or any projecting appendages on the body which would interfere with easy passage through the soil. On each segment are four pairs of **bristles,** or *setas*, which protrude from four small sacs in the body wall and are extended or retracted by special muscles. The bristles are used to anchor the worm firmly in its burrow, as can be readily discovered by trying to pull one out. But their main function is **locomotion.** The worm works its way along by extending the anterior part of the body, taking hold by means of the bristles, and by expansion of the body, then retracting the bristles of the posterior region and drawing up the posterior part of the body.

The lack of prominent sense organs on the head does not mean that the earthworm is insensitive to stimuli but only that there is no concentration of sensory cells into highly specialized organs at the anterior end. As in the nereis, cells sensitive to light, touch, and chemicals occur among the epithelial cells of the epidermis.

The absence of definite eyes, in an animal belonging to a phylum in which well-developed eyes are common, is not unusual. Almost all animals that live in complete darkness have degenerate eyes or no eyes at all; examples are burrowing forms like the earthworm or mole, cave animals like certain fish and crayfish, and nocturnal forms like many beetles.

The *light-sensitive cells* of the earthworm are absent from the ventral surface and are most abundant at the anterior and posterior ends, the regions most frequently exposed to the light. Thought to be organs of *touch*, probably because they occur all over the body, are groups of from thirty-five to forty-five cells, each with a hairlike process which projects through the cuticle covering the surface. Perhaps they are also sensitive to *chemicals* and to changes in *temperature*, stimuli to which earthworms respond. *Taste* cells probably occur in the mouth and pharynx, since the worms seem to show definite food "preferences" —neglecting cabbage if celery is also offered, and passing up celery if carrot leaves are available. The sense of *smell* is very feeble; and the worms are unresponsive to *sound*, which requires a complicated receiving apparatus not found in lower animals. More important for a subterranean animal is the ability to detect *vibrations* transmitted through solid objects. To these, earthworms are extremely responsive. It is said that one way to collect earthworms is to drive a stake into the ground and then move it back and forth, setting up vibrations in the ground, which cause the worms to emerge from their burrows.

The **central nervous system** is essentially the same as that of the nereis. A brain (suprapharyngeal ganglion) lying above the pharynx connects by two nerves with a large ganglion (subpharyngeal ganglion) lying below the pharynx. These two ganglia send nerves to the sensitive anterior segments and are considered to be the "higher centers." The brain is supposed to direct the movements in response to sensations of light and touch which it receives; but if it is removed, the behavior of the worm is affected little. After removal of the lower ganglion the worms no longer eat, and they cannot burrow in normal fashion. From this first ventral ganglion the double nerve cord runs to the posterior end of the body, enlarging in each segment to a double segmental ganglion from which nerves go to all parts of the segment. Each ganglion serves as a center which receives impulses coming from sensory cells in the skin and sends impulses that result in contraction of the muscles.

The ganglia co-ordinate the impulses so that the longitudinal muscles relax while the circular muscles contract, or the opposite. Without this arrangement the two sets of muscles might only counteract each other's activities, and no movement would result. The smooth muscular waves which pass down the body in the ordinary creeping movements of the earthworm are not controlled by the large ganglia at the anterior end, for almost any sizable piece of an earthworm will creep along as well as a whole worm. The co-ordination is thought to be achieved through impulses relayed from one segment to another by nerve cells in the cord which run from one ganglion to the next. Since there is

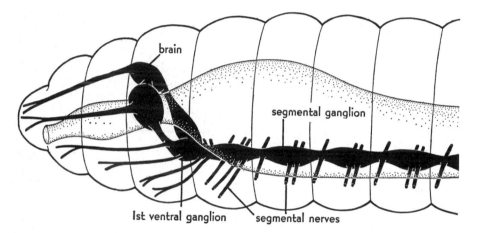

Anterior end of the earthworm, showing the **nervous system.** The digestive system is drawn as if transparent.

a certain amount of delay involved in the transfer from ganglion to ganglion in this chain-like succession of connecting fibers, these impulses travel very slowly. If measured as the speed of travel of the waves of thinning or thickening in the body of a moving worm, the rate is only about 1 inch per second. Besides the ordinary creeping movements earthworms can suddenly contract the whole body in response to strong stimulation of any region. If the anterior end is extended from the burrow and receives some unfavorable kind of stimulus, the longitudinal muscles contract as a whole, and the worm disappears into its burrow almost instantly. Such a response requires very rapid nervous transmission, and we do find certain "giant fibers" in the ventral nerve cord which pass over long distances or even throughout the length of the cord. The speed of transmission in these giant fibers has been estimated at 1.5 yards per second. The speed is about the same in the giant fibers of the nereis but may be almost 10 yards per second in some of its relatives. These figures seem very low when compared with the rate of nervous conduction in the motor nerves of man, in which impulses travel at about 100 yards per second.

The **digestive system** is differentiated into a number of regions, each with a special function. Food enters the mouth, is swallowed by the action of the muscular pharynx, and then passes through the narrow esophagus, which has on each side three swellings, the **calciferous glands.** These glands excrete calcium carbonate into the esophagus and in this way dispose of the excess calcium obtained from the various salts present in the food. The esophagus leads into a large, thin-walled sac, the **crop,** which apparently serves only for storage, since the food undergoes no important change and does not remain there very long. Behind the crop is another sac, the **gizzard,** with heavy muscular walls which (aided by mineral particles and very small stones swallowed by the worm) grind the food thoroughly. From the gizzard the food passes through the intestine, which continues practically uniformly to the anus. In the intestine the food is digested by juices from gland cells of the lining epithelium. The roof of the intestine dips downward as a ridge or fold (the typhlosole), which increases the digestive surface that comes in contact with the intestinal contents. The digested food is absorbed into the blood vessels of the intestinal wall, and from there distributed to the rest of the body.

The **circulatory system** is very similar to that of the nereis. A median contractile dorsal vessel, which lies on the digestive tube and accompanies it from one end of the body to the other, is the main collecting vessel. In it the blood flows forward, propelled by rhythmic peristaltic waves. A median noncontractile ventral vessel, suspended from the digestive tube by the ventral mesentery, is the main distributing vessel. In it the blood

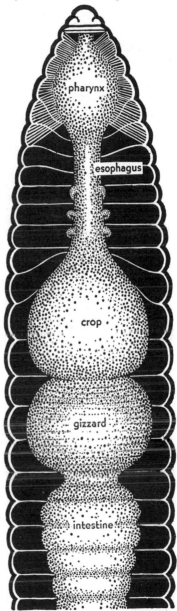

Digestive system of the earthworm. The three pairs of swellings on the esophagus are the *calciferous glands.*

flows backward and out into branches which supply the various organs. In almost every segment blood flows from the ventral to the dorsal vessel through capillary beds of the body wall, digestive tract, and nephridia. In the region of the esophagus the dorsal and ventral vessels are connected directly through five pairs of enlarged muscular transverse vessels, the **hearts,** which pump the blood through the ventral vessel. Valves in the dorsal vessel and hearts prevent the blood from backing up during irregular contractions.

Branches of the transverse segmental vessels supply blood to the capillary beds of the nephridia and body wall. This blood is then returned to the dorsal vessel. In segments 7–11 this is reversed; the blood flows downward directly from the dorsal to the ventral

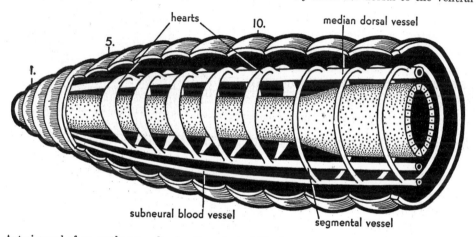

Anterior end of an earthworm, showing the principal blood vessels. Five pairs of **hearts** surround the esophagus.

vessel through the hearts. In front of the hearts the blood in the ventral vessel flows forward to the head; behind the hearts it flows backward and out into the transverse branches. The ventral vessel also sends segmentally repeated branches to the wall of the digestive tube, where the blood becomes loaded with absorbed food. From the intestinal wall the blood returns through paired segmental vessels to the dorsal vessel. Besides draining the body wall and nephridia, the transverse vessels carry blood directly from the subneural vessel to the dorsal vessel. The subneural vessel runs below the nerve cord and supplies it with blood. Two lateral neural vessels (not shown in the diagram) run one on each side of the nerve cord and send branches to the segmental nerves.

Earthworms are terrestrial animals, but they have not really solved the problems of land life; they have merely evaded them by restricting their activities to a burrowing life in damp soil, by emerging only at night, when the evaporating power of the air is low, and by retreating deep underground during hot, dry weather. Animals well adapted for land life have a heavy impermeable skin which prevents excessive drying, but it also prevents respiratory exchange through the skin. In such animals oxygen reaches the internal tissues by means of special respiratory devices, such as lungs. Earthworms, on the other hand, breathe in the same way as their aquatic ancestors. That is why they can live for months completely submerged in water, yet will die if dried for a time. The outermost layers of the earthworm are

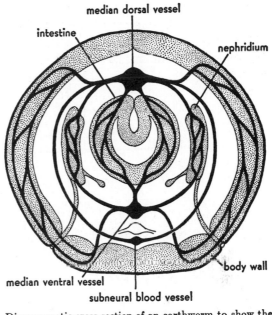

Diagrammatic cross-section of an earthworm to show the **main segmental blood vessels.**

thin and must be kept moist so that **respiratory exchange** can occur by diffusion through the general body surface, which is underlain by capillary networks. Moistening of the surface is accomplished by mucous glands which occur in the epidermis and also by the coelomic fluid which issues from *dorsal pores* located in the mid-dorsal line in the grooves between segments.

The **excretory system** is like that of the nereis, with a pair of excretory organs, or *nephridia*, in every segment (except the first three and the last). Each nephridium really occupies two segments, because it opens externally by a pore on the ventral surface and internally by a ciliated funnel which lies in the coelom of the segment anterior to the one containing the body of the nephridium and its external pore. The passage of fluid is caused not so much by the cilia lining the nephridial tube as by waves of contraction of the muscles in the wall of that portion of the nephridium which leads to the external pore.

The nephridia are not the only means of excretion in the earthworm. The coelomic lining surrounding the intestine and the main blood vessels is modified into special **chloragog cells.** Wastes extracted from the blood accumulate in the chloragog cells, which finally become detached and

float in the coelomic fluid. Some of the chloragog detritus is removed by
the nephridia. Some of it is engulfed by the ameboid cells of the fluid,
which finally wander into the tissues and disintegrate, leaving their wastes
as a deposit of pigment in the body wall.

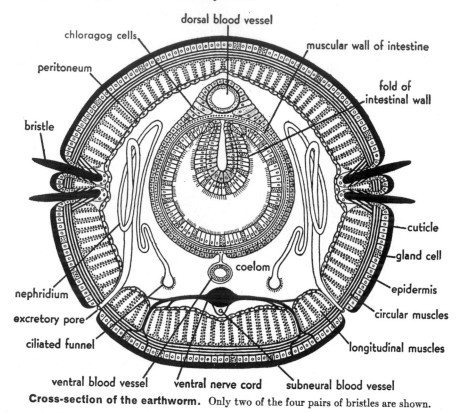

Cross-section of the earthworm. Only two of the four pairs of bristles are shown.

The pigment which thus accumulates in the body wall probably serves to shield the
underlying tissues from the light, particularly the ultra-violet, which is very harmful to
earthworms. One hour's exposure to strong sunlight causes complete paralysis in some
worms, and several hours' exposure is fatal. This is thought to explain the death of many
of the earthworms seen lying in shallow puddles after a rain. They have not been drowned
by the water, as many people suppose, for earthworms can live completely submerged in
water. However, during a rain the water that fills their burrows has filtered down through
the soil and therefore contains very little oxygen. This forces some of the worms to come
to the surface, where they are injured by the light and after a time can scarcely crawl.
They probably remain in the rain puddles because of the protection afforded by the
layer of water. Many of the dead worms seen after a rain were no doubt sick before-
hand, perhaps as the result of heavy infestation with parasites; their death has only been
hastened by the rain.

The organ systems of the earthworm described up to this point have shown little or no increase in division of labor among segments over the condition in the nereis. The nephridia are identical in all segments in which they occur, and the central nervous system is practically the same as that of the nereis. The circulatory and digestive systems of the earthworm show some increase in specialization. Certain of the transverse vessels are enlarged and modified as hearts; and there are (in addition to the pharynx, esophagus, and intestine, also present in the nereis) two separate regions: the crop for storage, and the gizzard for mechanical breakdown of food. However, the greatest specialization among earthworm segments is found in the reproductive system.

The complexity of the **reproductive system** is an adaptation to land life, where the naked sex cells cannot simply be discharged to the exterior, as in the aquatic annelids, but must in some way be protected from drying and other adverse conditions during the development of the young. Earthworms, unlike the nereis, are *hermaphroditic*, each individual having a complete male and female sexual apparatus. This is thought to be an adaptation to sedentary life, which provides relatively few contacts between individuals. Hermaphroditism makes possible two exchanges of sperms, instead of only one, for each meeting of two individuals.

The **sex organs** are located in the anterior end of the worm, each organ in a particular segment. The male sex cells are formed in two pairs of testes, located in segments 10 and 11, and each pair is inclosed within a testis sac that communicates with the sperm sac in which the sex cells undergo further development. The mature sperms pass back into the testis sacs, into the sperm funnels, and through the sperm ducts to the two male genital openings on the ventral surface of segment 15. Two pairs of small sacs, the sperm receptacles, in segments 9 and 10, open through pores to the ventral surface. During mating they receive the sperms from the other partner. The eggs are formed in a pair of ovaries in segment 13. As they attain maturity, they are shed from the free end of the ovary into the egg funnels situated on the posterior face of segment 13. These funnels lead into the oviducts, which open by two minute pores on the ventral surface of segment 14. The beginning of the oviduct has a lateral pouch, the egg sac, in which ripe eggs are stored. Behind the sex organs is a swollen ring, the **clitellum,** formed by the thickening of the surface epithelium, which contains great numbers of gland cells. These secrete mucus which, as we shall see, plays an important role in the protection of the developing embryos.

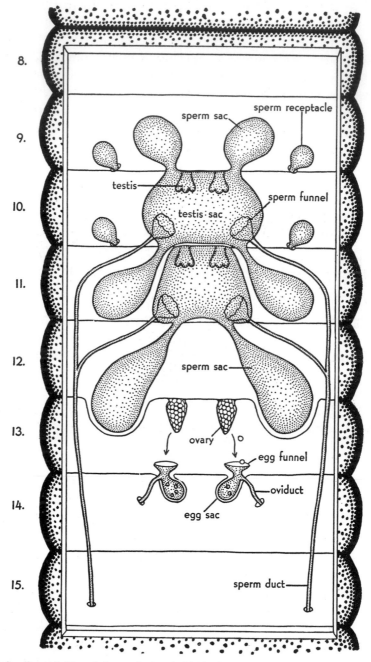

8.

9.

10.

11.

12.

13.

14.

15.

sperm sac

sperm receptacle

testis

sperm funnel

testis sac

sperm sac

ovary

egg funnel

oviduct

egg sac

sperm duct

The **reproductive system** of the earthworm is highly specialized. The testis sacs are drawn as if they were transparent, so that the testes and sperm funnels, located within them, can be seen. The sperm ducts actually run beneath the sperm sacs, but have been drawn out at the sides to show their connections clearly. The large sperm sacs in segment 12 press down on the coelomic partition, so that they appear to lie also in segment 13. This diagram is based on the structure of *Lumbricus terrestris;* in some other species, certain of the details are different.

The **mating process** is no simple shedding of the gametes, as in the aquatic annelids. At the sexual season, when the ground is wet following a rain, the worms may emerge and travel some distance over the surface before they mate; more often they merely protrude the anterior end and mate with a worm in an adjoining burrow. The two worms appose the ventral surfaces of their anterior ends, the heads pointing in opposite directions. The clitellum of one worm is opposite segments 9–11 of the other, and this is the region of most intimate attachment. Mucus is secreted until each worm becomes inclosed in a tubular mucous "slime

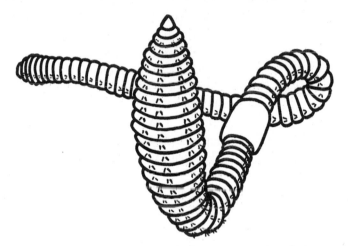

An "ant's-eye" view of an earthworm with the anterior end lifted to show the four **rows of bristles.** The thickening of the body wall in the region of the **clitellum** obscures the external segmentation and also the bristles of that region. The position of the clitellum is definite for each species of earthworm. In *Lumbricus terrestris* it extends over segments 31 or 32 to 37.

tube," which extends from segment 9 to the posterior edge of the clitellum. When the sperms issue from the male genital openings in segment 15, they are carried backward (in longitudinal grooves which are converted into tubes by the presence of the mucous sheath) to the openings of the sperm receptacles on segments 9 and 10 of the mating partner. Then the worms separate; the egg-laying and fertilization occur later.

The **egg-laying** starts when the gland cells of the clitellum secrete a mucous ring which glides forward over the body of the worm. As it passes the openings of the oviducts (segment 14), it receives several ripe eggs; and then, as it passes the more anterior openings of the sperm receptacles (segments 9 and 10), it receives sperms which were deposited there previously by another worm in the mating process. Fertilization of the eggs takes place within the mucous ring, which finally slips past the anterior tip of the worm and becomes closed at both ends to form a sealed capsule (sometimes called a "cocoon"). Within the capsule, which lies in the soil, the zygotes develop directly into young worms and then escape. As in other land animals (and in fresh-water forms as well), there is no free-swimming larval stage comparable with that of marine annelids.

POLYCHETES

THE marine bristle worms, such as the nereis, comprise the largest and most generalized class of annelids, the **Polychaeta,** named from the many bristles borne upon the parapods. They are among the most common animals of the seashore, some living under stones and in the mud in tubes or burrows, while others swim freely in the water, especially during the breeding season.

The **free-swimming polychetes** resemble the nereis in having a well-differentiated head with prominent sense organs, an eversible pharynx bearing horny jaws or teeth, and well-developed parapods. Many of these animals, like the nereis, live in burrows but can leave them and build new ones. Some of them have a special sexual phase, which in the case of the nereids is called a *heteronereis.* This modified sexual animal looks so different from the normal burrowing type that it can scarcely be recognized as belonging to the same species. The eyes are enlarged, the sensory projections shrunken, and the body differentiated into two distinct parts: an anterior region with unmodified parapods, and a posterior region with parapods that have enormous lobes and flattened, oar-shaped bristles. These changes are associated with the increased activity of the free-swimming sexual form, for the heteronereis does resemble those polychetes which are permanently free-swimming.

In some polychetes the sex cells are formed only in the posterior part of the worm. After undergoing changes in shape and color this part breaks off, rises to the surface and swims about, shedding the eggs or sperms. In the palolo worm of the South Pacific this takes place at a specific time— just at dawn one week after the November full moon. The sexual pieces rise to the surface in countless millions, and the appearance of the water at this time has been compared to vermicelli soup. Later it appears milky from the eggs and sperms that are discharged. The anterior part of the worm, which remains hidden in some crevice in the coral rock when the posterior piece breaks off, regenerates the missing parts. On the corresponding day of the next year, the regenerated posterior end, laden with sex cells, breaks away.

The natives of the Samoan and other islands are familiar with the habits of the palolos. They consider them a great delicacy and look forward to their breeding season. When the day arrives, they scoop them up in buckets and prepare a great feast, gorging themselves just as we do on Thanksgiving day, knowing that there will not be another treat like it until exactly the same day of the next year. Actually, there is a small "crop" of swarming palolos a week after the October full moon, but it is too small to interest the natives.

Since the sex cells of animals are capable of being fertilized for only a short time after they are released into sea water, the swarming habits of polychetes, which provide for the simultaneous release of eggs and sperms from a great many closely approximated individuals, are an adaptation for insuring the fertilization of the greatest possible number of eggs. In addition to the periodicity of the swarming, some polychetes have other devices which bring about the simultaneous extrusion of eggs and sperms. In the nereids the discharge of sperms is set off by a secretion from the swarming female. In the so-called "fire worms" of Bermuda the meeting of the sexes involves the exchange of light signals. The worms come to the surface to spawn each month a few days after the full moon at about an hour after sunset. The female appears first and circles about, emitting at intervals a greenish phosphorescent glow which is readily visible to observers on the beach. The smaller male then darts rapidly toward the female, emitting flashes of light as it goes. When the two sexes come close together, they burst, shedding the sex cells into the sea water. Then the spent worms, reduced to shreds of tissue, perish. It has been suggested that the phosphorescent flashes of spawning polychetes were the lights seen by Columbus on the night he approached the New World.

The true **tube-dwelling polychetes** rarely or never leave their tubes, which may be made of mucus hardened to a parchment-like material, of particles of sand or shells stuck together by a mucous secretion, or of lime laid down on a mucous framework. These worms have degenerate parapods; their heads, though reduced, are provided with long brilliantly colored tentacles which protrude from the opening of the tube. Some forms also have extensible gill filaments, full of circulating blood, which serve as respiratory organs. The anterior part of the digestive tube is not eversible, and there are no jaws. These worms feed on minute animals or plants which are carried toward the mouth by rows of cilia on the tentacles.

The **burrowing polychetes** have reduced heads and parapods, as in the tube-dwelling types. But they have an eversible pharynx, which in the com-

Giant earthworms are found in tropical regions, especially in Australia. *Megascolecides australis*, in the drawing, may be 11 feet long. It lives in burrows with volcano-shaped openings. (Modified after an old cut from Sterne)

mon lugworm *Arenicola*, is covered with minute papillas and, in addition to its use in feeding, serves as the chief organ of locomotion through the sand. Arenicola feeds like an earthworm, passing large quantities of sand through the digestive system to obtain the organic matter mixed with it. The castings can be seen on sandy beaches when the tide is out.

OLIGOCHETES

THE second largest class of annelids is the **Oligochaeta** ("few bristles"), of which more than four-fifths are the earthworms and the rest are mostly small or minute worms which occur in the soil or in fresh water. (This last type is extensively used as fish food and is familiar to breeders of pet fish. A hydra can be seen eating a fresh-water oligochete in the photographs in chapter 7.)

The oligochetes differ from polychetes in several important respects. There are no parapods, and the bristles emerge from pits in the body wall. Whereas the polychetes have separate sexes and the sex cells are budded off from the coelomic lining in numerous segments, the oligochetes are hermaphroditic and the sex cells are produced in special organs which occur only in certain segments. Because both groups have clearly marked external segmentation, share many annelid structures, and bear on each side of every segment two separate bundles of bristles which are moved by muscles and serve in locomotion, they are sometimes grouped together as the *Chaetopoda* ("bristle-footed"). These resemblances may be the result of descent from a common ancestral group which was more primitive than

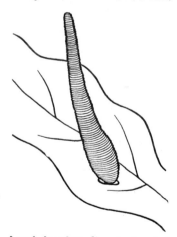

Land leeches live in tropical forests, attached to leaves by the posterior sucker, with the body held erect and prepared to fasten on to the first mammal that comes within reach. These leeches occur in such numbers that horses are sometimes driven wild by them and men suffer serious loss of blood. (After a photograph by Heinrich)

either polychetes or oligochetes. But if one of the groups is derived from the other, the polychetes are certainly the more primitive group and the oligochetes the derived group.

ARCHIANNELIDS

FORMERLY both polychetes and oligochetes were thought to have evolved from a supposedly primitive group of annelids, the class **Archiannelida** ("primitive annelids"). Now the best opinion considers these worms to be not primitive but *simplified* annelids, which have lost the external segmentation and in most cases also the parapods and bristles. Many of them are ciliated, a juvenile character retained from the trochophore larva.

LEECHES

THE leeches, class **Hirudinea**, have no bristles; and the external segmentation of the body does not correspond with the internal segments, of which there are fewer. The body is solid, the coelomic spaces being crowded out by the growth of connective tissue. There is a definite number of segments, for leeches do not add them on throughout life, as do other annelids. At each end of the body is a sucker, the posterior being much larger than the anterior, which has the mouth in its center. Despite all the modifications the leeches are closely related to the oligochetes and are probably derived from them. There is no character of leeches which is not present in at least some degree in some oligochete.

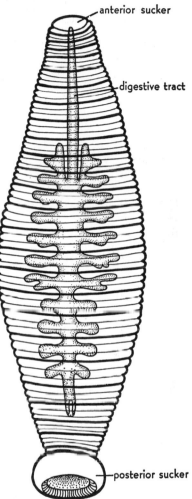

A leech. The digestive tract has large side pouches which greatly increase its capacity for blood. The surface of the leech is thrown into folds, of which only certain ones (indicated by heavy lines in the drawing) correspond to segments. This animal lives in the gill chambers of a fish. (After Hemingway)

Most leeches lead a semiparasitic life, sucking the blood of vertebrates, although some of them have lost this habit and feed on small animals. They show some of the adaptations for parasitism that were noted among flatworms, namely, the development of clinging organs, the suckers, and the extreme complication of the reproductive system. On the other hand, since they need to swim about to locate victims, they are less modified than flukes or tapeworms and have eyes. In sucking blood, a leech at-

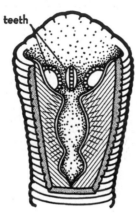

The **head of a leech**, cut away to show the three sawlike teeth with which it makes a wound. (Modified from Pfurtscheller)

The three teeth of a leech inflict a **Y**-shaped wound. (After Reibstein)

taches to some vertebrate by the posterior sucker, applies the anterior sucker to the skin, makes a wound, often with the aid of little jaws inside the mouth, fills its digestive tract with blood, and then drops off, remaining torpid while digesting the meal. Large blood meals are few and far between, but the digestive tract has lateral pouches which hold enough blood to last for months. The salivary glands of leeches manufacture a substance called *hirudin*, which prevents the coagulation of the blood while the leech is taking its meal. For this reason a wound made by a leech continues to bleed for a long time after the leech has detached itself.

Free-swimming polychete worm, *Nereis,* spends most of its time in a burrow in the mud or sand but swims actively at the surface during the mating season. Nereis is well known to fishermen and clam-diggers of the New England coast, who call it the "clam worm" because it commonly occurs where clams live. Sometimes the worms are found between the valves of an empty shell, hence the false belief that it preys on live clams. Actually, it eats much smaller animals. (Photo of living animal. Mount Desert Island, Maine)

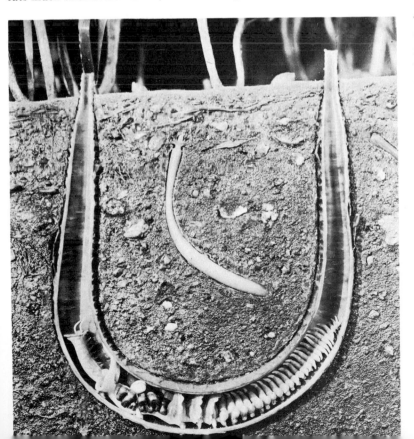

Tube-dwelling polychete worm, *Chaetopterus,* lives in burrows in the mud or sand along the Atlantic Coast. The U-shaped burrow, here shown in section, is lined with a tough, parchment like substance secreted by the worm. The animal has a delicate body and remains always in its well-protected burrow, feeding on small organisms brought in by the steady current of water that enters one end of the burrow and leaves through the other. The current, which also brings in oxygen and carries away wastes, is maintained by the flapping of the large, modified parapods on the mid-region of the body. The worm is luminescent, emitting a bluish-green light; but no one knows how this could be of any use to an animal that spends its life in a tube in the mud. (Photo of model. American Mus. of Nat. Hist.)

Fan worm, *Bispira* (*left*), and **peacock worms,** *Sabella* (*below*), are polychete worms that live in long tubes which they build in the sand, usually among rocks. Only the anterior end, a degenerate head, is extended from the tube. It bears long feathery gills which are respiratory and collect food by entangling small organisms in a layer of mucus and conveying them, by means of cilia, to the mouth. The gills are brightly colored, usually red or purple, and a group of these worms looks like a small patch of flowers—until one approaches closely and sees them whisk the tentacles into the tube with a lightning-like speed that immediately identifies them as animals. (Photos of living animals by Douglas P. Wilson. Plymouth, England)

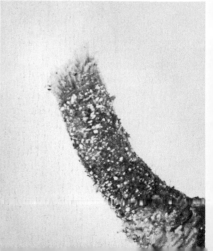

Feather worms, like their two relatives above, have eyespots on the gills and are very sensitive to changes in light. The shadow of one's hand passing over the extended worm (*left*) will cause it to pop back into its tube, as at *right*. (Photos of living animal. Pacific Grove, California)

Copulating earthworms usually do not leave their burrows but extend only the anterior end and mate with a neighboring worm. The animals oppose their ventral surfaces and exchange sperms, for, though every earthworm has both male and female sex organs, it does not fertilize itself. The actual fertilization occurs later at the time of egg-laying. (Photo of living animals made at night by L. Keinigsberg. Chicago)

Giant earthworms are found in tropical regions. They may be located by the gurgling sounds they make as they move underground. These men, in Australia, are extracting a long worm from its burrow. When extended, it may be up to 12 feet long, with as many as 500 segments. The only bird known to feed on this worm is the laughing kingfisher. (Globe photo)

Pulling an earthworm out of its burrow is not easy, as any one can find out by trying. The animal hangs on by inserting its bristles into the walls of the burrow. (Photo of living animal by L. Keinigsberg. Chicago)

Leech (*Placobdella*) taken from the naked skin at the base of the hind leg of a snapping turtle. Removed to an aquarium, it attached itself to the glass by means of the two suckers, the larger of which is at the posterior end, the smaller at the mouth end. Leeches are segmented worms, but the folds of the surface are more numerous than the internal segments. A good meal lasts several months, and during this time the blood is stored in stomach pouches, which can be seen through the body wall as dark bands. The animals are hermaphroditic, having both male and female sex organs. Large specimens are 3 or 4 inches long when partly extended. The ground color is deep olive green, with spots of brown. (Photo of living animal by S. T. Brooks, courtesy *Nature Magazine*)

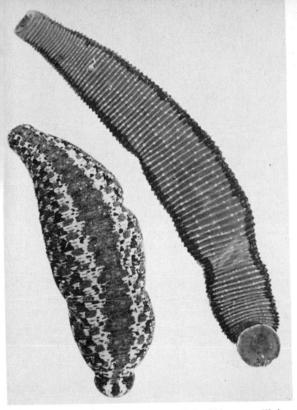

Medicinal leeches (*Hirudo medicinalis*) are still imported from Europe for the removal of black-and-blue spots, particularly around the eyes. These two were purchased in a Chicago drugstore. The leech can take three times its own weight in blood at a single meal and injects into the wound an anticoagulant. *Left,* upper surface of contracted worm. *Right,* lower surface of extended worm. (Photo by P. S. Tice)

Pond leeches are easily collected, usually unwillingly, by wading in a pond with bare feet. In addition to the well known "leechlike" method of movement, leeches can swim actively by undulations of the body. They feed on worms, aquatic insect larvas, and even other leeches. When sexually mature, they take a large blood meal, which may be slowly digested over as long as a year. (Photo by Cornelia Clarke)

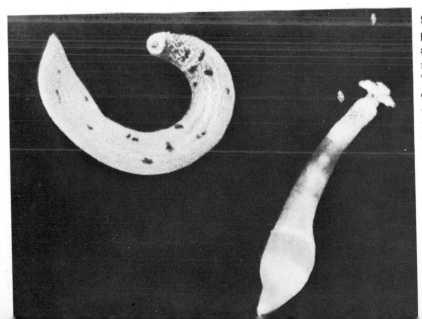

Sipunculans (Phylum **Sipuncula**) are marine wormlike animals which, though unsegmented, are related to annelids. They swallow great quantities of mud or sand, from which they extract organic material. When removed from their burrows, they alternately evert, *right,* and invert, *left,* the anterior end of the body, which has at its tip a circle of tentacles surrounding the mouth. A model of a sipunculan can be seen in the mud in the picture of *Chaetopterus.* (Photo of living animals. Pacific Grove, California)

1. *Urechis* (phylum **Echiura**) is wormlike, unsegmented and related to annelids. It lives in West Coast mud flats and can be dug up at low tides.

2. It inhabits a U-shaped burrow, whose exact position is determined by inserting a rubber tube into one opening and blowing until water spouts from the other.

3. Following the path of the rubber tube, the mud is rapidly shoveled away until the burrow is laid open and the pink, cylindrical worm is exposed.

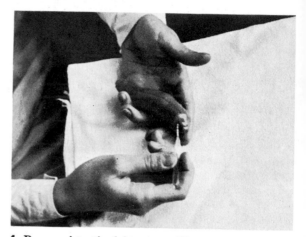

4. Removed to the laboratory, *Urechis* is an excellent source of eggs or sperms, which can be sucked up with a glass pipet inserted into the sexual openings.

5. Through the microscope, biologists observe the fertilization and development of the egg under normal and also under various experimental conditions.

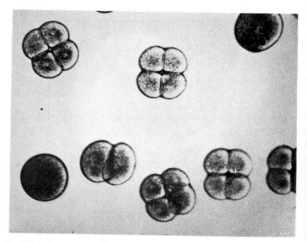

6. Highly magnified under the microscope, are one-, two-, and four-cell stages—practically indistinguishable from the early stages of most other animals.

CHAPTER 21

A MISSING LINK—PERIPATUS

IF WE could find an animal clearly intermediate in structure between two modern phyla, we would have good evidence that the two phyla are closely related. That such an animal has never been found is not surprising. Indeed, it would be more remarkable if the very form which at some remote time in the past gave rise to two stocks, now represented by two modern phyla, had also persisted unchanged through the ages. We have fossil records to show that certain species have remained unchanged for very long periods of time, but none are so old that they trace back to the time before all of the modern phyla had evolved. Therefore, we often speak of these missing ancestral forms as "missing links."

An animal that comes closer than any other to being the "missing link" between any two phyla is the **peripatus,** member of the small phylum **ONYCHOPHORA,** the name of which means "claw-bearing" and refers to the curved claws on the feet. The peripatus is a rare animal, found in moist places under logs in the tropical forests of Australia, Africa, Asia, South and Central America, and a few other regions. Its occurrence only in local regions in such widely separated parts of the world suggests that it was probably a more successful and widespread form in the past but is now gradually disappearing.

A peripatus looks much like a caterpillar, 2 or 3 inches long, with soft velvety skin and many pairs of legs. While many of its structures are like those of the phylum Arthropoda, to which the caterpillars belong, the peripatus also has many similarities to the annelids and, of course, some special features of its own. As already pointed out, neither the peripatus, as we find it now, nor any other living animal could be ancestral to any group as old as a phylum; but there is little doubt that the peripatus is a descendant from a line which branched off close to the primitive annelid-arthropod stock.

Unlike typical annelids and arthropods, the peripatus shows no external segmentation, though there is a pair of legs for each internal segment of the body. The legs end in claws, which superficially resemble those of arthropods, but differ from arthropod legs in that they are not divided into joints.

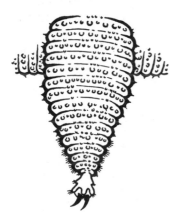

The outer covering is a thin **cuticle** like that of annelids, although it is ridged and covered with microscopic projections which give it a *velvety texture* unknown in other animals and which appear to prevent the cuticle from being readily wetted by water. Beneath the epidermis which secretes the cuticle are layers of muscles, as in annelids. The body wall of arthropods is somewhat different, having a heavy outer covering and no continuous layer of muscle.

Leg of a peripatus, showing the arthropod-like claws.

The peripatus usually comes out at night; and, though it has a pair of simple annelid-like eyes, it feels its way about by means of two sensory projections, or antennas, on the head. When attacked, it gives off a slimy secretion from a pair of glands which open on two projections, the oral papillas. It feeds on small insects and other animals by means of a pair of horny cutting jaws. Each of the three pairs of head appendages (antennas, oral papillas, and jaws) occurs on one of the three segments which compose the head. The fusion of segments, particularly at the head end, is characteristic of the most highly developed segmented animals; and the **three-segmented head** of the peripatus is thought to indicate a condition midway between that of annelids and arthropods, since the latter commonly have a six-segmented head.

The **internal anatomy** is a mixture of annelid-like and arthropod-like structures. The digestive tract is simple and not particularly distinctive. The circulatory system is like that of arthropods. A long contractile dorsal vessel, the heart, extends the length of the body and has along its sides a pair of openings for each segment of the body. As in arthropods, there are no definite vessels to return the blood to the heart. After leaving the vessels that carry blood away from the heart, the blood flows into large spaces in the tissues and finally collects in the space surrounding the heart, into which it enters through the paired heart-openings. The coelom is practically obliterated by the growth of connective tissue in which the blood spaces occur, and this is typical of arthropods. But the most arthropod-like character of all is the respiratory system, consisting of **air tubes** (tracheal tubes) which open from the external surface and extend throughout the body, piping air directly to the tissues. Although such structures occur nowhere else in the animal kingdom except in terrestrial arthropods, they are thought to have arisen independently in the two groups and not to be evidence that the Onychophora are related to arthropods.

The air-tubes of a peripatus differ from those of an arthropod in several important respects. In arthropods there are relatively few openings in the body wall, and they all have closing mechanisms. The openings lead into large air tubes which branch repeatedly, the branches decreasing in size and ramifying throughout the body. In the peripatus a large bundle of unbranched air tubes arises directly from each external opening and penetrates into the deep tissues of the body. The external openings are necessarily numerous and scattered over the body, and, as each is a small pit which lacks any kind of closing mechanism, the loss of water through this system of exposed air tubes is considerable. Experiments designed to test water loss under comparable conditions showed that a peripatus (*Peripatopsis*) lost water twice as rapidly as an earthworm, forty times as rapidly as a smooth-skinned caterpillar, and eighty times as rapidly as a cockroach. Since the cuticle of a peripatus, though very thin, is probably not readily permeable to water, the water loss is almost wholly through the system of air tubes.

The most annelid-like character is the **excretory system.** This consists of segmentally arranged pairs of coiled tubes which open by external pores at the bases of the legs. They resemble the excretory organs (nephridia) of annelids. The inner end of each organ opens into a very small coelomic sac, from which wastes are collected, presumably entering there by diffusion from the large internal blood space. Cilia in the tube sweep the wastes out through the external pore.

The **nervous system** is primitive. From the brain in the head run two widely separated ventral nerve cords which show small thickenings in each segment. Some annelids have widely separated cords, but even the primitive ones have segmental ganglia that are larger than those in the peripatus.

The **reproductive organs** are *ciliated*, as in annelids; cilia do not occur anywhere in arthropods. The eggs are fertilized within the body of the female. In some species eggs are laid, but in most forms the eggs develop within the female, and the young are born fully developed. As internal fertilization and development of the eggs are adjustments to land life and have evolved independently in terrestrial animals of many phyla, they have no special significance for the relationships of the peripatus.

THE existence of an animal with structures peculiar to two different phyla is a situation which follows naturally from what we know of the continuous nature of the process of organic evolution. But it creates difficulties in classification. The problem has been solved temporarily by placing the peripatus group in a phylum by itself. Still, there are some zoölogists who think that the animal is definitely an annelid and that its arthropod-like characters have arisen independently in the two groups. Others feel that these curious animals should be made one of the classes of the arthropods, with which group, they maintain, it belongs. That such a controversy exists makes the peripatus the best living candidate for the title of "missing link"; or, since it is not missing, perhaps we should call it a "connecting link." It suggests what the intermediate stage between two phyla might have been like, although the picture is much modified by the fact that the peripatus has undergone considerable evolution since the time it first branched from the primitive annelid-arthropod stock.

One may ask why the annelids and arthropods have gone on to become large and successful groups while the onychophorans may well be on the road to extinction. A conclusive answer cannot be given. The success or failure of a group depends upon a complex of many factors, some inherent in the structural development of the group, some determined by changes in climate, some quite accidental. The onychophorans have made many efficient adjustments to terrestrial life. Among these are a dry skin through which water is not easily lost, a type of metabolism that increases the production of water, mechanisms for decreasing the excretion of water, and a viviparous mode of reproduction. The exposed system of air tubes, however, makes impossible a close control over water loss, and it is quite possible that this is chiefly responsible for the apparent failure of an otherwise seemingly well-adapted group of animals. In the arthropods, and especially in the insects, we shall see a truly successful solution to the problems offered by the rigors of land life.

The peripatus is an animal much talked about but seldom seen, even by those who carefully search under every likely rotten log or mound of leaf mold in the tropical rain forests where the peripatus is known to live. Of extremely retiring habits, it comes out only at night to capture its small animal prey. When poked with a finger, both mother and offspring shot out a sticky material. The large female shown (about 5 inches long) gave birth to two young, but one ate the other. (Photo of living *Macroperipatus geayi*. Panama)

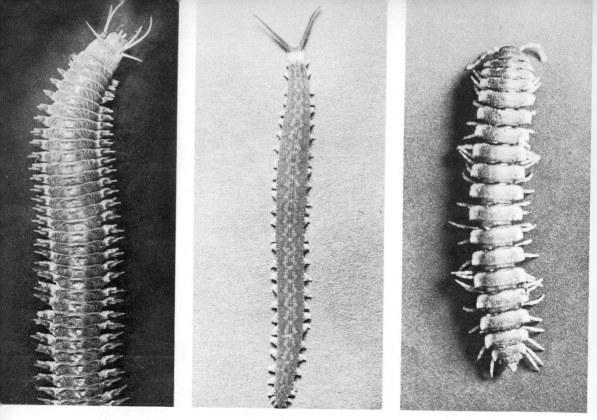

Nereis (Annelida)   **Peripatus** (Onychophora)   **Millipede** (Arthropoda)

The peripatus bears a general resemblance to both annelids and arthropods. Notice that in each of the three animals the appendages are all much alike and move in a series of waves. (Photos of living animals)

Tardigrade, *Hypsibius,* washed from moss close to a freshwater pond. To appreciate both common and scientific names of the phylum **TARDIGRADA** one need only watch through the microscope when the short, plump little "water bears" are slowly crawling about on plant debris with 4 pairs of stumpy clawed legs that can best be compared with those of onycophorans, which they resemble in other ways also. Tardigrades live in sea sediments and in lakes and ponds, but mostly inhabit temporarily wetted places on land such as mosses, lichens, garden walls, and cemetery urns. They are well adapted to withstand drying, even for years, and to become active again within a few hours of being wetted. Tardigrades feed mostly on plant juices, sucking them from plant cells pierced by two sharp stylets protruded from the mouth. (Photo of living animal about 0.75 mm long)

CHAPTER 22

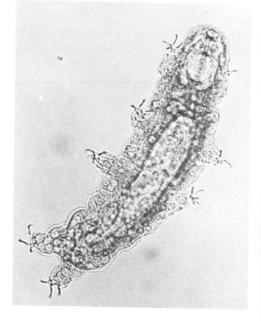

JOINTED-LEGGED ANIMALS

IN HUMAN society "success" is commonly expressed in terms of the number of dollars a man controls or the level of esteem which he occupies in the minds of his associates. But when we talk of the "biological success" of man or of any other animal as a species, we have in mind very different criteria.

The animal groups which we judge to have attained the greatest "biological success" are those which have the largest numbers of species and of individuals, occupying the widest stretches of territory and the greatest variety of habitats, consuming the largest amount and kinds of food, and most capable of defending themselves against their enemies. By these standards the phylum which occupies first place among the animals (vertebrate and invertebrate) is the phylum **ARTHROPODA.**

More species of arthropods have been described than of all other kinds of animals put together. Of the million or so known species of animals, over three-fourths are arthropods. The **major classes** of arthropods are the *crustaceans* (crayfish, lobsters, shrimps, crabs, water fleas, barnacles), the *centipedes*, the *millipedes*, the *arachnids* (spiders, scorpions, ticks, mites), and—by far the largest class of all—the *insects*. Only a few insects have been able to invade the ocean but the group is extremely abundant in fresh water and on land. In temperate regions insects cannot compete with the warm-blooded vertebrates during the winter; but in the tropics, where they suffer no handicap, they are dominant at all seasons.

Some arthropods are beneficial to man, providing food or some valuable service. Others do untold damage, destroying crops, undermining wooden buildings, and transmitting diseases. Parts of the world's most

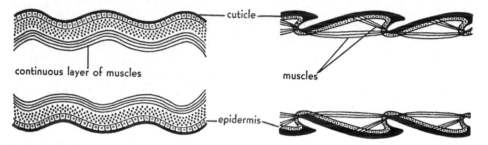

Body wall of annelid and arthropod contrasted. In annelids the cuticle is thin, and the epidermis is underlain by heavy layers of circular and longitudinal muscles. In arthropods the cuticle is heavy; the muscles occur in separate bundles; there is no continuous layer.

fertile regions are closed to man by the presence of disease-bearing arthropods. And where they do not exclude him altogether, they are man's chief competitors for food and shelter.

THE **arthropod body plan** may be roughly described as an elaboration and specialization of the segmented body plan of annelids. Primitive arthropods are composed of a series of similar segments bearing similar appendages. But in the higher types almost every segment of the body has a somewhat different structure and function. The outer layer, or **cuticle,** very thin in annelids, in arthropods is usually a heavy layer which serves as a *protective armor*. It is nonliving but is secreted by the underlying epithelium and is composed of several different substances, each of which contributes some useful property. Horny outer coverings occur in many groups of animals (for example, the covering of the obelia colony or the cuticle of annelids), but in no case are they used so effectively or produced into so great a variety of structures as in arthropods. Made of the cuticle, in whole or in part, are outer protective coverings, biting jaws, piercing beaks, grinding surfaces, lenses, tactile sense organs, sound-producing organs, walking legs, pincers, swimming paddles, mating organs, wings, and innumerable other structures found among the highly diversified insects. This horny material is to the arthropods what steel is to civilized man, and it is partly to the possession of this hard cuticle that the arthropods owe their success.

The surface of the cuticle is a thin waxy layer which makes it *waterproof*. Under this is a heavy layer composed of a protein and of **chitin,** a horny flexible substance which is the most characteristic component of the cuticle, if not the principal one, and which provides *elasticity*. Wherever the cuticle is relatively rigid, as it is over most of the surface of an arthropod, there is a third layer, which lies between the other two. It is formed by the infiltration of the upper part of the chitinous layer with the substance of the waxy layer and other hardening materials (which in crustaceans are

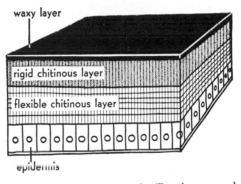

Body wall of an arthropod. (Based on several sources)

mostly lime salts). The middle layer is responsible for the *rigidity* of the cuticle, the property which makes it so effective as a protective armor. Since the hardening occurs in definitely limited areas, between which the cuticle remains as flexible membranes, or joints, the outer covering of arthropods provides *protection* without sacrificing *mobility*. This is what makes it so superior to the armors of such animals as the snails and clams, which have heavy cumbersome shells that limit movement.

Since the rigid cuticle furnishes a supporting framework for the tissues within and provides a surface for the attachment of muscles, it is appropriately called an *exoskeleton*. And though the chitin is responsible for the elasticity, rather than the rigidity, of the cuticle, in order to distinguish it from the external supports of other animals we usually call it a **chitinous exoskeleton.** In sharp contrast to this kind of framework is the endoskeleton of vertebrates, which lies on the inside and is surrounded by the soft fleshy parts. We can imagine how it might feel to be an arthropod by mentally putting on an iron suit of armor which adheres closely to the skin, and then thinking of our bones being eliminated and our muscles being attached instead to the iron armor.

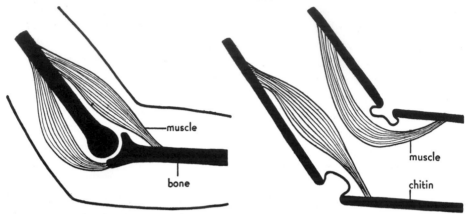

Diagram contrasting skeletal and muscular systems of vertebrate and arthropod. *Left*, part' of a vertebrate limb, showing that the bones lie internally and have muscles attached to their outer surfaces. *Right*, part of an arthropod limb, showing that the cuticle lies externally and has muscles attached to its inner surface.

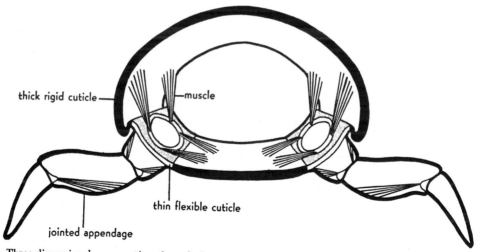

Three-dimensional cross-section through the **exoskeleton** of an arthropod to show that it consists of hardened plates, joined by more flexible membranes of cuticle, and serves as a place of attachment for muscles. (After Snodgrass)

To their chitinous exoskeleton arthropods owe their ability to live on land. **Land life** requires, among certain other adjustments, a relatively *impermeable outer covering* to prevent drying of the watery tissues within and a fairly *rigid framework* of some kind to support the soft tissues. In vertebrates the covering is furnished by scales or heavy skin, and the supporting framework is a bony skeleton. In arthropods the waterproof and rigid cuticle fills both requirements and enables the group to exploit the land with practically no serious competition from the other invertebrate groups, most of which are largely aquatic.

The name Arthropoda means "jointed legs" and refers to the most characteristic structures of arthropods. To distinguish them from the jointed appendages of vertebrates (for example, the arms and legs of man) we call them **chitinous jointed appendages.** The various appendages of arthropods have a specialized structure which adapts them to some particular function. This increases efficiency but sacrifices the versatility possessed by the more generalized hands of man.

LIKE the nereis, which has a pair of swimming flaps (parapods) on nearly every segment of the body, the arthropods also have, typically, a pair of appendages to every segment. In the embryos of both groups the appendages arise in a similar way from similar structures, and hence are said to be "homologous." The principle of **homology** is the basis of our scheme for determining animal relationships. Thus, if two animals have similar structures which develop in the same way from corresponding embryonic parts, the animals are judged to be closely related. The more similar the structures and their mode of origin, and the greater the number of such structures, the closer their relation. In other words, we assume that the homologous structures of two different animals have come, by a process of gradual modification, from the same or corresponding part of some remote common ancestor.

Not all structures which resemble each other indicate a common evolutionary origin of the animals which possess them. Many are only superficially alike, being adapted to the same environmental conditions; and such structures arise in entirely different ways in the embryos. They are similar in function but not in basic plan or mode of origin, and are said to be **analogous.** The wing of a bird and that of a bee are both used for flying, though one is made of feathers and the other of chitin and they do not develop in the same way. They are analogous but not homologous. On the other hand, the wing of a bee is homologous to that of a dragonfly or a cockroach. In all three insects the wing is essentially the same and arises from a corresponding part of the embryo. In this case the homologous organs, all used for flying, are also analogous. Sometimes homologous structures have different functions: the legs of a bee are used in walking, while those of a water beetle are adapted for swimming.

When corresponding structures in different segments of the *same* animal are considered, we say that they are **serially homologous.** One parapod of a nereis is serially homologous to any other. The front leg of a bee is serially homologous to the third leg, and both legs are serially homologous

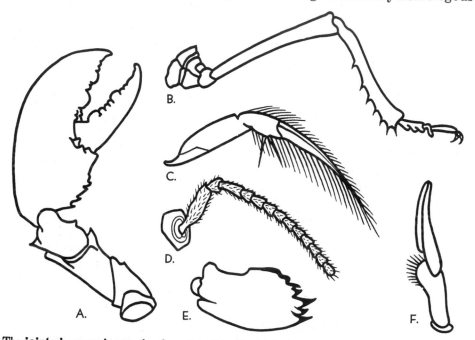

The **jointed appendages** of arthropods, originally chiefly for locomotion, have been modified for a great variety of functions, even in the same animal. They are all homologous but not analogous. **A,** pinching leg of a lobster. **B,** walking leg of a grasshopper. **C,** swimming leg of a water beetle. **D,** sensory antenna of a honeybee. **E,** chewing jaw of a cockroach. **F,** mating organ of a male lobster. (After various sources)

to the antennas and the jaws, which are modified segmental appendages and arise from corresponding parts of their respective segments. On the other hand, the eyes or wings arise in a different way and, therefore, are not homologous to the jaws, antennas, and legs.

IN THE insects and in many other arthropods the segments are grouped into **three body regions:** head, thorax, and abdomen. The heads of arthropods are roughly comparable in that they bear sense organs, the mouth, and feeding appendages. The thorax and abdomen correspond less closely in groups of arthropods. The thorax of a crustacean does not correspond to the thorax of an insect; they are similar only in that they both represent the middle region of the body. The head and thorax may be fused, as in the lobster, or the abdomen may be much reduced, as in crabs; but in most arthropods the total number of segments is much smaller than that of annelids. Further, adult arthropods have a fixed number of segments and do not add them on throughout life, as do most annelids. The same is true of the vertebrates, which have a fixed number of segments and a fusion of many of them. If we wish to generalize, we may say that primitive animals have a large and indefinite number of repeated but similar parts, while more specialized animals have a smaller and definite number of repeated parts with much division of labor among them, or they have the repeated parts fused into compact masses or organs. (For a specific example of this, see diagrams of the nervous systems on p. 248.)

In the **head** of an arthropod, each segment except the first typically bears a pair of jointed appendages, which are sensory or have

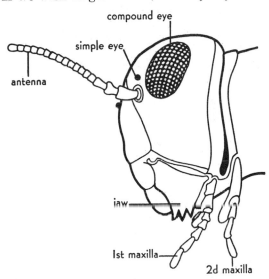

Head of an insect, showing structures characteristic of the heads of most arthropods. (Modified after Snodgrass)

to do with feeding. The segments are clearly visible in the embryo; but as development proceeds they fuse, so that, in the adult, segmentation of the head is indicated only by the presence of the several pairs of appendages. The head also bears a pair of compound eyes in the primitive arthropods, crustaceans and insects; the others have only simple eyes or clusters of simple eyes. Most insects have simple eyes in addition to the compound eyes.

Knowledge of the correspondence between appendages and segments comes from the study of embryology. The following description applies in general to the crustaceans, centipedes, millipedes, and insects but not to arachnids (see chap. 23). The first segment never has an appendage. The second bears a pair of feelers, or *antennas*. The third has a *second pair of antennas* in crustaceans but lacks a segmental appendage in insects. The

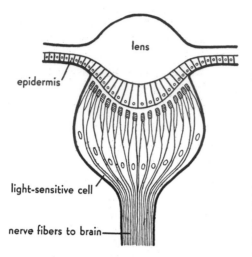

Simple eye of a spider. All the sensory cells have a single lens, made of cuticle and secreted by the underlying epidermis. (After Hentschel)

fourth has the *jaws*, which usually serve for biting but may be modified for other methods of feeding. The fifth and sixth segments each bear a pair of *maxillas*, accessory jaws which aid in feeding, particularly in handling the food and in holding it to the mouth. The mandibles and maxillas are referred to collectively as "mouth parts," and are frequently very highly modified.

The **eyes** of arthropods are composed of visual units, each of which is a bundle of cells consisting of two functional parts. The first is represented by *refractive bodies* which transmit the light rays and condense them upon the light-sensitive cells. The cuticle which covers the

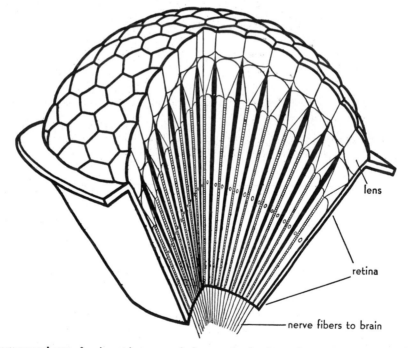

The **compound eye** of an insect is composed of many visual units, each with a small bundle of light-sensitive cells and a lens. The lens is made of cuticle and is secreted by specialized epidermal cells. The light-sensitive cells are also thought to be modified epidermal cells. (Partly after Hesse)

surface of the body is transparent and usually much thickened to form a lens over the surface of the eye; and there are one or more additional refractive structures within the eye. The second part, or *retina*, lies deeper and is composed of a layer of light-sensitive cells continued at their lower ends into nerve fibers which enter the central nervous system. A **simple eye** has a single light-condensing apparatus for all the sensory cells. A **compound eye** is composed of hundreds or thousands of units, each with its own light-condensing apparatus. This kind of eye is unique to arthropods and (except for the "camera" eyes of certain mollusks described on p. 203) is the most highly developed of invertebrate eyes. It does not give as sharp an image as the "camera" eye of man; perhaps the arthropod sees something a little worse than a newspaper photograph as it would look to us under a magnifying glass. However, some insects must have fairly good images, for they have been seen attempting to extract nectar from flowers on wall paper. In any case, arthropods react not so much to details in an image (as we do) as to *motion*. Since the movements of objects are recorded successively in every unit, the compound eye is admirably adapted for detecting the slightest movement of prey or enemy.

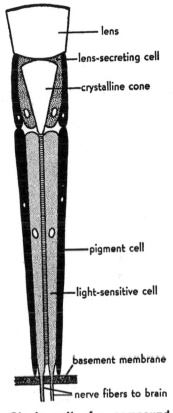

lens

lens-secreting cell

crystalline cone

pigment cell

light-sensitive cell

basement membrane

nerve fibers to brain

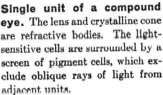

Single unit of a compound eye. The lens and crystalline cone are refractive bodies. The light-sensitive cells are surrounded by a screen of pigment cells, which exclude oblique rays of light from adjacent units.

Each unit of the compound eye is isolated optically from its neighbors by a screen of pigment cells, so that only a narrow band of parallel rays enters each unit; and it is thought that the image thrown on the retina is a **mosaic** composed of as many points of light as there are units. Most compound eyes are adapted to see in dim light by a migration of the pigment, leaving the sides of the visual units exposed. In this case each unit throws on the retina an image of a larger part of the visual field, and the adjacent images overlap somewhat. Such overlapping images are not as sharp but enable the animal to see in dimmer light, since they do not waste the light which enters obliquely and is therefore absorbed by the pigment in the mosaic type of vision.

The **thorax** of arthropods has different numbers of segments in the various groups. In the crustacea, which are more primitive, there are often numerous segments, with appendages for feeding and walking. In the insects, which are more specialized, the thorax is composed of three segments, each of which bears ventrally a pair of **legs** and dorsally, on the second and third segments each, a pair of **wings.** The thoracic appendages of arthropods are most often used for walking but may also serve other functions. In the lobster one pair is modified for grasping; and in the honeybee the legs, though used for walking, are highly modified for collecting pollen (see p. 292).

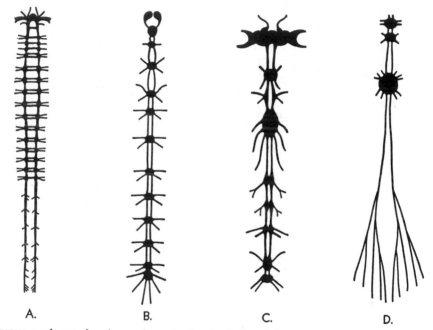

A.　　　　　　　B.　　　　　　　C.　　　　　　　D.

Nervous systems of various arthropods, showing fusion of the ganglia. **A,** primitive crustacean. **B,** caterpillar. **C,** honeybee. **D,** water bug. (After several sources)

The **abdomen** may or may not have appendages. In the crustaceans, such as the lobster, there is a pair of appendages on every abdominal segment; but in higher forms these have been lost, until in the insects there are practically no abdominal appendages homologous with the appendages of the other segments, except the egg-depositing structures on the most posterior segments.

In the nereis we saw a clearly segmental arrangement of parts, both external and internal—with only the beginnings of fusion and specialization at the head end. Every organ system had a representative in each segment which provided for the local needs of the segment. In the arthropods we have already seen modifications of the primitive external segmentation in the fusion of the head segments and the specialization of the various appendages. The **internal segmentation** is even more modified. Some of the organ systems consist of single large organs which serve the whole body, and segmentation is clearly apparent only in the repeated branches of the circulatory and respiratory systems and in the ganglia and segmental branches of the nervous system.

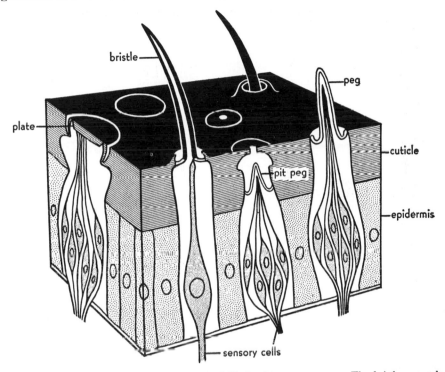

Portion of body wall of an insect to show several kinds of **sense organs.** The *bristle* responds to touch. The *peg* is made of thin cuticle and is therefore thought to be a receptor for taste or smell; the same is true of the *pit peg*, which lies in a cavity that opens to the surface through a pore. The *plate* (without a pore) may be a receptor for chemical stimuli, but its exact function is not known. (Combined from McIndoo and Snodgrass)

The **nervous system** consists of a dorsal brain which connects, by a ring of nervous tissue encircling the digestive tract, with the first ganglion of the ventral nerve cord. In primitive arthropods this system can hardly be distinguished from that of annelids. In higher arthropods there are all stages of condensation of the ganglia, reaching a peak in certain animals which have all the ganglia of the thorax and abdomen fused into one large mass.

One might suppose that an animal that lives incased in a nonliving cuticle as heavy as that of arthropods would be handicapped in establishing connections between the central nervous system and the external environment. On the contrary, the cellular and cuticular layers of the body wall of arthropods have been modified to form highly specialized **sense organs** of a variety greater than that found in any other phylum. The *eyes*, sensitive to light, and the *antennas*, sensitive to touch and to chemical stimuli, have already been mentioned. Some arthropods also have *balancing organs*, composed of sensory pits containing hard particles, and *auditory organs*, which have a flexible membrane stretched across an opening in the hard cuticle. In addition, the surface of the body is covered with a variety of *sensory bristles*, "hairs," spines, scales, and pits. The simplest of these is a bristle formed by a hollow outgrowth of the cuticle

Left, **closed circulatory system** of annelids; the blood is confined within blood vessels. *Right*, **open circulatory system** of an insect. The only blood vessel is a pulsating tube, the heart. Blood flows forward through the heart and then passes out the open anterior end into the tissue spaces and large blood cavities, eventually returning to the heart through paired openings in its sides. In the less specialized arthropods the vessels are more extensive, arteries leading from the heart to the main regions of the body.

and connected with a sensory cell which extends to its base. The bristle articulates with the cuticle, and any mechanical stimulus which moves the bristle sets up an impulse in the sensory cell with which it connects. Certain small and slender bristles which are not movable and have thin and permeable walls are assumed to be among the *receptors of chemical stimuli* (taste and smell).

The **coelom** of arthropods is practically obliterated. It appears in the embryo as a series of cavities in the mesoderm, but in the adult is represented chiefly by the cavities of the sex organs. The apparent body cavity of the adult is not a coelom at all but is a large "blood cavity" which forms part of the circulatory system.

The **circulatory system** has evolved in the direction of simplification. The *heart* is a pulsating tube which lies dorsally. The arteries lead not into capillaries but into the "blood cavities" in the mesenchyme throughout the body. In these spaces the blood bathes the various organs. Then it returns to a large cavity surrounding the heart and enters the heart through paired openings in its sides. Since the blood is not at all times confined within blood vessels, this type of system is called an **open circulatory system,** in contrast to the *closed* systems of such groups as annelids and vertebrates.

The great adaptability of the arthropod body wall is further emphasized by the structures concerned with **respiration.** Most terrestrial arthropods have a system of branching *air tubes*, formed by tubular ingrowths of the surface ectoderm. The ectoderm secretes an inner lining of cuticle, which strengthens the walls of the delicate tubes and prevents them from collapsing. Air enters and leaves the tubes through openings on the sides of the body and is piped directly to the tissues, partly or almost completely replacing the respiratory function of the circulatory system. Most aquatic arthropods breathe by means of *gills*, thin-walled extensions of the body wall through which carbon dioxide and oxygen pass readily.

The cuticle also forms an important part of the **digestive system.** The ectoderm turns in at the mouth and anus and lines the anterior and posterior regions of the digestive tube with the cuticle which it secretes. In the anterior region the cuticle may be produced into hard teeth for grinding up the food. In the insects and many other arthropods the anus serves also as the exit for nitrogenous wastes, since the **excretory organs** are tubules which open into the digestive tube.

Aquatic arthropods range in size from minute crustaceans like the "water fleas" to monster crabs measuring up to 12 feet across their long, spindly, outspread legs. Such large size is possible in the ocean, where the water supports most of the weight of the animal. But on land, legs with an exoskeleton thick enough to support such a load above the surface of the ground would be too heavy for much movement. Thus, terrestrial arthropods are limited, by their exoskeleton, to a relatively **small size,** which is not without its compensations if we are to judge from the success of these animals. Small size, combined with great development of the muscles, makes for active habits and easy escape from enemies. In addition, small size requires relatively little growth, and many forms develop from the egg to the sexually mature adult in a few days or weeks. Such a **short life-cycle** results in many generations in a year; and this means that such species may undergo *rapid evolution*, which explains, in part, the great numbers of species of terrestrial arthropods, particularly insects.

Like most invertebrates, arthropods lay large numbers of small **eggs;**

usually the young hatch from the egg in an immature state and must feed to obtain the necessary materials for further growth and differentiation. The *larvas* of aquatic forms are free-swimming and undergo a gradual change into the adult. In the most specialized insects, the larvas (caterpillars, grubs, etc.) are so different from the adult that it is not possible to have a gradual change. The larva surrounds itself with some kind of protective material and becomes transformed into the **pupa,** which is referred to as a "quiescent" stage—and so it is, from all external appearances. Internally, however, many important changes take place, for the larval tissues break down and become reorganized through the growth of certain cells which were set aside early. After a time the sexually mature adult emerges. Such radical changes from larva to adult are known as **metamorphosis** ("change in form"). (See chap. 24 for examples.)

In many insects which undergo metamorphosis there is a marked **division of labor among the different phases of the life-history.** In the butterfly, for example, the caterpillar has chewing jaws and feeds on leaves. It eats large quantities of food in a relatively short time and grows very rapidly. Thus, its role in the life-history is *feeding.* The pupa undergoes profound changes in structure, and its function may be said to be that of *transformation* and *differentiation.* The adult is a winged form which has no chewing jaws and can feed only by extracting nectar from flowers by means of a long sucking tube. Its ability to fly makes it important in *distribution* of the species, but perhaps its chief role is that of *reproduction,* for it does not grow and much of its food goes to produce the eggs. Moreover, some adults never feed at all but mate soon after emerging from their pupal cases; the females lay eggs and the adults die.

Specialization between different stages in the life-cycle may also involve more than one habitat. This enables one species to exploit two very different sources of energy, increasing the amount of energy available in any locality for the growth of that species. For example, the adult mosquito is a flying, terrestrial form that sucks blood, while the larva lives in fresh water and feeds on minute organic particles wafted into the mouth by special bristles.

Many of the more primitive insects have no metamorphosis or have only an **incomplete metamorphosis,** like that of the cockroach. The cockroach hatches from the egg as a young form which looks like a miniature adult except for some differences in general proportion and in the possession of only the beginnings of wings. (See photographs in chap. 24.) It leads a life like that of the adult and grows rapidly. Naturally, an animal with a hard outer covering cannot grow indefinitely without making some kind of readjustment. And the cockroach periodically sheds or **molts** the outermost layers of the cuticle. When the cuticle ruptures and is cast off, the animal already possesses a new "roomier" cuticle, which has formed beneath. But until the hardening substances are laid down, the newly shed cockroach has a light-colored and delicate cuticle, which is elastic and stretches to accommodate the animal. From molt to molt, as growth continues, there is a gradual increase in specialization until the fully mature form is attained. Most insects do not molt after the adult stage is reached; but crustaceans, centipedes, millipedes, and arachnids do.

THE phenomenon of **polymorphism,** as we saw it in coelenterates, was a division of labor among the structurally differentiated subindividuals of a colony. Arthropods are not colonial in the structural sense; that is, the different individuals are not physiologically connected, as in the obelia. But many insects show polymorphism in that different members of the species are structurally specialized for the performance of different functions. They live together as co-operating members of a **social colony.** The social insects most familiar to everyone are the highly evolved ants, bees, and wasps; but the lowly termites, relatives of the cockroaches, have one of the most interesting types of social structure. The termite *workers* are sterile; they build the nest, collect food, care for the king and queen, and raise the young. The *reproductives* are fertile and hatch as winged forms which fly out to establish new colonies. They mate; and the female, or *queen,* spends the rest of her life laying eggs. She cannot feed herself but is fed and cared for by the workers. The *soldiers* cannot feed themselves; they protect the colony from invaders. (For further details about the termites, see photographic section of chapter 24.) Social life in insects has the same advantages as the social life of man. One individual need not perform all the necessary labors, the various duties being distributed among different individuals specialized for the job.

In human society the individuals are not born anatomically suited to their various occupations but become trained physically and mentally to fit their particular jobs. Among arthropod societies the individuals are *structurally adapted* from the very start, and are so specialized that sometimes they cannot even perform such an ordinary activity as feeding.

The polymorphism of social arthropods extends also to their **behavior.**

The behavior patterns of a termite, for example, are established at the outset, little or no learning being necessary for the animal to take its place in the life of the colony. This is demonstrated every time a new colony is formed. The workers which hatch from the eggs laid by the queen have never seen the nest from which the queen came; yet they construct a nest exactly like it. Such complex inherited behavior is unlearned, or **instinctive.** Not all the behavior of social insects is instinctive. If a beehive is moved, the bees return first to the original spot; but when, afterward, they find the new location, they "learn" the new place.

Instinctive behavior is superior to learned behavior for animals, such as the insects, which live only a few days, weeks, or months, and can ill afford to spend time learning how to catch prey, eat, build a shelter, and lay eggs. Life is short, and there are many things to do. On the other hand, learned behavior has distinct advantages for animals which live a long time and have parental care and training. The human infant is helpless and would die if left without care. Years of training are a necessary preparation for an independent life in human society. However, men can learn new kinds of behavior and solve new kinds of problems throughout life, whereas the arthropod is more or less limited to the original set of instinctive reactions.

THROUGH the animal groups we have seen many *successive levels of structural differentiation.* Specialization within single cells, among cells, among tissues, and among organs we saw respectively in protozoans, sponges, coelenterates, and flatworms. The beginnings of segmental specialization were already apparent in annelids. Division of labor among different individuals and among the several stages in the life-history occurred in coelenterates. All of these specializations reach their extremes in the arthropods, which represent the peak of invertebrate evolution.

CHAPTER 23

THE LOBSTER AND OTHER ARTHROPODS

THE appendages of vertebrates are four in number; and though they show a variety of structure in adaptation to different methods of locomotion and to additional services which they may perform, such as digging or holding prey, they are primarily locomotory—with the notable exception of the fore limbs of man. The appendages of arthropods, however, are greater and more variable in number; and some of them have no locomotory function but serve as sense organs, jaws, mating organs, or respiratory structures. Further, in contrast to the versatile limbs of vertebrates, arthropod appendages are often specialized for a single function. Thus, to describe the appendages of an arthropod is to tell almost everything about the habits of the animal: where it lives, how it moves, and how it feeds.

Primitively, arthropods had along the whole length of the body a series of simple, flattened appendages which were all alike. Each served several functions: locomotory, food collecting, respiratory, and perhaps also sensory. Such a condition is found in no living arthropod, but something very much like it is seen in the extinct trilobites (chap. 27). Among living arthropods the closest approach to this occurs in certain crustaceans, the fairy shrimps, which have a specialization of appendages on the head and a loss of appendages on the posterior region but have on the trunk region a series of similar flattened appendages for swimming, food-collecting, and respiration. At the other extreme is an insect like the honeybee, in which every pair of appendages on the body is different. To bridge the gap between the fairy shrimp and the honeybee would be difficult if it were not that among the other arthropods we find almost every intermediate stage. The lobster, for example, has a pair of appendages on almost every segment. Those on the abdomen are simple swimming flaps which are almost all alike, but those of the head and thorax are highly diversified in structure and function. Further, in the development of the lobster embryo even the appendages of head and thorax appear first as simple, similar structures which only gradually become specialized and differentiated from each other. Thus, by tracing the development of the lobster appendages and homologizing the different parts of each, we are better able to understand how a simple, flattened swimming oar can, by gradual changes, become a chewing jaw or a sensory antenna.

THE LOBSTER

ASIDE from minor details, the lobster is so much like its fresh-water kin, the crayfish, that the description of the lobster applies, in general, to both animals.

The body of the lobster consists of twenty-one segments (or less, if certain segments which lack segmental appendages are not counted). The first fourteen are united into a large **cephalothorax,** which represents the combined head and thorax. The fusion is complete dorsally and at the sides, but the segmentation can still be recognized on the ventral surface. The **abdomen** consists of seven distinct segments, which are clearly marked externally.

The cuticle, secreted by the underlying epidermis, covers every part of the body, forming a jointed **exoskeleton** which is made particularly hard by an infiltration with calcium salts. The cuticle also furnishes some internal support to the cephalothorax by means of thin plates of cuticle secreted by infoldings of the epidermis. These plates increase the area for the attachment of muscles, and they protect important organs. Over the dorsal surface and sides of the cephalothorax the calcified cuticle forms a single large shield, the **carapace.** Over the abdomen it is folded between segments, allowing for flexibility.

The **least modified appendages** are those of the **abdomen.** Each consists of a *basal piece* (protopodite), which bears at its free end an *outer branch* (exopodite) and an *inner branch* (endopodite). The numbers of joints in the three pieces may vary, but the basic plan of this two-branched

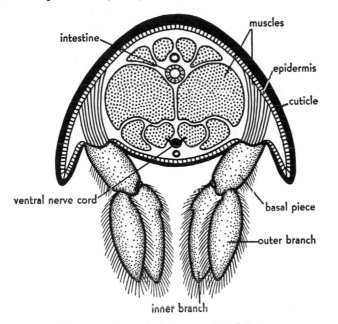

Cross-section of abdomen of the lobster.

(biramous) appendage is thought to be the fundamental plan of all crustacean appendages; and we find it throughout the group, both in highly specialized adults and in the simplest larvas. In the lobster the two-branched plan is obscured in the appendages of head and thorax by the presence of additional lobes or extensions on the basal piece or by the absence of the outer branch. In the lobster embryo, however, all the appendages arise as simple two-branched structures. (See figure on p. 265.)

The **head** of the lobster is fused with the thorax, but its component segments have been determined from careful studies of the embryology of appendages and other structures, especially the nervous system. On the first segment of the head there is a pair of **compound eyes** set on the ends of jointed, movable stalks. These are not serially homologous with the other appendages, since they arise in a different way. The second segment bears the **first antennas,** sensory structures which have two filaments. (While the antennas themselves are serially homologous with the rest of the appendages, their two-branched condition is not. For in the lobster embryo the first antennas remain single until long after the other appendages have become two-branched; and even when the larva emerges from the egg, the inner filament is represented by only a small bud from the base of what finally becomes the outer filament.) The **second antennas,** located on the third segment, have only one long filament; this is serially homologous with the inner branches of other appendages, and the outer branch is represented by a scalelike process. (From this point on, only the location and function of the appendages will be mentioned. All are paired and are serially homologous with each other; some bear extra processes or lack the outer branch, as can be seen in the diagram of the appendages.) The fourth head segment bears toothed **jaws** (mandibles) for crushing the food. On the fifth and sixth segments are the **first** and **second maxillas,** which pass food on to the mouth. The second maxilla is a thin, lobed plate and is chiefly respiratory, serving as a "bailer" for driving water out of the respiratory cavity.

The **thorax** has a pair of appendages on every segment. The first three bear the **first, second,** and **third maxillipeds.** These are somewhat sensory but serve chiefly to handle food, mincing it first and then passing it on to the mouth. Only the third is powerful enough to do much real chewing of the food, unless it is soft. In each the basal piece bears a thin flap (epipodite), to which, on the second and third maxillipeds, is attached a gill. The flaps separate and protect the gills. The fourth thoracic segment bears the large claws or **pinching legs** (chelipeds), used both in offense and defense. The next four segments have each a pair of **walking legs.** All five pairs of legs have attached to their bases a gill-separator and a gill. The walking movements of the legs move the gills and stir up the water in the respiratory cavity under the carapace. The pinching legs are not symmetrical in lobsters over $1\frac{1}{2}$ inches long. In the smallest lobsters both of them are slender and have sharp teeth; but as the animal grows, they gradually differentiate. One becomes larger than the other, and its teeth fuse into rounded tubercles; it is used for crushing. The other remains smaller and more slender, the teeth become still sharper, and it is used

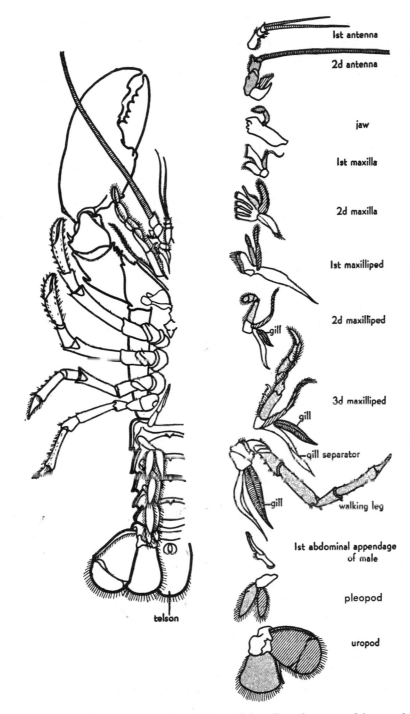

Appendages of the lobster show a marked division of labor. Some have extra lobes or other processes, and some lack the outer branch; but they all can be reduced to a common basic plan. The *inner branches* are stippled, the *outer branches* are shaded with diagonal lines, and the *basal piece* and its processes are left unshaded.

especially for seizing and tearing the prey. The first two pairs of walking legs also have small pincers which aid in seizing prey. The last pair of walking legs are used also for cleaning the abdominal appendages.

The **abdomen** has a pair of appendages on every segment except the last. Those on the first abdominal segment are different in the two sexes. In the male they are modified to form a troughlike structure used for transferring sperms in the mating process. In the female they are much reduced. The next four segments all bear similar two-branched append-ages, the **pleopods,** which serve in the female as a place of attachment for the eggs. The sixth abdominal appendages are the **uropods,** which resemble modified and enlarged pleopods. Together with the flattened last abdominal segment, the **telson,** they form a tail-fan, used in backward swimming.

The appendages of the lobster have been stressed for a number of rea-sons. They furnish a striking example of *specialization among appendages* of different segments and, in the case of the large pincers, between the right and left sides of the same segment. While the flattened, two-branched swimmerets are not very different from the appendages of the hypothetical arthropod ancestral type, on the same animal we find such specialized appendages as the jaws, which have a counterpart even in the most advanced insects. In the development of the lobster appendages we see how a series of originally similar parts can become gradually dif-ferentiated into highly specialized and dissimilar structures which, though no longer analogous, are still homologous.

The internal parts of the lobster with which some of us are familiar are the large (and very edible) abdominal **muscles.** These are segmentally arranged and include muscles for moving the swimmerets, extensor mus-cles for straightening the abdomen, and much larger flexor muscles, which furnish the major source of power for locomotion. For rapid movement the lobster flexes the abdomen ventrally and with such force that the whole animal shoots backward through the water. In the cephalothorax are numerous muscles for moving the appendages and certain organs. Most of the muscles of the lobster are of the striated type, characteristic of arthropods and vertebrates. **Striated muscles** contract very rapidly and are therefore well suited for moving the body and appendages. Both arth-ropods and vertebrates have unstriated muscles for organs such as the digestive tube and blood vessels, which undergo slow, rhythmic contrac-tions. Lower invertebrates possess the slower, unstriated type.

The **digestive system** of the lobster consists of three main regions, of which only the middle one has an endodermal lining. The anterior and posterior ends develop as tubular ingrowths of the ectodermal epithelium and so become lined with a cuticle which is continuous with the exoskeleton and is shed when the animal molts. Lobsters are scavengers, but they also catch live fish and dig for clams; and they have been seen to attack large gastropods, breaking off the heavy shell, piece by piece, to obtain the soft inner parts. The food is shredded by the maxillipeds and maxillas and then further crushed by the jaws before it enters the mouth. As if this were not enough, part of the stomach is specialized as a gizzard,

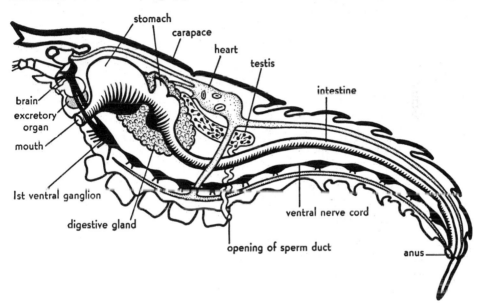

Internal anatomy of the lobster, which, like most highly specialized segmented animals, has single large organs (or a pair of organs), instead of small local representatives in each segment. Of the systems represented here, only the nervous system is clearly segmental. Segmental blood vessels of the abdomen, omitted here, are shown in the diagram of the circulatory system. (Based partly on Herrick)

which is lined with hard chitinous teeth and worked by numerous sets of muscles. In the stomach the food is pulverized, strained and sorted. The smallest particles are sent in a fluid stream to the large digestive glands for digestion and absorption; larger particles go in a steady current to the intestine; and the coarsest particles are returned to the grinding mechanism.

The anterior portion of the stomach is large and bulbous and serves chiefly for storage. The posterior part is mainly for sorting and straining. Between the two lies the grinding region, which reduces the food to minute particles. Since these are readily digested in the tubules of the digestive glands, the work of the intestine is less important than in the earthworm. This explains how a large animal like the lobster can get along with such a short uncoiled intestine.

The extensive respiratory surface needed to supply the demands of a large and active animal like the lobster is furnished by twenty pairs of

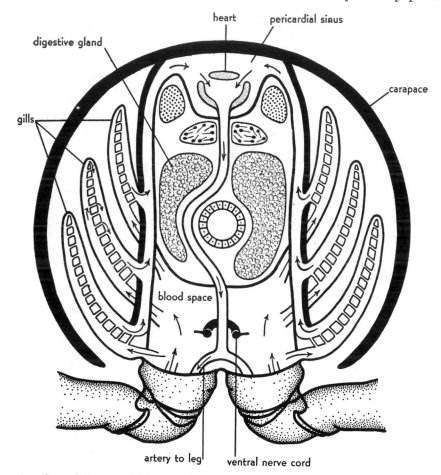

Cross-section of thorax of lobster to show relations of gill chambers to other organs and the path of the blood through some of the main blood channels.

gills, feathery expansions of the body wall, which are filled with blood channels. The gills are attached to the bases of the legs, the membranes between the legs, and the wall of the thorax. They lie on each side of the body in a cavity inclosed by the curving sides of the carapace. Water enters the cavity under the free edges of the carapace, passes upward and

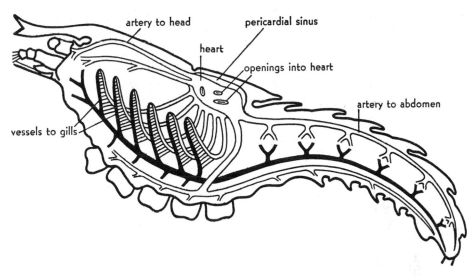

Circulatory system of the lobster, showing the main blood channels. Blood returning from the tissues goes through the gills before returning to the heart. (Modified after Gegenbaur)

forward over the gills, and is directed out anteriorly in a current maintained by the flattened plates of the second maxillas.

The **circulatory system** is an **open** one. The muscular heart lies dorsally in a chamber filled with blood. In the sides of the heart are three pairs of openings through which blood from the chamber enters the relaxed heart. When the heart contracts, valves prevent the blood from going out the openings; instead, it is driven into arteries which go to the tissues of the body. The smallest branches of the arteries open, not into veins, but into blood cavities in the tissues called **sinuses**. Blood returning from the tissues collects in a large ventral sinus and from there enters the gills, where it gives up carbon dioxide and takes up oxygen. Then it is returned, through a number of channels, to the large pericardial sinus which surrounds the heart.

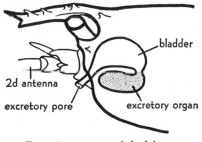

Excretory organ of the lobster.

The single pair of **excretory organs,** sometimes called the *green glands* because of their greenish color, consist each of a glandular sac and a coiled tube which opens into a muscular bladder. Wastes extracted from the blood are poured into

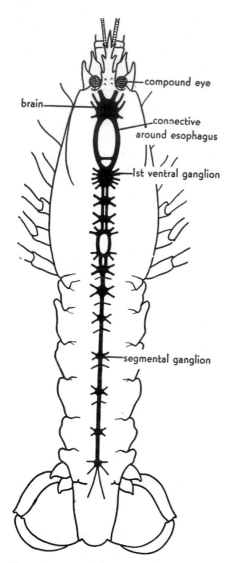

compound eye

brain

connective
around esophagus

1st ventral ganglion

segmental ganglion

The **nervous system** of the lobster is much
like that of the nereis, with a ring of tissue
around the esophagus and a ventral, double,
ganglionated cord.

the bladder and from there emptied
to the outside through a pore at the
base of the second antenna.

The general pattern of the **nervous
system** is like that of annelids. The
large brain is in the head near the eyes.
From it a pair of connectives pass ventrally, one on either side of the esophagus, and unite below the digestive
tract to form a double ganglion, the
first ventral ganglion, from which the
double nerve cord extends backward,
enlarging into paired ganglia in almost
every segment.

The most conspicuous **sense organs**
are the antennas and the compound
eyes. As the lobster is most active at
night, and even in the daytime lives at
depths where there is not enough light
for clear vision, the eyes are probably
secondary in importance to the *sensory
bristles* which are distributed all over
the surface of the antennas, body, and
appendages—from fifty thousand to
one hundred thousand of these bristles
occurring on the pincers and walking
legs alone. The bristles are of two types
—one sensitive to touch, and the other
to chemicals. Occupying the basal segment of each first antenna is a water-
filled sac which opens to the outside by
a fine pore. On the floor of the sac is
a ridge of sensory hairs, among which
are numerous fine sand grains. As any
movement of the lobster would sway
the hairs or cause the sand grains to roll over them, this structure is
thought to be a balancing organ.

To demonstrate the balancing function of this organ in shrimps, one investigator performed a very ingenious experiment. He obtained a shrimp that had just molted and

therefore had no sand grains in the sensory sac. He put the animal in filtered water and supplied it with iron filings. The shrimp picked up the filings and placed them in the sac. Then, when the investigator held a powerful electromagnet above the animal, it turned over on its back—apparently because the magnetic pull on the iron filings in the sac was greater than the opposing pull of gravity.

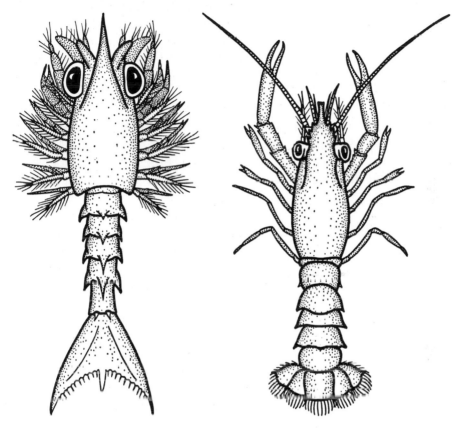

Left, the first **larval stage of the lobster,** is just over ⅓ inch long. The appendages are all two-branched, similar structures. The swimmerets at this stage are only small buds, and the larva swims about at the surface by the rowing action of the flattened, fringed outer branches of the limbs. *Right*, the fourth larval stage of the lobster is about ½ inch long and resembles a miniature lobster. Like the first stage, it swims at the surface, feeding on small organisms; but forward swimming is by means of the swimmerets. The outer branches of the legs are reduced and no longer visible; the inner branches are differentiated, though right and left large claws are still similar. (After Herrick)

The **reproductive system** consists of a pair of ovaries or testes, which lie in the dorsal part of the body and from which a pair of ducts leads to the external openings at the bases of the third legs in females, fifth legs in the male. The sexes can be distinguished by the position of these sex openings as well as by the structure of the first abdominal appendages. In the sex act the male deposits sperms near the female pores, and the eggs are fertilized as they emerge. They are fastened, by a sticky secretion, to the swimmerets of the female and are kept well aerated by the movements of the swimmerets.

The young lobster hatches from the egg as a free-swimming **larva** and goes through a series of changes before it comes to resemble the adult. The young crayfish, like the young of most fresh-water animals, hatches as a juvenile form which is much like the adult except in size.

CRUSTACEANS

THE name Crustacea was originally used to designate an animal having a hard but flexible "crust," as contrasted with one having a hard but brittle shell like that of oysters or clams. Since nearly all arthropods have a hard, flexible exoskeleton, we now use more distinctive criteria for assigning an animal to the class **Crustacea,** of which the lobster is a member. Crustaceans may be roughly distinguished as arthropods which breathe by means of gills and have two pairs of antennas. The lobsters and crabs are giants among crustaceans; most kinds are small animals, under half an inch in length.

Although the most primitive crustaceans, such as certain branchiopods, now live in fresh water, the earliest crustaceans certainly lived in *salt water*, and the class is still predominantly marine. This is not surprising when we consider that the crustaceans as a group are the most primitive living arthropods, and that the ocean, being the easiest place to live in, requires the fewest adjustments on the part of its inhabitants. Because of their tremendous volume, the seas provide relatively constant salt content, oxygen content, and temperature throughout the year. The salt concentration of animal tissues is much closer to that of sea water than to fresh water. Besides, sea water is buoyant, offering greater support. Crustaceans are so abundant in the ocean that they have been called "the insects of the sea," and there is hardly any way of life in the sea not followed by some member of this diversified class.

Among the most highly modified crustaceans are the **barnacles,** sessile marine animals which live attached to rocks, wooden pilings, ships, and the bodies of many animals. If you are surprised to find them among the arthropods, you are like most laymen, who assume them to be mollusks because of their thick calcareous shells. Early zoölogists, too, classified them with the mollusks until their true relationships were discovered through a study of their development. The young larva which hatches from the egg is free-swimming. In the possession of three pairs of appendages and in other characters it resembles the **nauplius larva,** characteristic of crustaceans. After swimming about for a time, it undergoes changes and then settles on some solid object, becoming attached by the head end. In spite of their extreme modifications, adult barnacles can be recognized as arthropods by their chitinous jointed appendages, which are two-branched, as in other crustaceans, and are heavily fringed with bristles. The appendages are thrust out of the shell and sweep through the water like a casting net, entrapping small animals and organic fragments. With few exceptions barnacles are hermaphroditic; and, as in many other groups of animals, this is thought to be

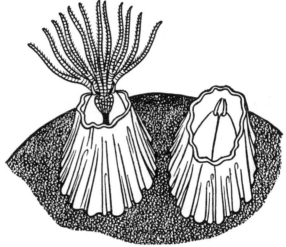

Barnacles are common marine crustaceans which live permanently attached. One is shown extended, the other withdrawn into its shell. About 2 × natural size.

associated with their sessile life, which prevents contacts between individuals. Another curious marine crustacean closely related to the barnacles is *Sacculina,* which has a free-swimming nauplius but in the adult stage fastens onto a crab and sends rootlike processes into every part of the host's body, parasitizing it so completely as to seriously affect its whole physiology and arrest its growth.

The crustaceans have done almost as well in *fresh water* when we consider that this medium is a more difficult one for all groups of animals and that, compared with the ocean, which is sometimes described as a "thick soup" because of the abundance of its animal life, fresh water hardly rates the title of "thin lemonade." To invade the rivers that connect directly with the ocean, a crustacean not only must become adjusted to the lowered salt content but must be able to maintain itself against the downstream current. This is not so difficult for the adults, but their small and fragile larvas are easily swept downstream and back into the ocean. No doubt this has been a factor in preventing some invertebrate groups

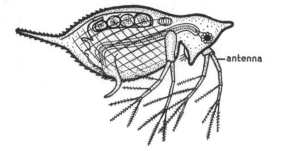

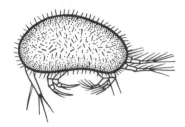

—antenna

Daphnia is the most common fresh-water representative of the cladocerans. This group belongs to the subclass Entomostraca, which includes the smaller crustaceans, most of which are under a quarter of an inch in length.

Ostracods, another group of entomostracans, live in fresh water and in the ocean. Though the body lacks segmentation, crustacean characteristics are evident in the jointed appendages and, in some, in the nauplius larva.

from ever establishing themselves in fresh water. The ones that are successful usually suppress the free-swimming larval stages, and the young hatch as miniature adults. Further, bodies of fresh water are subject to violent fluctuations of temperature, and small ponds dry completely in the summer and freeze solid in the winter. Crustaceans have become adapted to these rigorous conditions by the development of thick-shelled eggs which resist drying and freezing.

Adaption to *land* life is a still more difficult step. Temperature fluctuations are even more extreme, drying is a constant threat, and breathing mechanisms must be adapted to air respiration. A few crustaceans, some crabs and "wood lice," are fairly successful on land—but only because they avoid certain problems by living in moist places. For truly successful land forms we must look to the other classes of arthropods.

ARACHNIDS

THE class **Arachnida** includes the spiders, scorpions, mites and ticks, harvestmen ("daddy longlegs"), and a few minor groups. No other class of animals is less loved by most people. There is some basis for this dislike, in that scorpions and some spiders can inject a poison which produces painful, though usually not serious, results in man; some mites are parasites in human skin; and some ticks suck human blood and spread disease. But relatively few people in large cities have ever had a single unpleasant experience with an arachnid. The sinister reputation of a group like the spiders, which do little harm and some good (by killing

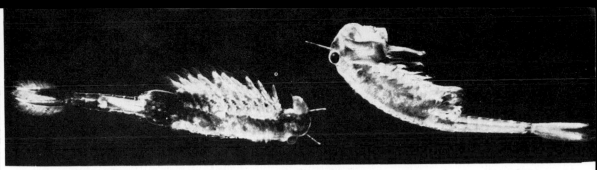

Fairy shrimps are not really shrimps but belong to the most primitive group of crustaceans. These (*Eubranchipus*) live in fresh water; their relatives, the brine shrimps, live in salt lakes. They row themselves about, on their backs, by means of numerous, similar, flattened appendages. (Photo by Peltier. Courtesy *Nature Mag.*)

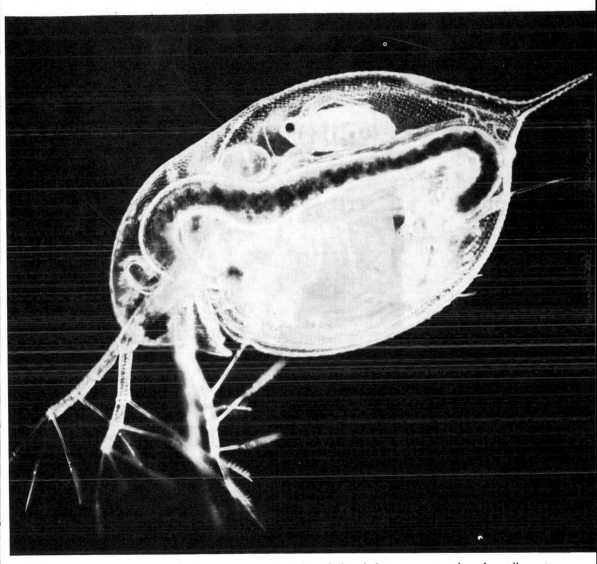

Daphnia is the most common fresh-water representative of the cladocerans, an order of small crustaceans which have a compact body often inclosed in a bivalved carapace. They swim by rapid jerks of the large two-branched second antennas. Daphnia is about $\frac{1}{16}$ inch long, but most cladocerans are smaller. This specimen is a mature female; an unborn Daphnia can be seen inside the brood pouch. (Photo of living animal by P. S. Tice)

Ostracods are minute crustaceans, mostly fresh-water forms. The animal lives incased in a bivalved carapace. When the valves are open, as shown here, it can protrude the appendages which propel it through the water. Actual size, about $\frac{1}{25}$ inch long. (Photo of living animal by P. S. Tice)

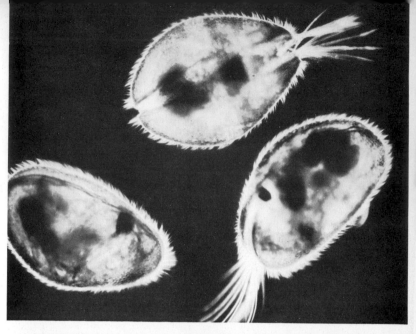

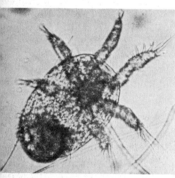

Nauplius larva is first stage after hatching of many crustaceans. (Photo of living animal. Pacific Grove, California)

Marine copepods are crustaceans which occur in countless millions. (Photo of living animal, $\frac{1}{20}$ in long. Pacific Grove, California)

Cyclops, so named from the single median eye, is about $\frac{1}{10}$ inch long and is the most familiar fresh-water copepod. Mature females usually have two groups of eggs attached to the body. Most members of the order Copepoda are marine. Copepods are very abundant and form an important part of the food of fishes. (Photo of living animal by P. S. Tice)

Stalked barnacles live in the open ocean. Each has a flexible stalk from which hangs the main portion of the body, inclosed in a series of calcareous plates. From between the plates are extended the long bristle-fringed appendages with which they fish for their food. The buoy-making barnacle, *Lepas fascicularis*, secretes a frothy substance at the base of the stalk that keeps it afloat. Shown here are three individuals sharing a common float. Natural size. (Photo of living barnacles by D. P. Wilson. Plymouth, England)

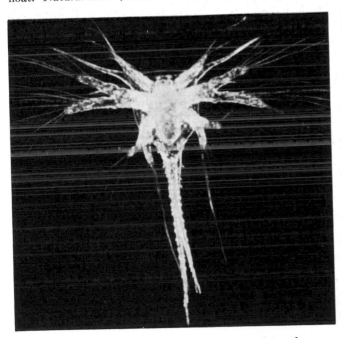

The nauplius larva of a barnacle, *Lepas*, confirms the crustacean affinities of the sessile adult barnacle. The nauplius swims about for a time and then settles down, head first, on a solid object. (Photo of living nauplius by D. P. Wilson. Plymouth, England)

Stalked barnacles on tooth of a whale. Stalked barnacles live attached to almost anything that floats or swims in the ocean: certain animals, floating logs, ships and empty bottles. (Photo of preserved specimen)

Acorn barnacles live attached to rocks on every rocky seashore. From the top of the cone-shaped calcareous shell they rhythmically extend and retract the fringed appendages, straining from the water small organisms. When the tide is out, the appendages are withdrawn, and the opening of the shell is closed against loss of moisture. (Photo of living barnacles made at 1/10,000 second. Mount Desert Island, Maine)

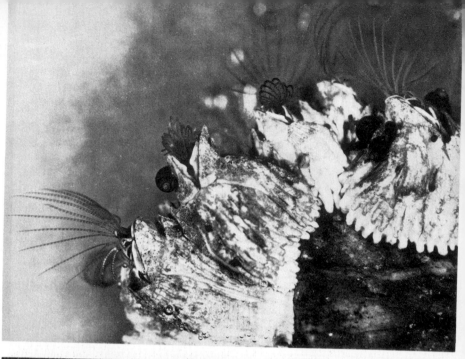

Model of a section through an acorn barnacle, *Balanus*, to show the animal with appendages withdrawn as when disturbed or when the tide is out. Barnacles have been described as animals which sit on their heads and kick food into their mouths. (Photo, courtesy American Museum Nat. Hist.)

Barnacles may live on almost any hard object in the water on which the free-swimming nauplius larva happens to settle down. This old lobster is covered with barnacles, but young lobsters manage to keep free of them. (Photo of living animal by F. Schensky. Helgoland)

Land isopods, also called "wood lice" or "sow bugs," are among the few successful land crustaceans. They are found under logs and stones and feed on decaying vegetation. Because their delicate gill-like breathing organs must be kept moist, they live mostly in damp places. The young develop in a brood pouch. (Photo by Cornelia Clarke)

Aquatic isopods ("legs all alike") live in fresh and salt water. These are *Limnoria lignorum*, marine wood-boring forms, shown here in their burrows. Tiny animals, only ⅛ inch long, they occur in large numbers and cause wholesale destruction of wooden pilings which support wharves. (Photo courtesy *Nature Magazine*)

Amphipods are crustaceans flattened from side to side. Some amphipods live in fresh water, but most are marine. The "beach fleas," so called because they jump about, live a more or less terrestrial life on ocean beaches. (Photo courtesy *Nature Mag.*)

The fresh-water shrimp, *Palae-monetes*, is a true shrimp, very similar to certain of its marine relatives. Actual length, 1½ inches. Collected at Wolf Lake, Indiana. (Photo by P. S. Tice)

Marine shrimp, *Stenopus*, which shows the general tendency of animals in tropical waters toward bizarre shapes and brilliant colors. This shrimp is white with bands of iridescent blue-green, red, orange, and purple. About natural size. (Photo of living animal. Bermuda)

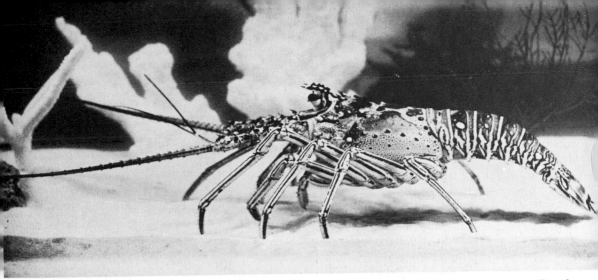

The spiny lobster (*Panulirus*) has no large pincers and would seem to be an easy animal to approach. But the body is covered with a formidable array of spines, and the spiny antennas are large and deal vicious, tearing blows. The flesh is delicious, and the animals are eaten extensively. In the U.S. they occur in the warm waters off Florida and off the coast of southern California. (Photo of living animal, courtesy Shedd Aquarium)

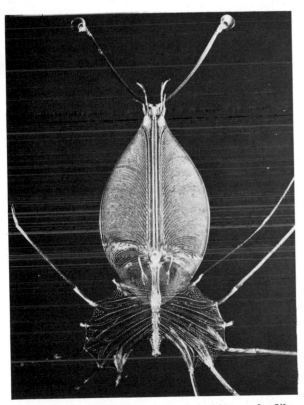

The larva of the spiny lobster is a bizarre, leaflike, transparent animal with eyes at the ends of long stalks. The head and thoracic appendages are present, but the abdomen is underdeveloped. After several molts, the animal comes to look like the adult. (Photo by William Beebe)

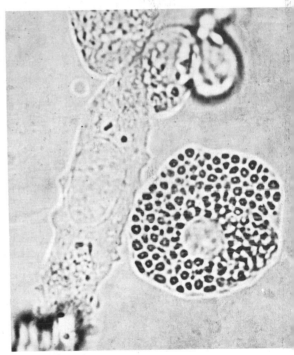

Blood cells from the spiny lobster can be kept alive for a considerable time outside the body. Two are shown highly magnified; the cell on the right was only 1/1,250 inch in diameter and moved about actively in ameboid fashion. The blood of larger crustaceans clots readily and is bluish from the presence of hemocyanin, a copper-protein compound which carries oxygen. Hemoglobin, found in man, occurs in some of the smaller crustaceans. (Photo of living cells. Bermuda)

The lobster *Homarus*, is mostly dark green when alive; but when boiled, like this one and like millions of others every year, turns bright red. About half an hour after this picture was taken this lobster was reduced to an empty exoskeleton.

Crayfishes live in streams and ponds, and are very similar to their marine relative, the lobster. They are scavengers, feeding upon decayed organic matter, and also catch small fish. Like shrimps and lobsters, they can walk forward slowly but in escaping enemies shoot backward by suddenly contracting the powerful abdominal muscles. (Photo of living animal, courtesy Shedd Aquarium)

Swamp crayfish *Cambarus*, beside the "chimney" which surrounds its burrow. The burrows are 1–3 feet deep, have at the bottom a water-filled cavity, and are built in swamps and meadows, often far from a stream. (Photo by C. Clarke)

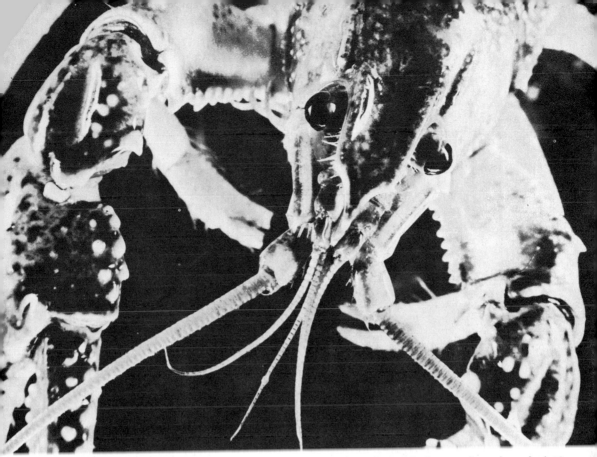

Head of crayfish, showing the stalked eyes and jointed antennas, the first pair short and two-branched, the second pair long and single. The two pairs of antennas are a distinguishing characteristic of crustaceans. Insects, millipedes, and centipedes have one pair; arachnids have none at all. (Photo by P. S. Tice)

The eggs of the crayfish develop while attached to the pleopods (*left*). They hatch into young (*right*) that look like miniature adults and cling for a time to the pleopods (Photos by Cornelia Clarke)

Fiddler crab is so named from the way in which the male brandishes the huge left pincer—extending it out and then suddenly drawing it in. The "fiddling" becomes especially vigorous in the presence of a female. The female has two small pincers of equal size. These crabs are scavengers and live in burrows on sandy ocean beaches. (Photo, courtesy Am. Mus. Nat. Hist.)

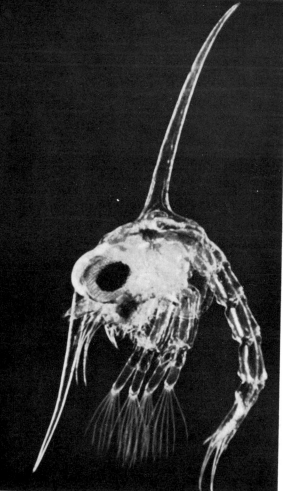

Beach crab in mouth of burrow on sandy beach. Sometimes called "ghost crab" because of the way in which it appears and then suddenly disappears, it is one of the most active of crabs. It burrows above high-water mark and is very successfully adapted to a semiterrestrial life. (Photo of living crab. Panama)

Larva of the velvet swimming crab, *Portunus puber,* is typical of the free-swimming larvas of crabs, though they differ in such details as the proportions of the long spines, which increase the surface of the larva and help to keep it afloat. The naturalists who first saw a crab larva thought it was a special kind of adult animal and named it a "zoea." Now we use the name to indicate a particular stage in the development of the larger crustaceans. The zoea of the velvet swimming crab molts several times, each time growing a little larger. Shown is the second zoeal stage. (Photo of living larva by D. P. Wilson. Plymouth, England)

Land crab has not made a complete transition to terrestrial life. It must still pass through its earliest stages in the sea. Land crabs are common in the tropics, but it is startling even to a biologist to lift a rotting log in the tropical forest and find hiding under it a crab. (Photo of living crab. Panama)

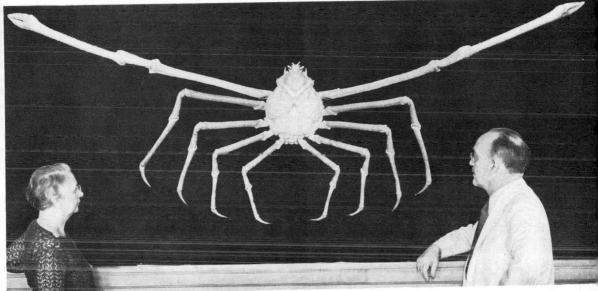

A giant crab from the waters off Japan. As in the fiddler and all true crabs, the abdomen is reduced and bent under onto the ventral surface; it cannot be seen from above. (Photo courtesy Buffalo Museum of Science)

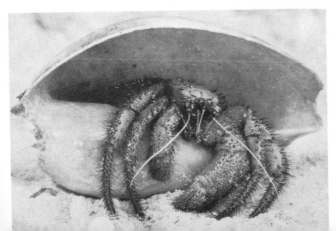

The hermit crab is not a true crab, for it has a long abdomen, which is soft and spirally coiled and is inserted into the empty shell of some marine gastropod. The abdominal appendages are atrophied, but one pair has hooks for holding on to the inside of the shell. The cephalothorax can be protruded, and the animal walks about carrying its shell, into which it retreats at any sign of danger. (Photo by Otho Webb. Australia)

The king crab, *Limulus*, often called the "horseshoe crab" because of its shape, is no crab at all, but a primitive marine animal sometimes put in the same group with scorpions and spiders, which it resembles in many details. It lives along our Atlantic Coast, scooping its way through the sand or mud as it hunts for the bivalves and worms, especially nereids, on which it feeds. (Photo by L. W. Brownell)

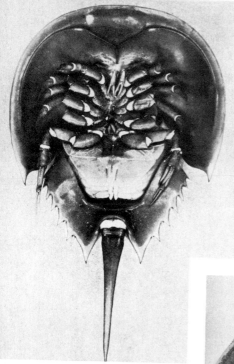

The scorpion is an arachnid common in our Southwest. It hides in crevices during the day and comes out at night to hunt spiders and insects, which it catches with the large pincers, stings to death with a poison injected by the curved spine at the tip of the tail, and then sucks dry. The sting of a scorpion is painful but not dangerous to human adults, though it sometimes proves fatal to children. (Photo by P. S. Tice)

Underside of Limulus shows walking legs and other appendages that lie under the protective hood. To the flat abdominal plates are attached the leaflike gills.

Tarantulas are large spiders which hide during the day in the cracks of trees and under logs, stones, or debris, and at night come out to stalk their prey. They are common in our South and Southwest, where they reach a length of 2 inches; their bite is painful but not dangerous. Most people insist they are revolted by the long legs and hairiness, but no one on record has ever objected to these same characteristics in a Russian wolfhound. (Photo by Lee Passmore)

A tropical tarantula, shown here about life size, is a giant among spiders. Some South American tarantulas have a 7-inch span and can catch small birds. (Photo of living animal. Panama)

The trap-door spider lives in a silk-lined burrow and waits, just beneath the hinged trap door that closes the burrow, for passing prey. Here the spider is pouncing on a sow bug (a small land crustacean). Trap doors, hinged with silk, guard the entrance to the burrow. When closed, the door matches the surroundings perfectly. The size of the door is an indication of the size of its occupant. (Photos by Lee Passmore)

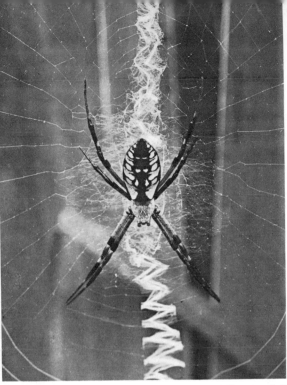

A garden spider spins a web and waits "patiently" until a small animal becomes entangled. Then it rushes out, seizes the prey, injects a poison and a digestive juice, and sucks up the tissue fluids. (Photo by Cornelia Clarke)

Regeneration of a missing leg occurs if the spider is young and growing. The adult trap-door spider (*above*) has regenerated the right second walking leg. The young spider (*below*) is a miniature of the adult. (Photos by Lee Passmore)

Just after molting the trap-door spider (*left*) has a white, delicate cuticle and is helpless; but in a day or so the cuticle hardens and darkens. The old cuticle (*right*) was first loosened by a molting fluid; then it split along the sides and was shed. (Photo by Lee Passmore)

The black-widow spider (*Latrodectus mactans*) is one of the two really dangerous spiders in the United States, and, though common in the South and Southwest, has been reported throughout the U.S. The male (*left*), as in most species of spiders, is small and harmless. The female (*right*) is ½ inch long, black, and has a red mark shaped like an hourglass on the ventral side of the abdomen. The venom is very poisonous, and the bite is followed by pains and fever. Victims usually recover after two weeks, but fatalities occur.

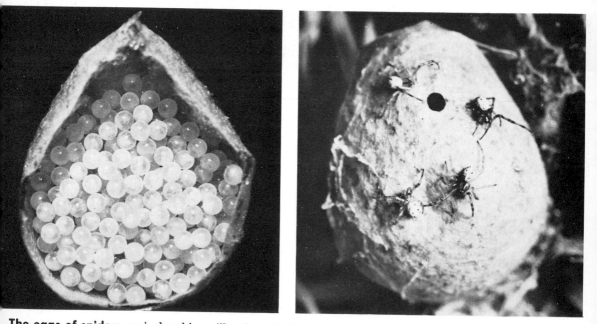

The eggs of spiders are inclosed in a silken bag which is hung from the web or some solid object or is carried about by the female. *Left:* Opened egg sac of black widow. *Right:* The young spiders, just emerged from the egg sac, look like miniature adults. (Photos on this page by Lee Passmore)

A spider is distinguished externally from an insect by its four pairs of walking legs; a body composed of two regions, cephalothorax (head-thorax) and abdomen; a lack of antennas; only simple eyes; and in the possession of two pairs of appendages about the mouth. These are the *cheliceras*, shown in this huntsman spider beneath the eyes as broad appendages with fanglike tips and *pedipalps*—leglike appendages on either side of the cheliceras. Insects have three pairs of legs, three body regions, two antennas, two compound eyes, different mouth parts, and usually two pairs of wings. (Photo by Otho Webb. Australia)

The eight simple eyes of a spider, (*left*) are usually arranged in two rows. The central lens is surrounded by a ring of dark pigment. (Photo by P. S. Tice)

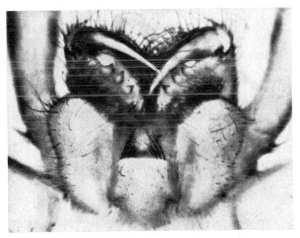

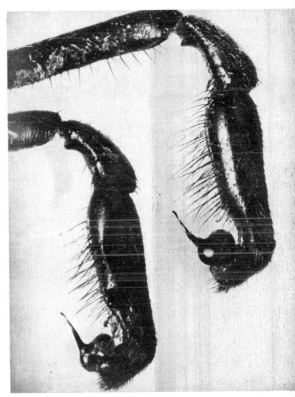

Underside of spider's head shows fanglike cheliceras which bite prey and inject a poison through openings near the tips. Just below them are the "jaws" of the spider—two flat plates, each topped by a tuft of dark hairs. They are extensions from the bases of the pedipalps and can be brought together to hold prey and to squeeze it when the spider is sucking up its fluids. (Photo by P. S. Tice)

The pedipalps are chiefly sensory, but in male spiders have their tips modified to assist in copulation. The males have been seen to spin a silken net on which they deposit sperms which issue from openings on the underside of the abdomen. The sperms are picked up by the pedipalps, stored in bulbs at their tips, and later transferred to the female at the time of mating. (Photo by Lee Passmore)

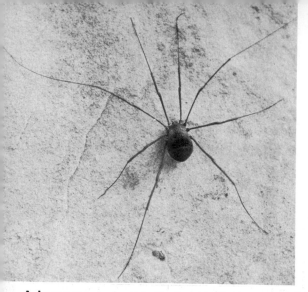

A harvestman (or "daddy longlegs") is so called because it is seen most often at harvest time. It looks like a long-legged spider but belongs to another group of arachnids, the phalangids, because, among other things, the cephalothorax and abdomen are broadly united and there are no silk glands. It feeds mostly on small insects. (Photo by Cornelia Clarke)

A mite, as the name implies, is small. This one (shown from below) found living in an open field, is ¼ inch long—quite sizable as mites go. Many of the parasitic types, like those that live in the oil glands and hair follicles of the human face, are only 1/50 inch long. There are four pairs of legs, as in all arachnids; the body is ovoid and all in one piece. (Photo by P. S. Tice)

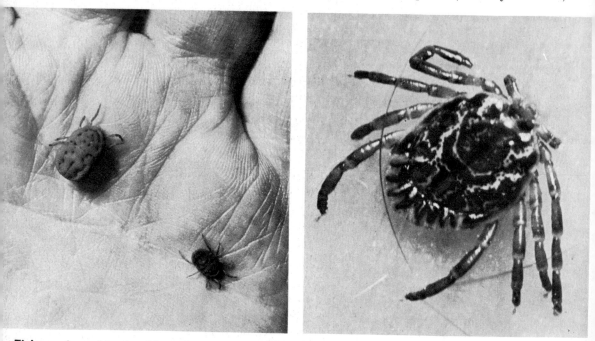

Ticks are large, blood-sucking mites, some of which parasitize man and his domestic animals and transmit to them serious diseases. Every year they cause millions of dollars of damage to cattle alone. The two shown at the left were kept for five years without food in the U.S. Public Health Service Laboratory. During this period of starvation they maintained within their bodies the organisms that cause relapsing fever when transmitted to man by the bite of the tick. These ticks are most common in our southwestern states. (Photo from Science Service). *Right:* Wood tick with beak inserted in human skin. (Photo of living animal. Panama)

he house centipede, *Scutigera*, differs from other ntipedes in having long, delicate legs and compound yes. (Typical centipedes have two clumps of simple yes.) It lives in damp places in houses, usually in the asement. It does no harm and preys on cockroaches nd other insects. Slightly larger than natural size. Photo by Cornelia Clarke)

Typical centipedes have shorter, stouter legs. The long trunk has similar segments, as in millipedes, but is flattened and has only one pair of legs to a segment. Centipede means "hundred-legged," but *Scolopendra*, shown here, has 21 pairs of legs and the number ranges from 15 pairs as in Scutigera to 173 pairs in the geophilids. (Photo by Otho Webb. Australia)

The poison claws of the centipede (shown here in a view of the underside of the head, enlarged 4 times) are not jaws but modified appendages of the first body segment. They are curved, hollow organs, perforated at their tips, which inject a poison that rapidly paralyzes prey such as insects, slugs, worms, and even lizards and mice. This 6-inch Bermuda centipede inflicts a bite that may keep a man in bed with a fever for several days. Some tropical centipedes are over a foot long. (Photo by P. S. Tice)

Millipedes are usually found among decaying leaves and under logs in moist shady woods or under stones in gardens. They feed mostly on decaying plants but sometimes eat living roots, becoming garden pests. Their long, cylindrical bodies consist of a head followed by a series of similar segments, almost all of which bear two pairs of legs. Millipede means "thousand-legged," and, though this is an exaggeration, some common ones do have 115 pairs. The eggs are laid in a nest made of earth (*right*) and are guarded by the mother. (Photo of living animal)

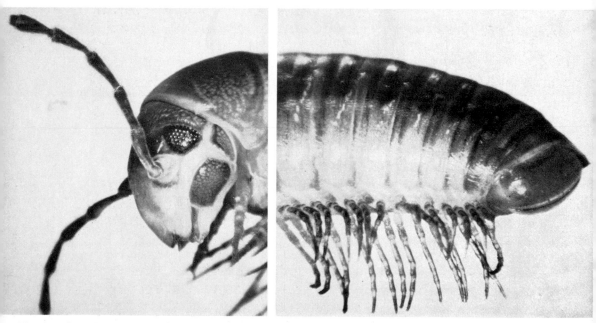

The head (*left*) of a millipede is similar in most respects to that of a centipede or an insect. It has paired antennas and chewing jaws, and a pair of accessory jaws (maxillas). There are two eyes, one of which can be seen just above the base of the antenna. It appears superficially like the compound eye of insects but is only a clump of simple eyes set close together. The posterior end of the animal (*right*) shows the terminal anus and a number of typical segments, cylindrical in shape and each with two pairs of legs. (Photos by P. S. Tice)

insects undesirable to man), is based on nothing more than a vague fear of animals which have long legs, run rapidly, live in dark places, and catch their prey in webs or other traps.

Hardly any description will fit all the orders; but, in general, arachnids are terrestrial arthropods which have the body divided into two main regions: a **cephalothorax** bearing six pairs of appendages, of which four of the pairs are walking legs, and an **abdomen** which has no locomotory appendages, though it may have some other kind. The *four pairs of walking legs* usually serve as a convenient, if superficial, way of distinguishing arachnids from insects, which have only three pairs. But the difference between the groups is much more deep-seated. Arachnids differ from crustaceans and insects in having *no compound eyes*, only simple ones. And they are even more clearly marked off from crustaceans, centipedes, millipedes, and insects by the nature of the segmental appendages on the head. Arachnids have *no antennas*, the function of these organs being served by an abundance of sensory bristles or "hairs" with which the body and particularly the appendages are covered. Also, they have *no true jaws* homologous with those of crustaceans and most other arthropods. None of the arachnid appendages are completely specialized for chewing, but on the basal segments of one or more of them are sharp biting processes. Many primitive arthropods have such chewing processes on the bases of the appendages, and it is thought that from such structures came the more specialized jaws of crustaceans or insects. In front of the mouth (on the third segment) arachnids have a pair of **cheliceras,** appendages which may take the form of pincers or of sharp, fanglike claws. Behind the mouth (on the fourth segment) is a pair of **pedipalps,** leglike appendages that serve a sensory function, as in spiders, or are used for seizing prey, as in scorpions. Among the various arachnids either the cheliceras or the pedipalps are the important weapons of offense, but never both in the same animal.

The spiders are by far the largest and most widely distributed order of arachnids. A generalized description of a spider, though applying in many respects only to this one group, will give some further idea of arachnid structure and habit.

IN A **spider** the cephalothorax is covered by a shield, the carapace, on which are set the simple eyes, usually eight in number. The cheliceras are sharp and pointed and are used for capturing and then paralyzing the prey by injecting a poison. Ducts from a pair of poison glands lead through the cheliceras and open near their perforated tips. The pedipalps look like legs but are sensory, and their basal joints have jawlike processes which hold and compress the prey. In the male the pedipalps are modified for transferring the sperms to the female. The four pairs of walking legs end in curved claws. The abdomen shows no external evidence of segmentation and has no appendages except the **spinnerets,** of which there are usually three pairs. The spinnerets are finger-like organs which have at their tips a battery of minute spinning tubes (sometimes a hundred or more on each spinneret), from which the fluid silk issues, and then hardens as it comes in contact with the air. The spinning tubes connect with several kinds of silk glands which produce different kinds of silk for spinning various parts of the web, making a protective cocoon for the eggs, binding the prey, etc. Some of the tubes produce not silk but a sticky fluid which makes the threads of the web adhesive.

When an insect or other small animal becomes entangled in the web, the spider apparently feels the tugging, for it hurries to the scene, seizes the struggling animal, and, holding it between the jawlike processes on the pedipalps, injects a poison. If the prey is large and formidable, the spider may use a more indirect method, first binding its victim with silk. The mouth of the spider is too small to swallow solid food; instead, the animal injects a digestive fluid through the wound made by the bite of the cheliceras. The predigested, liquefied tissues of the prey are then sucked up by means of a muscular **sucking stomach** (aided by the squeezing action of the pedipalps and the sucking action of the pharynx). Beyond the sucking stomach the digestive tract gives off several pairs of pouches, which increase the digestive and absorptive surface, and a large **digestive gland,** which branches extensively and occupies most of the spider's abdomen. This gland is the main organ of digestion and is capable of taking up very large quantities of food at one time, storing it, and then gradually absorbing it. This enables spiders to go for long periods without taking food (though they must have water quite often).

The circulatory system is open, as in other arthropods. The heart lies dorsally in a large sinus and receives blood through openings in its sides. The excretory system, as in most other arthropods, consists of tubules which open into the intestine. In many spiders there are also excretory sacs which open near the bases of the legs. The respiratory organs are of two types. The **lung book** is an air-filled sac which communicates with the external air through a slitlike opening. Attached to the walls of the sac is a series of leaflike folds of the body wall. These "leaves," which have suggested the name of the organ, are held apart by supports so that air can circulate freely between them. The spaces within the leaves are filled with blood and communicate with the blood sinuses of the abdomen. Thus the leaves of the lung book are simply another device for exposing a large amount of respiratory surface to the air. The **air tubes** of spiders receive air through openings on the abdomen and convey it to the tissues. The smallest tubes usually do not branch extensively, as in insects. The air tubes are not thought to be homologous in the two groups, for arachnids and insects probably did not have a common ancestor which lived on land. Some spiders have only lung books, and some have only air tubes, but most have one pair of lung books and one pair of openings to air tubes.

HORSESHOE CRABS

THE horseshoe crabs are not crabs at all but primitive marine arthropods, the only living representatives of the class **Palcostracha.** There are five living species of horseshoe crabs, all of them usually placed in the single genus *Limulus*. These animals are often referred to as "living fossils" because they have changed so little from the earliest fossil representatives of the group. No one can say with any certainty why they have been able, with "no modern improvements," to survive in competition with more highly developed aquatic arthropods. Perhaps their success results from a combination of unobtrusive habits and a heavy hoodlike carapace which forms a complete roof over the body and all the appendages.

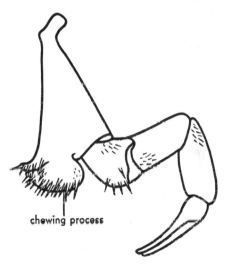

chewing process

First walking leg of *Limulus*, showing the **chewing process** on its base.

Horseshoe crabs live in shallow water along sandy or muddy shores. They can swim through the water by flapping the appendages on the abdomen, but spend most of the time burrowing in the sand or mud for the worms and mollusks on which they feed.

Aside from their interest as archaic forms, the horseshoe crabs have attracted attention because they are clearly related to arachnids and, though somewhat specialized, give us some idea of what the ancestral aquatic arachnid may have been like. The horseshoe crabs were, in fact, once classified as one of the orders of the class Arachnida, their chief differences being associated with their aquatic life. Attached to the flattened abdominal appendages are the **gill books,** groups of thin plates in which blood circulates; they are similar in plan to the lung books of terrestrial arachnids, but the two kinds of respiratory organs have different orientations and origins. As in arachnids, the body is divided into cephalothorax and abdomen; and the thorax has six pairs of appendages, of which the first is a pair of pinching cheliceras and the other five pairs

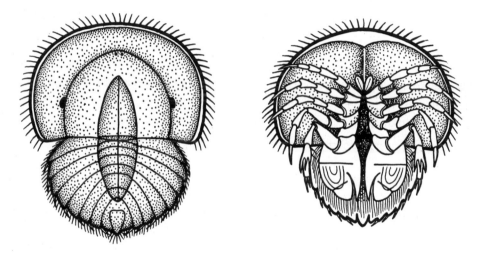

The free-swimming **larva of Limulus.** *Left,* dorsal view; *right,* ventral view. (After Kingsley)

are walking legs. The first four pairs of walking legs have on their bases spiny processes for masticating the food.

The eggs hatch into free-swimming larvas which lack the long tail of the adult.

CENTIPEDES

THE centipedes ("hundred-legged") form the class **Chilopoda.** The members are land arthropods which are flattened dorsoventrally and have a distinct head, followed by numerous similar body segments. The appendages of the first body segment (seventh segment of the animal) are modified as a pair of **poison claws.** These have perforated tips through which a poisonous secretion, from a pair of glands, can be injected into the prey. Each of the other body segments, except the last two, has a pair of walking legs. Some centipedes have as high as one hundred and seventy-three pairs of legs, and others have only fifteen, but thirty-five is probably an average number. The animals run very rapidly, and the numerous legs apparently work in perfect co-ordination.

The head, as in most arthropods, bears appendages which are homologous with those of a lobster or insect but resemble more closely those of an insect. There is a single pair of antennas; the jaws have no sensory process or palp; the first maxillas have two lobes; and the second maxillas are usually fused together, as in insects. There are two groups of simple eyes on the head; but other parts of the body must be sensitive to light, for some centipedes react negatively to a bright light when the eyes are completely covered with heavy paint.

In other respects they are much like insects (described in chap. 24). The digestive tract is a straight, simple tube. The excretory system consists of tubules opening into the hind portion of the gut. Oxygen is supplied to the tissues through branched air tubes which lead from a pair of openings in every segment. The circulatory system is slightly more elaborate than in insects, having a pair of arteries to every segment.

Lithobius, a common centipede, lays its eggs in the ground. The young hatch with only seven segments, and the rest are added later. During growth the animal sheds its exoskeleton frequently.

Centipedes are found in moist situations under the bark of decaying logs and under stones. They are carnivorous, feeding upon soft insects such as cockroaches, plant lice, and silverfish; they also eat earthworms and slugs. They hunt only in the dark and are probably guided in their movements mostly by touch, to which they are very sensitive. They seldom come to rest unless the body is in contact with some solid object on at least two sides. This is adaptive, since it keeps these animals under cover, where they are safe from enemies and from drying.

Though terrestrial, such a centipede as *Lithobius* can survive many hours completely immersed in water but will die in a few hours in an uncovered dish of dry earth. That their habit of keeping under cover is a positive reaction to contact as well as a negative response to light (which centipedes avoid when possible) can be shown by a simple experiment. A *Lithobius* placed in a glass dish will run about ceaselessly; but if some narrow, transparent glass tubing is placed in the dish, the animal will soon come to rest in the tubing, which affords a maximum of contact with the surface of the centipede.

MILLIPEDES

THE name millipede means "thousand-legged"; and though this is a gross exaggeration, millipedes do have very large numbers of legs—in fact, twice as many as a centipede of about the same length, since there are two pairs to each adult abdominal segment. The technical name of the group is class **Diplopoda,** which refers to the double-legged situation. A millipede embryo has only one pair of legs to every body segment, each innervated by a segmental ganglion. In the adult the first four (thoracic) segments remain single, but the other (abdominal) segments fuse in pairs, so that each adult ring represents two embryonic segments and has two pairs of legs.

The head has the same appendages as in centipedes except that the first maxillas, which appear in the embryo, do not persist to the adult stage. The eyes superficially resemble compound eyes, but each is only a clump of many simple eyes set closely together. The internal anatomy resembles that of centipedes.

In contrast to the centipedes, which are all carnivorous, most millipedes are herbivorous, and in spite of their larger number of legs run much more slowly.

CHAPTER 24

THE GRASSHOPPER AND OTHER INSECTS

MOTORING across the hot, dry, desolate regions of the West, one is surprised not by the scarcity of human beings but by the fact that anyone at all should choose to live in the Mojave Desert, when life is much more pleasant in Los Angeles, only a short distance away. The answer is that men, like all animals, compete with the other members of their species for food and shelter. And since the most desirable regions are usually filled to capacity, those individuals who can adapt themselves to the conditions of the less favorable places have the advantage of a relatively unexploited environment.

The most successful animal groups are those in which the members do not all live in one region in competition with each other but have spread out to every corner of the world that will support their mode of life. The problems of why particular animals live where they do, how they reached there from the places where they first evolved, and why they do not occupy certain other regions which appear to be suitable for them, belong to a major subdivision of biology which we call *animal geography* and which we cannot adequately consider here. We can only mention, in passing, that arthropods, more than any other group of higher animals, are **widely distributed.** They occur in all seas, in all bodies of fresh water, and in every land habitat. Of the many factors that make for wide distribution, one of the most important is small size; and arthropods are mostly small animals which can be carried from one place to another by water currents, by the wind, on floating debris, and on the bodies of other animals. Equally important is mobility; of all land arthropods, the most widely distributed are the winged insects, some of which are not stopped in their migrations even by barriers such as mountains or large bodies of water.

Ease of distribution is by no means the secret of insect success. For there is little advantage to animals in moving to new places if they cannot adapt themselves to the conditions of life there. Since the insect body plan does lend itself readily to structural specialization for almost any mode of life, insects have been able to spread from their original centers of distribution and are now the dominant invertebrates of tropical rain forest, temperate forest, prairie, plain, desert, and tundra. But even this is not enough. If all the insects in an oak woods were grass-eaters, they would be competing with each other for the limited amount of grass available. Instead, they show such a diversity of habits that they occupy every **niche**—and, in so doing, tap every source of energy available in that community. In the oak woods we would find insects living as predacious carnivores, herbivores, suckers of plant juices, suckers of vertebrate blood, pollen-gatherers, nectar-gatherers, scavengers, parasites of plants, and parasites of animals. Even within each of these major categories they do not all compete with each other. Some herbivores eat grass, some leaves of trees, some woody stems. Among leaf-eaters, different ones may specialize on particular species of plants. Among wood-eaters, some eat bark, some sapwood, and others only decayed wood. Of this last group, different ones are limited to wood in a particular stage of decay.

Thus we see that animals meet competition either by excelling their neighbors in one of the more typical modes of life, by becoming adapted to a relatively unexploited environment, or by becoming specialized to exploit some source of energy not available to the other members of a crowded community. This tendency of animals to spread out into every available niche in any habitat is called **adaptive radiation.** A few of the lower phyla, in which the simple body plan does not allow radical modifications, show practically no radiation and are limited to a single way of life. For example, all sponges are sessile forms which feed by drawing through the body a current of water from which they strain microscopic particles; all bryozoans and all brachiopods are attached forms which feed by ciliary currents. But this is not the rule. Most phyla, particularly the higher ones, which have complex body plans, show radiation not only

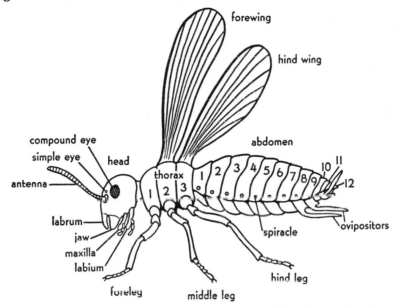

Diagram of a **typical insect.** Most insects have fewer abdominal segments, owing to loss or fusion at the posterior end. (After Snodgrass)

within the phylum as a whole but also within a single class or even an order. In categories smaller than this there is little significant variation; for while the different genera of a family may live in widely separated parts of the world, where they do not come in competition with each other, usually they fill corresponding roles in their communities.

Some variation within the family does occur, as in the case of the bark-beetle family (Scolytidae). This group is composed mostly of wood-eaters which excavate tunnels just under the bark, but some of the members dig through the solid wood and then use the tunnels to grow fungi, on which they feed. This can hardly be classed as adaptive radiation, by which we imply the penetration of a group into practically all the major roles in any region.

Of all animals, arthropods best illustrate adaptive radiation. It is clearly shown among crustaceans but is carried to the greatest extremes in the **insects.** The name "insect" comes from a Latin word meaning "incised" and refers to the fact that insects generally have a sharp division between the head and thorax and between the thorax and abdomen. The head of the adult insect is all in one piece, but the six segments can be seen in the embryo, and some of them are indicated in the adult by the paired appendages. The thorax, which has undergone less specialization and fusion than the head, has three segments, each bearing a pair of appendages, usually walking legs. The second and third thoracic segments each bear a pair of wings (except in certain primitive and degenerate forms). The abdomen is the least specialized region of insects, being composed of relatively similar segments, generally without appendages, unless we consider certain structures at the posterior end of the abdomen to be modified segmental appendages. The external anatomy of insects varies so much that it is less satisfactory to generalize than to describe a particular insect such as a grasshopper, which is not too specialized and is therefore usually considered a fairly typical representative of its class.

THE GRASSHOPPER

THE **head** of the grasshopper has two compound eyes and three simple ones (ocelli). The *compound eyes* are similar in structure to those of the lobster; and while they are not on stalks, they have a broad field of vision because they occupy a relatively large area and curve around the sides of the head. They occur on the first head segment, which has no segmental appendages. The second segment bears a pair of long, jointed sensory *antennas*, homologous to the first antennas of the lobster. The third segment bears the *upper lip* (labrum), which is not serially homologous to the other head appendages but is a secondary growth. The fourth has a pair of toothed horny *jaws* (mandibles). The fifth is indicated by a pair of accessory jaws, or *maxillas*, each with a jointed sensory palp. The sixth bears the *lower lip* (labium), which has on each side a sensory palp. The labium of the adult appears as a single plate, but in the embryo it arises as a pair of structures which later fuse in the mid-line. Thus, each half of the labium is homologous to one of the second maxillas of the lobster.

The **thorax** is partly covered dorsally by a chitinous shield and by the wings; but the three segments, with a pair of legs on each, are clearly visible at the sides. The *legs* are composed of a characteristic series of joints (as shown in the drawing) and end in two curved claws between which is a fleshy pad that aids in clinging to surfaces. The first two pairs

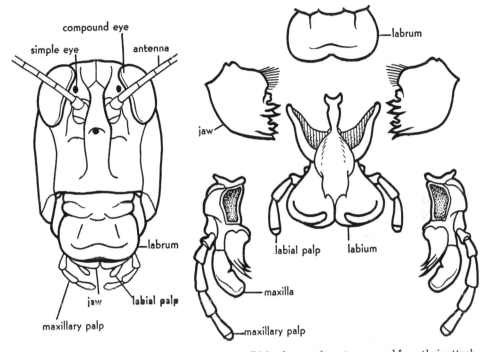

Left, front view of the **head of a grasshopper.** *Right*, the mouth parts removed from their attachments. (After Snodgrass)

of legs are typical walking legs. The third is specialized for jumping; it has one joint, the femur, which contains muscles for jumping, enlarged out of proportion to the others. The *two pairs of wings* are different from each

other. The forewings are narrow and hardened, and they serve as a cover for the hind wings. Both pairs are active in flying. The hind wings are quite broad when in flight, but when not in use are folded like a fan and fit under the first pair. This arrangement is a specializa-

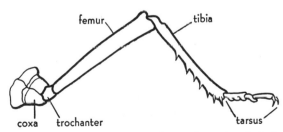

The **middle leg** of the grasshopper shows, in relatively unspecialized form, the parts of a typical insect leg. (After Snodgrass)

tion; in most generalized insects the two pairs of wings are more alike in size and in texture. The wings are made of cuticle and are stiffened by thickenings called *veins*.

The **abdomen** has no appendages except those at the posterior end, which are associated with mating and egg-laying. The abdomen contains much of the machinery of the body, since the head is small and the thorax is nearly filled with the muscles that move the legs and wings.

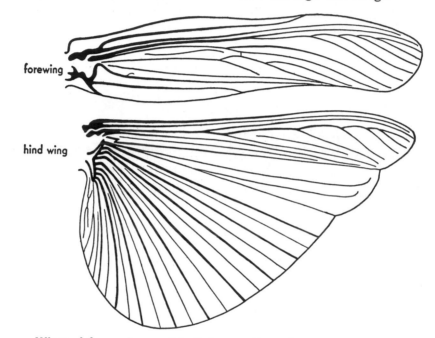

Wings of the grasshopper. Only the larger veins are shown. (After Snodgrass)

The **digestive tract** consists of three parts: fore-gut, mid-gut, and hind-gut. The *fore-gut* starts at the mouth, receives a secretion from the salivary glands, and runs on as a narrow esophagus, which leads to the *crop*, a large, thin-walled sac in the thorax. On the inner walls of the crop are transverse ridges armed with rows of spines which probably serve to cut the food into shreds. The crop is mainly a storage sac which enables the grasshopper to eat a large quantity at one time and afterward digest it leisurely. From the crop the food passes into a muscular *gizzard*, lined with chitinous teeth. At the posterior end of the gizzard is a valve, which prevents the food from passing into the stomach before it is thoroughly ground and also prevents food in the stomach from being regurgitated.

Digestion probably begins in the crop, for the food entering that organ is already mixed with salivary secretion, and it also receives some digestive juices which pass anteriorly from the stomach. Since the whole fore-gut is lined with cuticle, little, if any, absorption of food occurs there. The *mid-gut*, or stomach, which lies mainly in the abdomen, has no cuticular lining and serves as the main organ of digestion and absorption. Opening into the anterior end of the stomach are six pairs of pouches; one pouch of each pair extends anteriorly from the point of attachment, and the other posteriorly. These pouches secrete a digestive juice and also aid in absorption. The junction of the stomach with the *hind-gut*, or intestine, is marked by the attachment of long excretory tubules. The intestine is lined

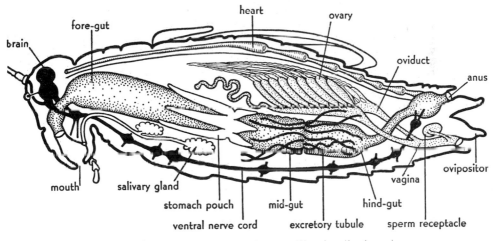

Internal anatomy of the grasshopper. (Based on Snodgrass)

with cuticle. It receives the waste materials of digestion and the nitrogenous excretions of the excretory tubules.

The **excretory system** consists of a number of tubules (called "Malpighian tubules," from the name of their discoverer) which lie in the blood sinuses and from the blood extract nitrogenous wastes. The wastes, in the form of crystals of uric acid, are poured into the hind-gut and leave the grasshopper, by way of the anus, as dry excretions. Dry wastes are characteristic of small land animals, which have a limited supply of water.

Air for **respiration** is not distributed by the circulatory system but is piped through branching *air tubes* (tracheal tubes) which form a definite system of longitudinal and transverse main trunks from which smaller branches ramify to all parts of the body. The air tubes lead from paired openings which lie at the sides of the abdomen and thorax in the thin membrane between segments. The openings, or *spiracles*, of which there are ten pairs, are guarded by hairs to keep out dirt, and by a valve which can be opened or closed to regulate the flow of air. Their closure also aids in decreasing the evaporation of water. The air tubes are prevented from collapsing by means of spiral thickenings in their walls. They branch

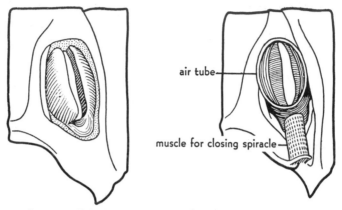

Spiracle of grasshopper. *Left*, the spiracle of the second thoracic segment, as seen from the out-side. It consists of a vertical slit guarded by two valve-like lips. *Right*, the same viewed from the inside to show the muscle which pulls the two lips together to close the spiracle. (After Snodgrass)

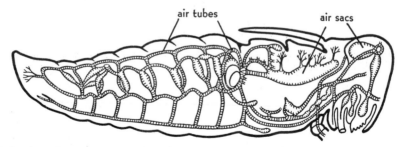

Respiratory system of a grasshopper, showing some of the main air tubes and air sacs. (After Vinal)

freely and become so small (1 micron, or 1/25,000 of an inch, in diameter) that the finest ones (air capillaries, or tracheoles) are made of a single cell; the smallest tubes have no spiral thickening and are usually filled with fluid. Oxygen in the air of the larger tubes dissolves in this fluid and passes, by diffusion, to the tissues. Here and there the system widens into large air sacs. The air is moved chiefly by diffusion; but muscular breath-ing movements, which alternately compress the air sacs and then allow them to expand, aid in changing the air. The greater the muscular activity, the greater the pumping action on the air sacs and the better the circulation of air. In the grasshopper the first four spiracles open into one set of air tubes only at inspiration, and the remaining six pairs are open only upon expiration; this facilitates the flow of air. The larger air tubes are impermeable to water (thus preventing water loss) but freely permeable to oxygen, which dissolves in the blood and has to travel only a very short distance to the tissues. There are no respiratory pigments for carrying oxygen. Carbon dioxide leaves by the reverse route but may also escape through the thin parts of the body surface.

The system of air tubes, as it occurs in the grasshopper and other insects, is one of the factors which limits the size of these animals. Since the air must travel mostly by diffusion, it could not reach the interior of a large animal fast enough to support the degree of activity displayed by insects. The other main factor in limiting the size is the chitinous exoskeleton, which in a larger animal would make flying more difficult.

The blood vessels of the **circulatory system** are much less extensive than in the lobster. In fact, there is only one vessel, the long contractile *dorsal vessel*, composed of the tubular *heart* which pumps the blood forward, and its anterior extension, the *aorta*. In each segment through which it passes, the heart is dilated into a chamber perforated on each side by a slitlike opening through which blood enters. The aorta carries the blood into the head and there ends abruptly. The blood flows out into spaces among the tissues and makes its way back into the thorax, where it bathes the thoracic muscles. From there it enters the abdomen and bathes the various organs, absorbing food from the stomach and giving up wastes to the excretory tubules. Then it returns to the heart. The course of the blood is really more definite than this brief sketch has intimated. There are partitions which deflect the blood so that it enters one side of each leg and emerges on the other. In the abdomen there are two large horizontal partitions which aid in directing its course, and the dorsal one separates the cavity containing the heart from that in which the other viscera lie. In an open system of this kind, where the blood flows among the tissues instead of in definite vessels, the rate of flow is relatively low. However, this is no disadvantage, since the distribution of oxygen has been taken over largely by the system of air tubes. The blood serves mainly as the distributing and collecting medium for food and wastes, but it has other functions. It acts as a reservoir for food; and when under pressure, it aids in hatching from the egg, in molting, and in the expansion of the wings. Besides, it contains cells which ingest bacteria and wall off parasites.

The **nervous system** is a ventral, double, ganglionated cord like that of the lobster. The embryo has a ganglion for each segment, including six in the head; but in the adult some of the ganglia are fused so that there are only two in the head, three in the thorax, and five in the abdomen. Still, this is a fairly generalized system, as compared with that of some insects, which have all the thoracic and abdominal ganglia fused into one mass. The *brain* lies above the esophagus and between the eyes. It is joined to the first ventral ganglion by a pair of nerves which encircle the gut. The brain has no centers for co-ordinating muscular activity; after removal of the brain the animal can walk, jump, or fly. As in lower invertebrates, the brain serves chiefly as a sensory relay which receives stimuli from the sense organs on the head and, in response to these stimuli, directs the movements of the body. It also exerts an inhibiting influence, for a grasshopper without a brain responds to the slightest stimulus by jumping or flying—a very unadaptive kind of behavior. And even in the absence of any external stimulation, the animal displays an incessant activity of the palps and legs. The *first ventral ganglion* controls the movements of the mouth parts and plays some role in maintaining balance. The *segmental ganglia* are connected and co-ordinated by

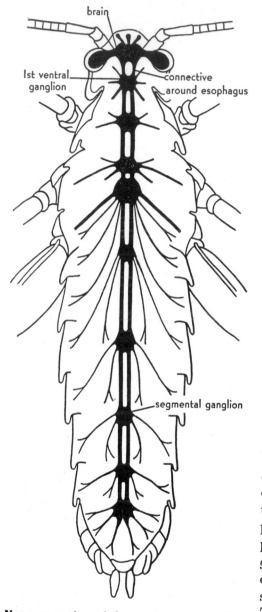

Nervous system of the grasshopper. (Modified after Snodgrass)

nerves which run in the cords, but each is an almost completely independent center in control of the movements of its respective segment (or segments) and appendages. In some insects these movements have been shown to continue in segments which have been severed from the rest of the body. An isolated

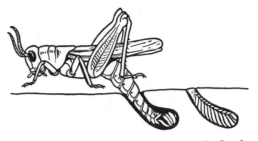

Grasshopper laying eggs in the ground. On the right is a completed batch of eggs. (After Walton)

thorax is capable of walking by itself, and an isolated abdominal segment performs breathing movements.

The **reproductive system** of the male grasshopper is a pair of *testes*, which discharge sperms into a *sperm duct*. The duct receives secretions (a fluid in which sperms are conveyed to the female) from glands and then opens near the posterior end of the body. In the female there is a conspicuous set of stout appendages near the posterior end of the abdomen. These are used for digging a hole in the ground in which to lay the eggs.

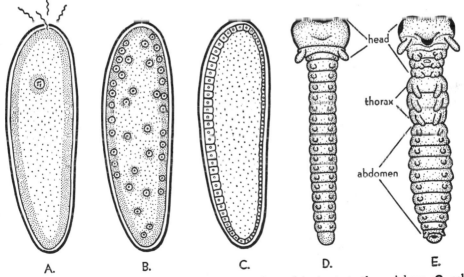

A.   B.   C.   D.   E.

Development of an insect. A, fertilization. **B,** the nuclei migrate to the periphery. **C,** cell walls appear between the nuclei, resulting in a single-layered embryo which corresponds to the blastula of other animals. **D,** when segmentation first appears, the head is composed of only three segments. **E,** later, the first three body segments are added to the head, forming a six-segmented head, which in the adult shows little trace of the original segmentation. (After Snodgrass)

Gradual metamorphosis of a grasshopper from first nymphal stage to adult. Between any two successive stages the animal molts. The first nymph has a relatively large head and no wings.

Internally there is a pair of *ovaries*, with *oviducts*. The two oviducts converge into a *vagina*, which opens into a *genital chamber*. In the sex act the male introduces sperms into a special sac in the female, the *sperm receptacle*, in which they are stored until the time for egg-laying. The mature eggs pass down the oviduct and, contrary to the usual procedure in animals, the shell and yolk are put on before fertilization. A small pore is left, however, through which the sperm enters. As the eggs pass into the genital chamber, they are fertilized by sperms ejected from the sperm receptacle.

The **development** of insects is rather different from that of most animals—chiefly because of the large amount of yolk, which provides enough food to enable the young to develop sufficiently to be able to feed. Instead of dividing into two cells and then into four, and so on to form a blastula, the zygote nucleus divides many times without the division of the cytoplasm. The nuclei then move to the periphery, and cell walls appear between them, thus forming a layer of cells. Subsequently, cells pass inward until there are roughly three layers. The outermost layer is the ectoderm; the endoderm and mesoderm are interior to it. Segments eventually appear as little hollow blocks of mesoderm. The mid-gut develops from endoderm. The fore-gut and hind-gut develop from infoldings of the ectoderm. Since the ectoderm alone secretes cuticle, this explains the presence of a lining of cuticle in the fore-gut and the hind-gut. A coelom forms as a series of pairs of hollow sacs in the mesoderm, one pair for each body segment. These coelomic sacs later break down and do not form the adult body cavity.

The young grasshopper, known as a **nymph,** hatches from the egg in a form which resembles the adult except in the relatively large size of the head, as compared with other parts, and in the lack of wings and reproductive organs. It feeds upon vegetation and grows rapidly; but since the chitinous exoskeleton cannot stretch very much, the animal must molt at intervals. **Molting** is a complex process which involves several steps.

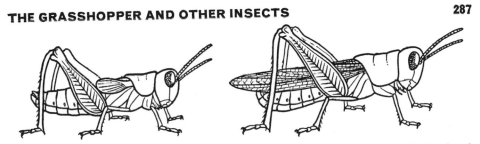

In the adult the head is smaller in proportion to the rest of the body, and the wings are fully developed. (Combined from several sources)

The outside layer is separated from that beneath by the secretion of a fluid from certain skin glands. This fluid dissolves part of the exoskeleton, which is absorbed by the epidermis and is presumably used to make the new cuticle. There is usually a weak spot in the skeleton down the middle of the back which breaks and provides an opening for the nymph to escape. The cuticle is finally ruptured, not by the increased size of the animal due to growth, which merely stretches the membranes between the segments, but by muscular contractions which build up great pressures on the skin. To aid in increasing temporarily the size of the body, the insect swallows air and closes the spiracles. The newly emerged insect is soft and white; and since it is in a precarious condition, it usually retires to a safe place until the soft cuticle hardens and darkens. With each successive molt, of which there are five in most grasshoppers, differentiation continues, so that the final molt results in the adult grasshopper.

VARIATIONS IN INSECT STRUCTURE

TO HAVE observed insects at all is to have noticed that they vary tremendously—from flattened, crawling cockroaches, which feed on scraps of food, to flying butterflies, which suck nectar from flowers, and swimming beetles, which chase animal prey. These differences are mostly in external structures; internally, insects are more alike.

Variations in the **digestive tract** are related mostly to what the animals eat. In the

"Garden fleas" are primitive **wingless insects** (order Collembola) which never have had any winged ancestors.

cockroach, which feeds on solid food, the gizzard is well developed, and its lining is armed with hard plates and spines. Insects which suck juices have no gizzard. In the honeybee the nectar is sucked up into a honey stomach, which corresponds to the crop of the grasshopper. The region between the honey stomach and the stomach (corresponding to the gizzard of the grasshopper) is a valve which prevents the food in the honey stomach, designed for storage in the hive, from going into the stomach.

Most insects obtain air for **respiration** through a system of air tubes; but some, like most of the tiny collembolans, have no air tubes and breathe through the body surface. Many insect larvas, which live as parasites in the fluids and tissues of their hosts, are equipped with well-developed systems of air tubes but must obtain oxygen by diffusion through the thin body wall. In many aquatic nymphs and larvas the spiracles do not function, and oxygen diffuses into the body through gill-like expansions of the body wall which contain air tubes (tracheal gills). Some aquatic larvas have thin-walled expansions of the body wall or extensions of the hind-gut which do not contain air tubes. In the absence of certain information we assume that oxygen diffuses into blood contained in these structures (blood gills) and so finally reaches the tissues.

There are also differences in the ganglia of the **nervous system.** In practically all insects there are two nervous masses in the head, but the number in the thorax and abdomen are more variable and depend upon the degree of fusion of the ganglia in these regions. The most extreme case of fusion is that of certain fly larvas, which have the entire ventral cord, including the first ventral ganglion, consolidated into a single mass.

The **excretory tubules** vary in number from two to over a hundred, but they all function in much the same way.

The essential parts of the **reproductive organs** of insects are as described for the grasshopper. Hermaphroditism is known to occur in one species (a cottony-cushion scale insect) in which the females are able to fertilize their own eggs. Unfertilized eggs give rise to males, which are rare. The males have been seen to mate with the females, but whether they can fertilize the eggs is not known. In some species no males have ever been found, and the eggs laid by the female develop parthenogenetically (without fertilization). In the vast majority of insects, sperms are stored in a receptacle of the female at the time of mating, and the eggs are usually fertilized, as they issue from the oviduct, at the time of laying.

No matter how conservative they may be in their internal anatomy, the insects, as a group, show the most radical modifications and the greatest amount of external variation known for any class of animals. There are, of course, the easily observed differences in shape of body, color, and size. But among the variations which are most important in adapting the animals to their different ways of life are those of the sense organs and appendages. From an ancestral arthropod with numerous segments each bearing a pair of appendages which were all alike and primarily locomotory in function, has been derived a vast array of insects which show a reduction and fixation of the number of segments, a loss of appendages on the abdomen, and a specialization of the appendages of the anterior part of the body into a series of structures which are, at least on the head, all different from each other. Moreover, in the different insect groups corresponding appendages have been modified in a great number of ways to fit the various animals to their particular niches.

Insects generally agree in having as sense organs a pair of compound **eyes,** three simple eyes, and a pair of antennas. In addition, the mouth parts may bear jointed sensory projections, the palps, and the body is clothed with a variety of sensory hairs, scales, pits, etc. There may also be special organs of smell or hearing.

The simple eyes probably do not form images but act to increase the sensitivity of the brain to light stimuli from the compound eyes. For, if the three simple eyes are painted over with an opaque substance, the insect does not react to light as rapidly as if the simple eyes were not covered. If the large eyes are covered, the insect does not respond to light. In insects which learn readily, color vision can be demonstrated. In one type of experiment a table is put near a beehive, and on the table are placed cards of different colors. On each card is set a glass vessel filled with water, and sugar is added to the water in one vessel—say the one on the blue card. In its excursions a bee finds the sugar water and, while busily feeding, is marked with paint, so that it can be recognized. After the bee has made several trips between the table and the hive, the sugar water is switched to the yellow card. The bee then returns to the blue card as before, even if the card is moved to another position on the table, showing that the bee is reacting to color and not to position or odor. Similarly, a bee can be trained to respond to yellow or to ultra-violet. Bees trained to red or black cannot discriminate between these two colors, or between them and dark gray. However, some insects do respond to red. If bees are trained to visit blue, and then blue is replaced by gray on which is set a yellow card, the bees now respond to the gray as if it were blue, apparently because, as in the case of man, blue and yellow are complementary colors and gray, set next to yellow, appears blue.

The **antennas** may be very long, as in crickets, cockroaches, and katydids. They are the chief organs of **touch,** for when the antennas are removed cockroaches can no longer be trained to turn right or left (see p. 296). The touch receptors of the antenna are the fine hairs with which it is clothed. The hairs are stiff and are joined by a very delicate cuticle at their bases to the rest of the antenna. The antennas also bear organs for the sense of smell.

Experiments which demonstrate the sense of smell are similar to those on color vision. First, it is necessary to determine if the insects react to odors. Sugar water is placed in small boxes, and, after bees have found them and are making trips to and from the hive, the box is substituted by one just like it, also containing sugar water, but sprinkled inside with flower extract. After the bees have made sufficient trips to get used to the scent, several new unscented boxes are placed beside a new scented one. When the bees return for more sugar, they buzz about the openings of the boxes but finally go inside the scented one. Further, when they are trained to go to one odor—say rose—they will not go to another, such as lavender. That the sense organs are on the antennas is shown by removing parts or all of the antennas from bees trained to certain scented boxes. When the last eight segments are removed from each antenna, the bees cannot distinguish odors. That this result is not due to the shock of the operation is proved by a control experiment in which some bees are first trained to visit blue boxes for sugar water. Then their antennas are removed, and it is found that they still return to the correct boxes. It may be noted, however, that with their antennas removed the bees have difficulty in entering the boxes probably because of the partial loss of their tactile sense.

In some insects, as in dragonflies, to which sight is more important, the eyes occupy nearly the whole head, while the antennas are relatively minute. Scattered over the bodies of most insects are numerous **tactile hairs.** Other sense organs may occupy rather unusual places. For example, the **sound-perceiving organs** of the grasshopper are on the sides of the abdomen just above the base of the third legs, while those of the katydid are near the upper end of the tibia of the first pair of legs. **Taste organs,** which enable the insect to distinguish sweet, salty, sour, and bitter, occur not only in the mouth but also on the antennas, palps, and feet. But just which of the several kinds of sense organs are the taste organs is not known.

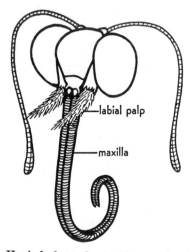
—labial palp

—maxilla

Head of a butterfly, showing **sucking mouth parts.**

The grasshopper, representing a fairly generalized group of insects, has *biting* **mouth parts.** These are the most primitive kind, and they are present also in beetles and in many other orders of insects. Two other main types of mouth parts are common: *sucking* mouth parts, as in butterflies; and *piercing and sucking,* as in the cicadas. In most butterflies the jaws are rudimentary, and the two maxillas are greatly elon-

gated, each forming a half-tube, so that
when they are held together they form
the long sucking proboscis of the adult,
through which liquids are pumped up
by the mouth pump in the forepart of
the head. The proboscis is extended
only when the insect is feeding; when
not in use, it is coiled under the head.
The piercing beak of the cicada consists of the mandibles, maxillas, and
labium, all thought to be homologous to
the mouth parts of the grasshopper.
The labium is a long tube but is not inserted into the food. It serves only as a
sheath for the other mouth parts, being
grooved on its dorsal surface to form a

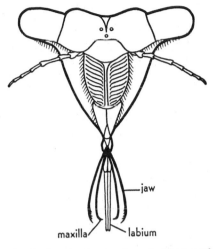

Head of a cicada, showing **piercing and sucking mouth parts.**

channel in which lie the mandibles and maxillas, which do the piercing.
The mandibles are long, fine stylets with minute teeth at the end. The maxillas are similar but hooked at the tips; each is crescent-shaped in cross-
section, and the two are fastened together by interlocking grooves and
ridges to form a channel through which the food is drawn up by the sucking action of a muscular pharynx. Biting flies, mosquitoes, fleas, and bedbugs have the mouth parts adapted in various ways for sucking blood.
Other insects have still other modifications of the mouth parts, such as
the sponging tongue of the housefly.

The thoracic **legs** of insects are modified in a variety of ways, but all are
composed of the same basic parts (figured and named in the diagram of
the grasshopper leg). Land forms have walking legs, with terminal pads
and claws, as in the grasshopper, for clinging to vegetation or other objects. Houseflies have sticky pads at the tips which enable them to walk
up smooth vertical surfaces, such as glass. Water beetles have flattened
legs, fringed with bristles, for swimming. But the legs may serve other
functions besides locomotion. The walking legs of the **honeybee** are modified for collecting food. Each is highly specialized and quite different from
the others, so that, together, they constitute a complete set of tools for
collecting and manipulating the pollen upon which the bee feeds.

The **first leg** has many branched feathery hairs for collecting pollen. Along one edge
of the inner surface of the tibia is a fringe of short, stiff hairs which form an *eye brush*
used to clean the compound eyes. The large first joint of the tarsus is covered with long,

unbranched hairs, forming a *pollen brush* for collecting the pollen grains that become caught among the hairs of the fore part of the body when the bee visits flowers. This first joint also has a semicircular notch, lined with a comblike row of bristles and known as the *antenna comb*. The antenna is cleaned of pollen by drawing it through the notch. As it is pulled through, it is held in place by a spur on the end of the tibia, which fits against the tarsal notch. Comb and spur together are called the *antenna-cleaner*. The **middle leg** is the least specialized of the three. The large first tarsal joint is wide and flat and covered with stiff hairs which form a brush for removing pollen from the first legs and

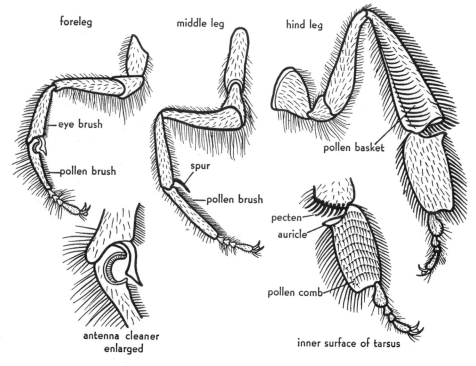

Legs of the honeybee.

the thorax. On the lower end of the tibia is a spine, the *spur*, for removing the plates of wax from the wax glands on the ventral side of the abdomen. The **hind leg** is the most specialized, being fitted to carry the load of pollen. Rows of *pollen combs* on the inner surface of the very large and flattened first tarsal joint scrape the pollen from the second legs and posterior part of the abdomen. A series of stout spines, the *pecten*, on the lower end of the tibia removes the pollen from the combs of the opposite leg, and it falls on the *auricle*, a flattened plate on the upper end of the first tarsal joint. The leg is then flexed slightly, so that the auricle is pressed against the end surface of the tibia, compressing the pollen and pushing it onto the outer surface of the tibia and into the *pollen basket*. The pollen basket is formed by a concavity in the tibia, which has, along both edges, long hairs that curve outward. Pollen clings together and to the basket hairs because it is

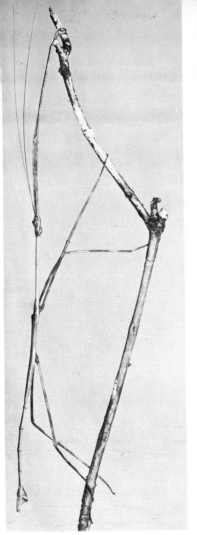

Insects vary in shape and color more than any other class of animals. Some are compressed from side to side, like the katydid (*above*). Others are compressed dorsoventrally, like the beetle (*bottom, left*). The walking-stick (*left*) is greatly elongated, while the Colorado potato beetle (*right*) has a compact body. Many insects resemble their surroundings. Some look strikingly like the green leaves, dried leaves, twigs, or bark on which they live. How much of this is accidental and how much adaptive, has long been a hotly debated subject. In any case, the katydid, which lives in dense foliage in Brazil, is green and leaflike, while the beetle (*bottom, left*) which lives under bark in the Malaysian Region is brown in color. The Kallima butterfly of India (*bottom, right*), when at rest looks very much like a dead leaf. However, many insects, like the potato beetle, have striking surface patterns which have no obvious resemblance to their habitat. (Walking-stick, living, Panama; potato beetle, courtesy U.S. Bur. Entomol.; other photos on this page, courtesy *Nature Magazine*)

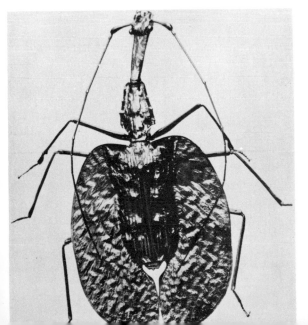

The legs of insects are six in number and are borne on the thorax, the middle region of the body. All are composed of the same five joints and usually end in two curved claws. Some have sticky pads which enable the insect to walk on smooth or on vertical surfaces. The least specialized legs are simple walking legs, all three pairs much alike, as in this darkling beetle, *Eleodes*. (Photo of living animal. Calif.)

The food-collecting legs of the honeybee are also used for walking. Pollen clings to hairs on legs and body and is transferred to "baskets" on the hind legs, here shown well loaded. (Photo by Cornelia Clarke)

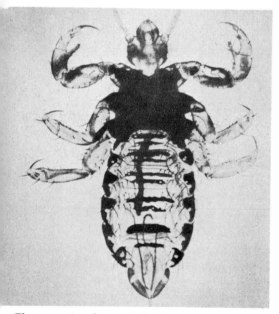

The grasping legs of the body louse, *Pediculus humanus*, enable the animal to cling to hairs on the body of man, or more often to his clothing. With its piercing and sucking mouth parts the louse sucks the blood of its host. The danger lies not in the bite itself but in the fact that body lice transmit the deadly typhus and other fevers. The true lice, all of which suck mammalian blood, comprise the order ANOPLURA. They lack wings and metamorphosis—not primitively but by secondary loss. (Photo, courtesy Army Medical Museum)

The swimming legs of the water-scavenger beetle are flattened and fringed with bristles. Though they live an aquatic life, these beetles have no gills but breathe by coming to the surface at intervals to obtain air, which they carry down with them as a film on the undersurface of the body. (Photo by Cornelia Clarke)

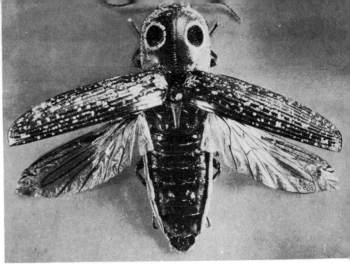

The wings of insects are typically two pairs of membranous appendages of the thorax. They are stiffened by thickenings, called "veins." In relatively unspecialized insects, like the dragonfly, veins are very fine and numerous and form a network. (Photo of living animal. Panama)

The thickened forewings of beetles, usually called "wing covers," form a protective armor over the body and the membranous hind wings. The elaterid beetle shown here has wing covers and hind wings spread out as if in flight. (Photo by P. S. Tice)

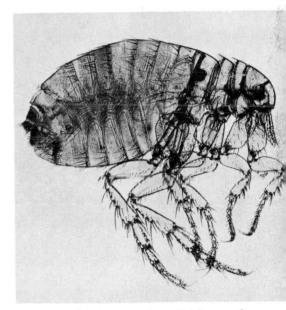

Absence of wings may be primitive, or the result of secondary loss as in the dog flea. The fleas (order SIPHONAPTERA) are small, wingless, insects with piercing and sucking mouth parts and complete metamorphosis. (Photo, courtesy U.S. Army Medical Museum)

One pair of wings is found in the crane fly and other members of the order Diptera. Behind the wings is a pair of stalked knobs, the halteres, which are thought to assist in some way in maintaining balance during flight. In this relatively specialized insect the wing veins are large, and few in number. (Photo of living animal. Chicago)

The mouth parts of insects consist of the same basic parts but are modified for various methods of feeding. In the stag beetle (*right*) the jaws are large and formidable, especially in the male, and are used not for chewing but for defense or holding the female during mating. These beetles are said to feed on honeydew and exudations from plants, which they gather up with their flexible, hairy labium. (Photo by P. S. Tice)

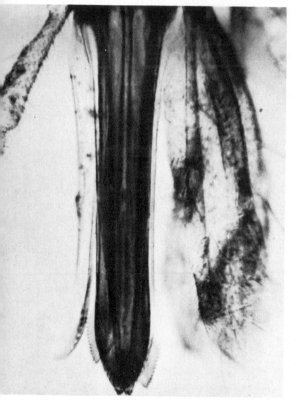

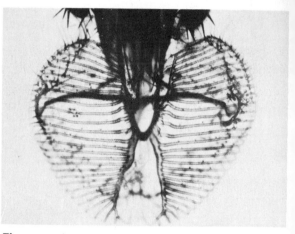

The sponging tongue of the housefly is the expanded, two-lobed tip of the labium. The food is sucked up through the opening at the end of the cleft between the two lobes. (Photo, courtesy *Nature Magazine*)

Piercing and sucking mouth parts occur in insects that feed on blood or on plant juices. The elongated parts are modified as a tube through which food is drawn up, or are stiff and sharp and used to make the wound. These last are usually fine stylets or flat blades, which may have saw edges, as in the beak, shown here, of a "punkie" or "sand fly." As in moths, butterflies, and houseflies, the sucking action is provided by a muscular "pump" in the head. (Photo, courtesy *Nature Magazine*)

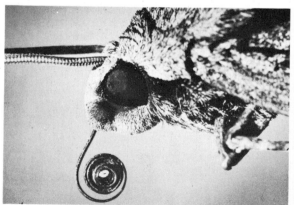

The sucking tube of moths and butterflies consists of the two elongated maxillas, each forming half the tube. (Photo, courtesy General Biological Supply House)

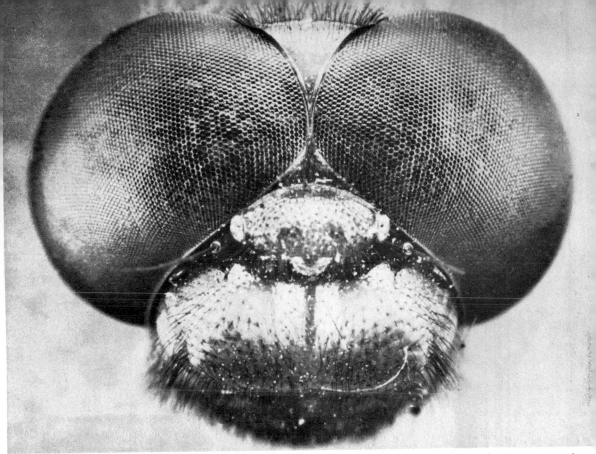

The compound eyes of the dragonfly occupy most of the head and are composed of nearly 30,000 separate units. The antennas, which look like mere bristles, one beneath each eye, apparently play a minor role. In the housefly the eyes have only 4,000 units and in some ants there are only 50, while some nocturnal insects have no compound eyes at all and rely chiefly on their well-developed antennas. (Photo by P. S. Tice)

The antennas of the Cecropia moth are large and branched and are important sense organs for this nocturnal animal. (Photo by P. S. Tice)

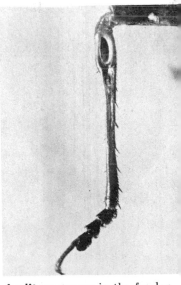

Auditory organ in the foreleg of a katydid. (Photo by C. Clarke)

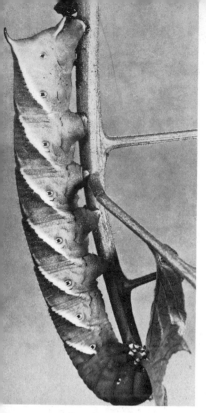

Insects breathe by means of a system of air tubes which communicate with the outside through openings, the spiracles, located along the sides of the body, not only in adults but also in many immature forms such as the larva (*left*) and the pupa (*right*) of a sphinx moth. Air tubes branch repeatedly, ramifying to all parts of the body and carrying oxygen directly to almost every cell. (Photos of larva and pupa are of living forms. Pupa by Cornelia Clarke)

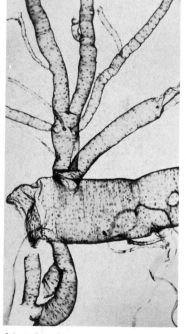

Spiracles usually have hairs to exclude dust and two lips which can be brought together to close the opening. (Photo courtesy General Biological Supply House)

Air tubes highly magnified, show thickenings of cuticular lining which keep walls of tubes from collapsing. (Photo courtesy Gen. Biol. Supply House)

Aquatic larva, *Corydalis*, breathes by diffusion of gases through air tubes contained within thin-walled tracheal gills. (Photo by P. S. Tice)

Spring-tails (order COLLEM-BOLA) are so called from their possession of a springing mechanism. Formed from a pair of abdominal appendages the spring (seen in released position *below*) is held in place under the abdomen by a catch formed from another pair of abdominal appendages (*as above*). Collembolans are very minute insects, primitively wingless, and lacking in metamorphosis. They have simple chewing mouth parts and a group of simple eyes on each side of the head. They live mostly on the ground under stones, rotting wood, and leaves. Some are found on plants. (Photos by P. S. Tice)

The most primitive insects are the members of the order THYSANURA ("tassel-tail"). Most thysanurans have at the tail end a "tassel" of two or three many-jointed filaments, and from these has come the common name "bristle-tails." All are wingless, and it is thought that they never evolved any wings. Another primitive character is the absence of metamorphosis. All have simple chewing mouth parts. Most bristle-tails are found on the ground, under stones, rotting wood, or leaves. The firebrat, shown here, frequents warm places such as fireplaces and steam pipes. Its relative, the silverfish, looks like the firebrat except for the dark markings. Silverfish live in cool damp places such as basements, but can often be seen in washbasins and tubs. They eat the starch in clothes and the paste in book bindings. (Actual size of firebrat ⅜ inch. Photo of living animal by P. S. Tice)

The cockroach is chiefly an inhabitant of warm parts of the world. In temperate North America there are a few native species, and these usually live in fields and forests. The large American cockroach (*right*) is native to tropical America but is common in our southern states, and in hot weather it flies indoors and out. (Photo of living animal. Panama)

The Oriental cockroach, originally from Asia, is about an inch long, blackish-brown in color, and easy to distinguish from the smaller, light-colored "Croton-bug," which was introduced from Europe. The oriental cockroach, especially the female, has reduced wings and cannot fly. On the left is an adult female with an egg case protruding from the body. On the right is an egg case cut open to show the eggs. Cockroaches have chewing mouth parts and are destructive to all sorts of organic material, particularly food. (Photos by Lee Passmore)

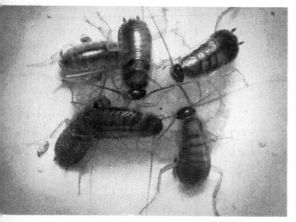

Newly hatched nymphs resemble adults except in size and wing development. Like all orthopterans, cockroaches have gradual metamorphosis. The nymph lives much like the adult and makes a gradual transition to adult form. (Photo by Lee Passmore)

Newly molted nymph lies helpless beside the empty exoskeleton which it has just cast off. The new epidermis is soft and elastic. Soon it will become hardened and later will be cast off by the constantly growing nymph. (Photo of living animal. Panama)

Grasshopper   **Katydid**

Praying Mantis

The order ORTHOPTERA includes cockroaches and the other insects on this page. Members have typically two pairs of wings, (the first pair thickened and acting as covers for the second pair, which are folded fan-like when not in flight), chewing mouth parts and gradual metamorphosis, the eggs hatching into nymphs. (Photos by L. W. Brownell)

Named for the way in which it holds the large forelegs while waiting quietly for its insect prey to approach. (Photo of living animal. Panama)

Crickets. *Above:* Field cricket, a solitary insect that lives in fields or about houses, where it betrays its presence by chirping. (Photo by L. W. Brownell.) *Below:* Tree crickets, whose high-pitched trill is one of the most conspicuous insect songs of summer nights. (Photo by Cornelia Clarke)

Walking-stick insect looks like a twig or a vine. Presumably, this aids in escaping detection by enemies. (Photo of living animal. Panama)

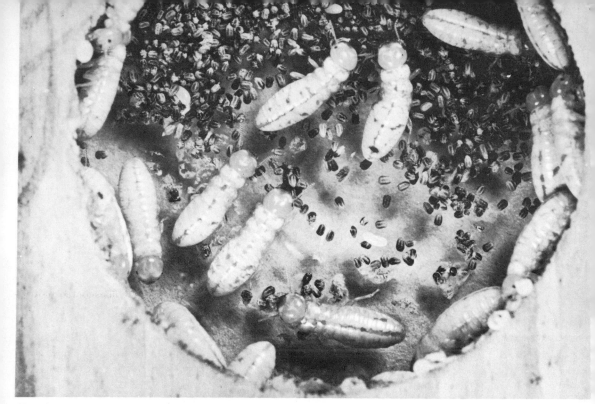

Termites (order ISOPTERA) are social insects. They have biting mouth parts and gradual metamorphosis. Here is a part of a colony of *Kalotermes* maintained, for observation, in a hole in a piece of plywood between two sheets of glass. Protected against drying and mechanical injury, as they would be in their nest in nature, the nymphs, which in this genus of termites do the work of the colony, are busily eating wood, which they digest with the aid of intestinal protozoans they harbor. Their chamber contains fecal pellets, among which can be seen several eggs. (At left is a millimeter scale. Colony of *Kalotermes* from Florida lent by V. Dropkin)

Termite castes include two or three reproductive castes, a large-jawed soldier caste with protective functions, certain special castes, and almost always a worker caste that excavates tunnels, chews wood, feeds and cleans the other castes, cleans and repairs the nest, and attends the young. Each caste consists of male and female individuals. *Left:* Soldier and winged reproductive (same scale as above). The latter leave the colony, mate, and start new colonies. (Photo of living *Kalotermes*. Florida.) *Right:* the largest termite queen known, *Macrotermes natalensis;* actual size. The queen lays one egg per second during a lifetime of about 30 years. Only the abdomen is greatly enlarged. (Specimen lent by A. E. Emerson)

Nasute soldiers, which substitute for the usual large-jawed type of soldier, are found in certain species. The jaws are small, but the head is a noselike snout from the tip of which a sticky fluid can be squirted to gum up the enemy. A few seconds before this picture was taken, the nest had been poked, making an opening about the size of a dime, and promptly bringing forth this array of nasutes ready to do battle. A few nonsnouted individuals can be seen; these are workers, who repaired the damage. (Photo of living *Nasutitermes ephratae.* Panama) .

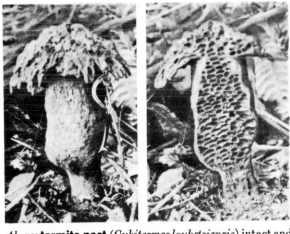

Above: **termite nest** (*Cubitermes loubetsiensis*) intact and cut open to expose elaborate system of chambers. The caplike roof is neatly adapted to shed rain in the tropical rain forest. (Photo by H. O. Lang)

Right: **Nest,** or termitarium, of an Australian species, *Amitermes meridionalis* showing the large size reached by some nests. (Photo by G. F. Hill)

The dobson fly, *Corydalis* (along with other insects on this page), belongs to the order Neuroptera, characterized by 4 wings with many veins and crossveins, biting mouth parts, and complete metamorphosis. (Photo, courtesy Amer. Mus. Nat. Hist.)

The dobson fly larva, known as the "hellgrammite" to fishermen who use it as bait, lives in rapid streams, under stones, and catches larvas and naiads of other insects. It breathes by tracheal gills. After three years it goes on land to pupate. (Photo by P. S. Tice)

Lacewing fly. Larva (*lower left*) known as "aphis lion," lives on vegetation and sucks the juices of plant lice and other small insects. Pupa (*upper left*) is emerging from its cocoon. (Photo by C. Clarke)

Ant lion. *Upper left:* Larva which digs a pit and lies buried with only the jaws protruding, ready to seize any ant or small insect that falls in. The pupa (*lower right*) is shown in opened pupal case. (Living, 3×nat. size)

Mayflies are delicate animals which emerge from streams, ponds, and lakes in thousands during the summer. They belong to the order EPHEMERIDA; and, as this name suggests, they live but a single day, during which time they molt twice, mate, and the female lays eggs in the water. Since the adults lack fully developed mouth parts, they never feed. The metamorphosis is incomplete, the egg hatching directly into an aquatic naiad, which feeds on plant material. Some naiads require one to three years to get ready for their brief adult life. (Photo, courtesy *Nature Magazine*)

1. The dragonfly naiad lives at the bottom of ponds, lakes, and streams, where it preys on small animals. Notice the developing wings and the large compound eyes. (Photo by L. W. Brownell)

2. When the naiad matures, it crawls up a plant that extends above the water. The skin splits down the head and thorax, and the adult emerges. (This photo and the two below, courtesy *Nature Magazine*)

3. The newly emerged adult has soft, limp wings which gradually expand as blood is pumped into them. When hardened (*right*), they make the dragonfly one of the fastest of all flying insects.

4. The adult dragonfly catches small flying insects. It belongs to the order ODONATA, characterized by four membranous, finely netted wings, chewing mouth parts and incomplete metamorphosis.

A bug to the layman is almost any kind of insect; but to the entomologist, only the members of the order HEMIPTERA are true bugs. Hemiptera means "half wings" and refers to the fact that the basal half of the front wings is thickened, the posterior half membranous. The second pair of wings is membranous and folded beneath the first. The mouth parts are modified for piercing and sucking, and the metamorphosis is gradual. The southern green-plant bug (*left*), sucks the sap of garden vegetables. (Photo, courtesy U.S. Bur. Entomol.) *Right:* Wingless stink-bug nymphs and eggs. Stink bugs emit a fetid odor which remains on anything they visit, and this explains the bad taste of occasional berries in a box of otherwise good ones. (Photo by Cornelia Clarke)

Giant water bug, also called the "electric light bug," is one of the largest insects, with a length of 4 inches in some species. It lives in water, swimming rapidly with its flattened hind legs, and feeding on fish, snails, and insects. It catches prey with its modified front legs and kills them with its piercing and sucking beak. The mating occurs under water, the larger female cementing the eggs all over the back of the male (*left*). The nymphs hatch (*right*), and are said to indulge in cannibalism, the older eating the younger. (Photos by Lee Passmore)

Tree-hoppers feed on plant juices. The thorax is prolonged backward over the abdomen and often has a grotesque shape as in the thorn tree-hopper (*above*), which resembles the thorns of the shrub on which it is resting. (Photo, courtesy *Nature Magazine*)

The order HOMOPTERA includes the insects on this page and many others such as cicadas and scale insects. Members typically have four wings, piercing and sucking mouth parts, and gradual metamorphosis. Above are shown some young nymphs and mature forms of a tropical leaf-hopper being "herded" by some larger ants, which lick the sweet exudations of their homopteran "cows." Ants attend in the same way many tree-hoppers and aphids. (Photo of living animals. Panama)

Mealy bugs, so called because they are covered with a powdery excretion, are common pests on trees and in greenhouses. The females are usually wingless. *Pseudococcus citri* is a pest of orange trees in the South. (Photo by Cornelia Clarke)

Plant lice, or aphids, are tiny, usually green, pear-shaped insects that infest almost every kind of plant. They insert their beaks into stems and leaves and suck the juices. *Above:* Wingless female; *below:* Winged female of the pea aphid. (Photo, U.S. Bur. of Ent.)

2. The eggs, imbedded in a twig, hatch in six weeks into nymphs, which fall to the ground and bury themselves. They suck the juices of the roots of forest and fruit trees. (Photo by Cornelia Clarke)

1. The cicada is one of the best-known homopterans, partly because of its large size and the loud shrill song of the male, but also because certain species have interesting breeding habits which have long attracted attention. The periodical cicada, *Tibicina septendecim*, is popularly known as the "seventeen-year locust"; but this is a misnomer, as locusts are properly migratory grasshoppers. Shown here is a female laying eggs in slits which she has made in a twig. (Photo by J. C. Tobias)

4. Adult emerges through a slit down the back of the nymphal covering. (Photo by P. Knight, courtesy *Nature Magazine*)

3. The nymphs emerge from the ground after seventeen years and crawl up on tree trunks. Since a whole brood emerges at one time, the empty nymphal coverings may later be seen by the thousands, clinging to the bark of trees. (Photo by C. Clarke)

5. Newly emerged adult with wings partly expanded. Adults mate and soon die. (Photo by Knight, courtesy *Nature Magazine*)

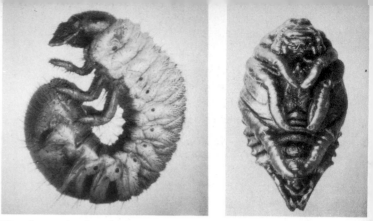

The Japanese beetle is typical of the order COLEOPTERA ("sheath wings"), members of which have the forewings thickened as wing covers that protect the membranous hind wings folded beneath them. They have chewing mouth parts and complete metamorphosis. *Middle:* The larva of the Japanese beetle, called a grub, lives in the ground, feeding on the roots of grasses. *Right:* Pupa. Japanese beetles do untold damage to fruit trees. (Photos, courtesy, U.S. Bureau of Entomology)

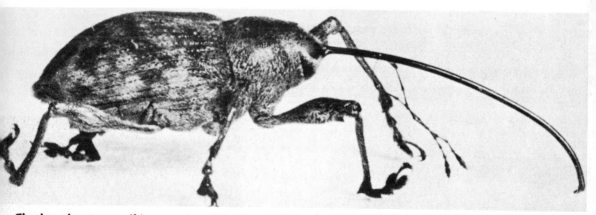

The hazel-nut weevil is a member of the snout beetles, which have the head prolonged into a snout, provided at its tip with a pair of jaws with which it bores into plants. (Photo, courtesy *Nature Magazine*)

Hercules beetles show the striking difference that may occur between the two sexes of the same species of insect. The male has a large horn on the head and another longer one on the thorax. The female is smaller and has no horns. The name is derived from the large size of the beetles, which are shown here about life size. Some of them are even larger. (Photo, courtesy *Nature Magazine*)

The rhinoceros beetle, a voracious herbivore, excavates burrows in palms. The three horns are on the thorax, which overhangs the small head. In spite of their heavy chitinous exoskeleton, beetles are good flyers. On the whole, the group is very successful; almost 20,000 species are known in the United States alone. (Photo by P. S. Tice)

Whirligig beetles swim on the surface of quiet ponds, feeding on small insects. The eyes are divided, the upper half for seeing in air and the lower half for seeing in water. (Photo by Cornelia Clarke)

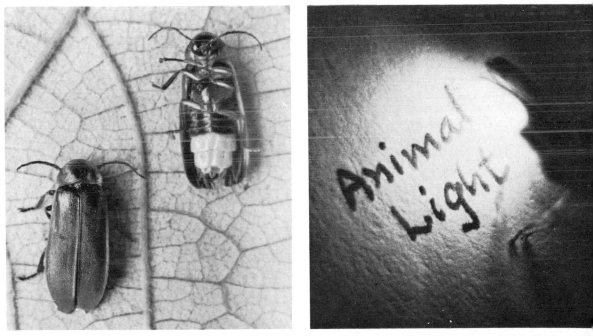

Fireflies are not flies but beetles; the wingless females and the larvas are called "glowworms." On the lower side of the abdomen are the light organs, which flash intermittently. That flashes enable the males to find their mates is shown by the fact that a female in a perforated opaque box does not attract males while one in a corked glass bottle does. (Photo by C. Clarke) *Right:* Photo made with the light of a firefly whose abdomen can be discerned toward the right. About 3 × natural size. (For further details on animal light see text)

Butterflies and moths are members of the order LEPIDOPTERA ("scaly wings"), characterized by four large wings covered with overlapping scales. They have sucking mouth parts and complete metamorphosis. The pictures on this page show the development of a typical member, the Monarch butterfly. (Photos by Cornelia Clarke) *Left:* The egg (shown greatly enlarged) is laid on milkweed plants. *Right:* The larva is a caterpillar with chewing jaws; it creeps about on plants, feeding almost continuously.

The larva changes into a pupa, hanging from a leaf. From *left to right* the larval covering is shed, revealing the pupa already formed beneath.

Above: **The pupa** (called a "chrysalis") appears to be "resting"; but inside, the whole structure of the insect is being reorganized. *Right:* **The newly emerged adult** clings to the old chrysalis covering while the body and wings expand and dry.

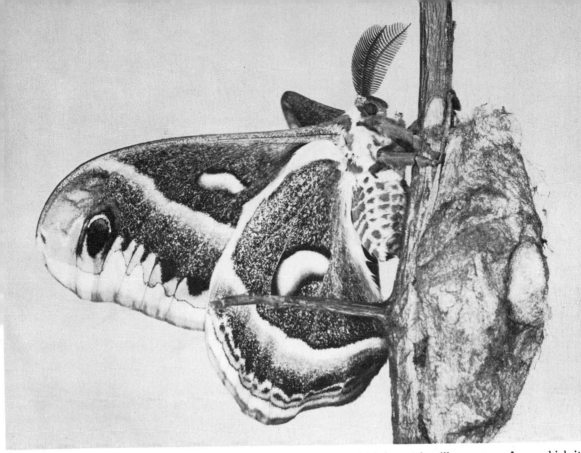

The Cecropia moth, with wings fully expanded, rests on a twig which bears the silken cocoon from which it has emerged. Moths are roughly distinguished from butterflies by their nocturnal habits, threadlike or feathery antennas, and their manner of holding the wings when at rest; the Cecropia is exceptional in holding the wings more like a butterfly; the tiger moth (*below*) has a more typical position. Butterflies are active during the day, have knobbed antennas, and hold the wings vertically when at rest. (Photo of living animal. Chicago)

Cocoon of a Cecropia moth, cut open. The outlines of the antennas and wings can be seen through the pupal covering. (Photo by L. Keinigsberg)

The tiger moth rests with the wings held rooflike over the body, the characteristic position of most moths. (Photo by L. W. Brownell)

Flies comprise the order DIPTERA ("two wings"). Included also are flies that are commonly called mosquitoes, gnats, and midges. Except for a few wingless forms, dipterans have a single pair of wings. All have either sucking, or piercing and sucking, mouth parts. Metamorphosis is complete. Dipterans are among the most obnoxious of insect pests, and many carry serious diseases. *Right:* A large tropical tabanid or horse-fly. (Photo of living animal. Panama)

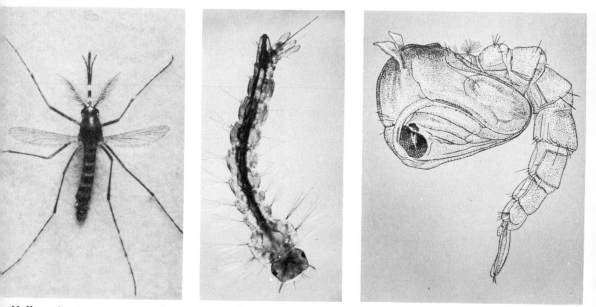

Yellow-fever mosquito (*Aëdes aegypti*). The male (*left*), feeds on plant juices. The female sucks mammalian blood and lays eggs in the water. The aquatic larva (*center*), called a "wriggler," breathes air through a spiracle at the posterior end. The pupa (*right*), shown by a drawing, also is aquatic and breathes through two tubes that look like horns. (Photos, courtesy U.S. Bureau of Entomology)

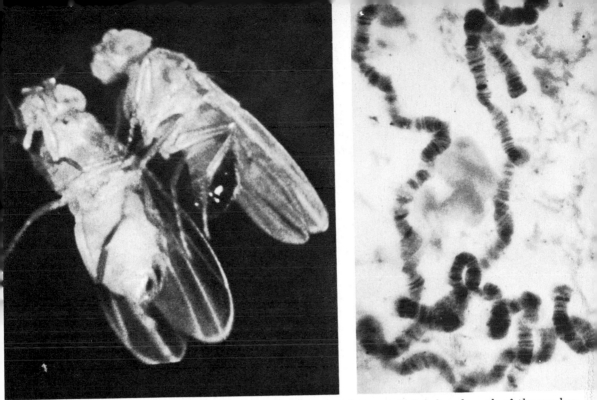

The fruit fly, *Drosophila melanogaster,* has furnished the material upon which is based much of the modern theory of genetics, the science of heredity. Fruit flies are found living about fermenting fruit. They can be reared in the laboratory in tremendous numbers with little effort and expense. The life-cycle, from egg to egg, may be completed in ten days. A single pair of flies, like the pair above (female, *left;* male, *right*) seen in ventral view, will produce hundreds of offspring. In this fly there are only four pairs of chromosomes. The chromosomes

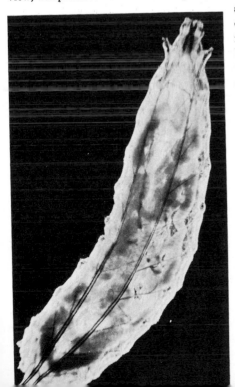

are bodies easily seen in the cell during cell division and gamete formation and which are the bearers of the hereditary units, the genes. The small number of chromosomes in Drosophila simplifies the interpretation of breeding results and the correlation of these results with evidence obtained by looking at the chromosomes themselves. However, certain details of chromosome behavior and structure were obscure until the giant chromosomes (*above, right*) of the salivary glands of the larva were studied. In these chromosomes there can be seen a constant pattern of light and dark bands. There is evidence that particular bands correspond to the position of particular genes. *Left,* larva; *right,* pupa of the fruit fly. (Photos by P. S. Tice)

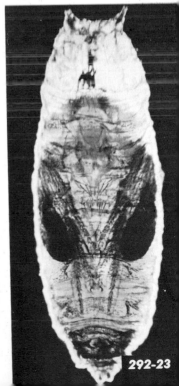

The bumblebee is one of the larger bees. It plays an important role in the fertilization of certain flowers, such as red clover, in which the nectar can be reached most easily by bees with long tongues. Bees, ants, wasps, and a great variety of other insects are all members of the order HYMENOPTERA, characterized by four wings with very few veins, mouth parts adapted for chewing and sucking, and complete metamorphosis. Bees differ from almost all other nest-building hymenoptera in provisioning their nest with pollen and honey instead of animal food. (Photo, courtesy U.S. Bureau of Entomology)

Honeybee worker with pollen grains clinging to the long, branched hairs on body and legs. As in ants, the worker is a sterile female; but, unlike ants, bee workers are winged. The elongated mouth parts are modified for sucking nectar, there are also chewing jaws for manipulating wax. (Photo by Cornelia Clarke)

A swarm of honeybees includes workers, drones (males), and queen. All individuals of a colony are offspring of the queen, who may lay over a million eggs during her lifetime. At any one time, a hive may contain from 50,000 to 80,000 bees, almost all workers. (Photo, courtesy U.S. Bureau of Entomology)

Ants are the most numerous of all terrestrial animals of comparable size. They are the most successful and widespread of insects in a time that is often called the "age of insects." Most are scavengers, taking to their nests any kind of organic matter. Some of the more primitive ants, like the large-jawed ponerine ant (*left*), are carnivorous. Actual size 1 inch long. (Photo of living animal. Panama)

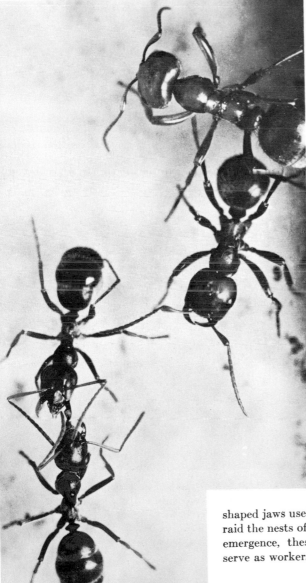

Ant larvas, pupas, and newly emerged adults are cared for by the worker caste of the highly specialized and closely integrated social colony in which all ants live. (Photo by Cornelia Clarke)

Slave-making ants (two at the right with sickle-shaped jaws useful only for fighting) cannot feed themselves. They raid the nests of another species of ant and carry off the pupas. On emergence, these adopted slaves (two at left with biting jaws) serve as workers in the colony. (Photo of living animals. Illinois)

Army ants occur in our South, but only in the tropics do they form colonies numbering millions of individuals or search for their prey in vast armies that overrun everything in their path. Army ants are nomadic, moving from place to place every few days, and each time building their temporary but remarkable nest out of the living bodies of the ants themselves. The nest, or bivouac, usually built under an overhanging log as in the photo at *upper left* is complete with runways leading to chambers where they care for and feed their young. *Right:* Army ants raid a wasp nest while the wasps wait helplessly near by. *Lower left:* Trail at night with workers transporting the larvae as the colony packs up and moves on. (Photos of living animals. Panama)

Leaf-cutter ants cut pieces of leaves from plants and carry them to their nest. There the pieces are chewed up and formed into moist balls on which are grown the fungus that is their food. Leaf-cutters occur in warmer parts of the U.S., but are best known in the American tropics. They can defoliate a 15-foot tree in one day. The underground nest is large and has trails leading out in all directions. Shown above is a portion of a trail with workers carrying pieces of leaves that average 5 times their own weight. Note "hitchhiker." Two large-headed soldiers stand guard. (Photo of living animals. Panama)

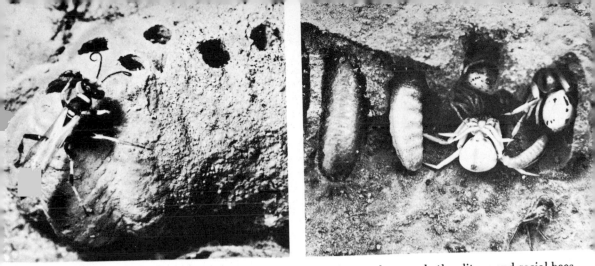

The mud-dauber wasp is solitary; but there are social wasps, just as there are both solitary and social bees. The mud-dauber builds a few cells of moist earth, fills them with live, paralyzed spiders, and lays an egg in each. *Right:* The nest cut open to expose four cells containing (*from left to right*), a pupa, a larva beginning to pupate, and two cells filled with spiders, one of which has a small larva clinging to its leg. (Photo by Lee Passmore)

Hornet's nest cut open to show the cells in which are reared the developing young. The whole structure is made of bits of weather-worn or decayed wood chewed up with saliva to form a kind of paper. Hornets have a social organization much like that of bees; the young are fed on insects brought to them by the winged workers. (Photo by Cornelia Clarke)

Oak apples are growths induced on oak trees by the larva of a hymenopteran known as a "gall wasp." The species of parasite can be determined by the type of growth induced. The larva feeds on the green tissues, and the plant responds by overgrowth. After pupation the insects emerge, mate, and lay eggs on new oak leaves. (Photo by L. W. Brownell)

A giant wasp is *Pepsis formosa*, the well-known "tarantula hawk" of our Southwest. The wings show several features characteristic of the wings of Hymenoptera. The venation is reduced; and the hind wings, which are smaller than the front pair, have on their anterior margin a row of hooks which fit into a groove on the posterior margin of the forewings. This holds the wings together in flying. Wasps have a long slender "waist" or stalk joining the abdomen to the thorax. (Photos on this page by Lee Passmore)

Many wasps provision their nests with spiders; and so it is not surprising that *Pepsis*, being a very large wasp, can capture tarantulas, paralyzing them with a poison injected by the sting on the end of the abdomen (*upper right*). Sometimes the wasp loses the struggle and is eaten by the tarantula; but, when Pepsis is successful, she places the tarantula in a burrow in the ground, lays her egg on its body, and closes the burrow. When the larva hatches, it feeds on the tarantula. Then it pupates; and, when the adult emerges, it digs its way out.

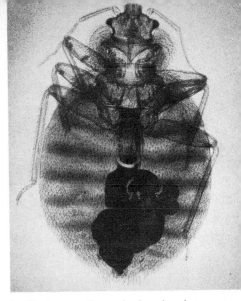

Insects spread human diseases such as malaria, yellow fever, elephantiasis, African sleeping-sickness, typhus, and many others; they are thus indirectly responsible for more deaths than any other agency, including war. Means of control are well known in most cases but are not applied on a wide enough scale because, among other things, men spend too much time fighting each other instead of their real enemies, the disease-carrying insects. Shown here is the malaria mosquito, *Anopheles*, sucking blood by inserting its piercing mouth parts through the skin. It can be recognized by the oblique position in which it holds the body while biting. (Photo from *Science Service*)

Bedbugs are flat, wingless hemipterans that live in human dwellings. They hide during the day in cracks in the furniture or floors and emerge at night to suck blood from their sleeping hosts. (Photo, courtesy U.S. Army Medical Museum)

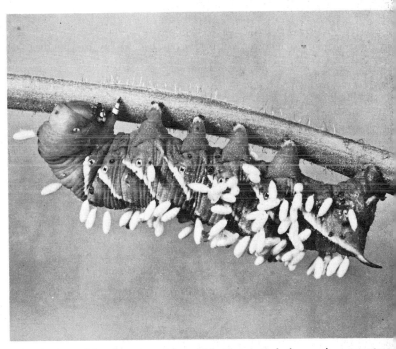

Botfly larvas attached to the lining of the stomach of a horse. The eggs are laid on the skin, licked off, and swallowed by the horse. In the stomach they hatch into larvas, which attach themselves and, when present in large numbers, cause indigestion. Finally, the larvas pass out with the feces, pupate, and emerge as adult flies. (Photo, courtesy U.S. Bureau of Entomology)

Insects parasitize insects, and so aid in the control of many insect pests. This sphinx-moth caterpillar is covered with cocoons of a braconid fly (hymenopteran), which have developed from eggs laid, by means of a long ovipositor, beneath the skin of the host. The larvas fed on the tissues of the caterpillar, emerged through the skin, spun their cocoons, and pupated. The exhausted caterpillar will soon die. (Photo of living animal)

Insects destroy man's plant food to the extent of hundreds of millions of dollars' worth every year. *Left:* A field of corn; *right:* The same field after an attack by migratory grasshoppers. (Photos, U.S. Bur. Entomol.)

Damage by termites (*Reticulotermes flavipes*) to book in library at Van Buren, Arkansas. (Photo, courtesy U.S. Bureau of Entomology)

Cotton boll weevil at work. One of the most destructive insects, this snout beetle attacks only cotton. (Photo, courtesy U.S. Bureau of Entomology)

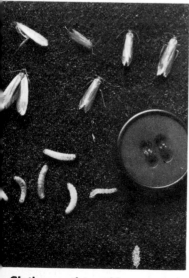

Clothes moths and larvas. Only the larvas eat woolen fabrics. (Photo, U.S. Bur. Entomology)

Grapefruit crop destroyed in Florida by larvas of Mediterranean fruit fly. (Photo, U.S. Bur. of Entomol.)

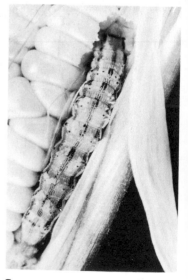

Corn-ear worm is a caterpillar that feeds on the tips of sweet corn. (Photo by P. S. Tice)

Insects are beneficial to man in many ways. The honeybee, *Apis mellifica* (shown here), and many other insects pollinate flowers while collecting nectar. Without this service our orchards would produce little fruit and many crops could not be grown. Unfortunately, the sprays used to kill pests on fruit trees also poison the honeybees and other nectar-gatherers. Honeybees are about fifty times as valuable for cross-pollination in orchards and fields as they are for the honey they produce. Honeybees are domesticated animals, living in colonies of about 60,000 in man-made hives. But they occasionally return to a wild state. (Photo of living animal)

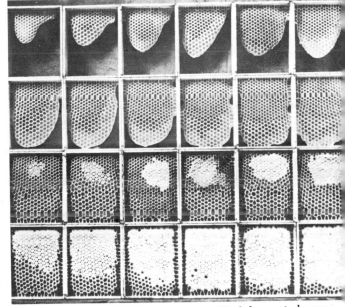

Honey, several hundred thousand tons of it, is harvested every year in the U. S. Honeybee workers lap up nectar of flowers and store it in a honey-stomach. When they return to the hive, it is disgorged into the wax cells. One pound of honey represents about 20,000 bee-trips from flowers to hive. (Photo, courtesy U.S. Bureau of Entomology)

Insects are food for man in many parts of the world. American Indians eat grasshoppers, crickets, and ants. African natives eat termites and fly maggots. What food is attractive is merely a matter of custom. Grasshoppers are grain-eaters and clean; their relatives, the lobsters, are scavengers of the sea bottom. *Left:* Toasted caterpillars of the Pandora moth are a delicacy in some parts of Mexico. (Photo, *Nature Magazine*)

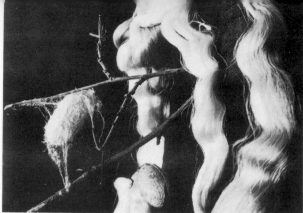

Silk is produced by the caterpillar of the silkworm moth, *Bombyx mori*. The caterpillars are fed mostly on mulberry leaves. When ready to pupate, they exude, from special glands, the fluid silk, which on contact with air hardens into a thread that is wound around the caterpillar to form the cocoon. The silk thread is obtained by killing the pupating moth with heat, and then unwinding the half-mile-long thread that forms a single cocoon. About 25,000 cocoons make a pound of silk. (Photos by C. Clarke)

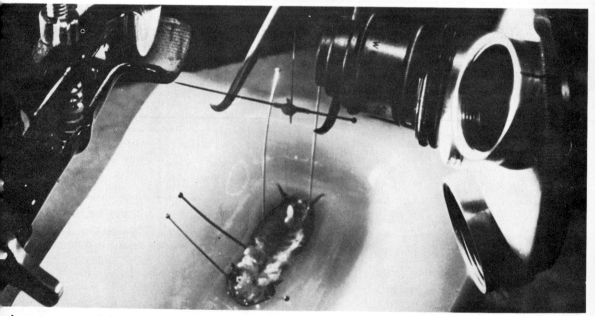

Insects are useful as experimental subjects for scientific investigations. Grasshoppers and cockroaches are among the most used. *Above*: The effects of nicotine on the heart beat of the cockroach is being studied. The movements of a fine glass thread connected to the heart by a hair are magnified through a microscope and recorded photographically. Cockroaches have also been used in growth and learning experiments. (Photo from Science Service)

moist with secretions from the mouth. When the baskets are loaded, the bee returns to the hive and deposits the pollen in special wax cells. Properly combined with sugar and other substances, the pollen mixture becomes "beebread." This provides a source of protein for both adults and larvas.

The **wings** of insects are flattened, two-layered expansions of the body wall and at first consist of the same parts: cuticle and epidermis. Later the two opposing layers meet and the inner ends of their cells unite, except along the channels in which lie nerves, air tubes, and blood spaces. In later stages the epidermal cells condense along these channels, forming the "veins" of the wing. The epidermal cells finally degenerate, and the adult wing is almost completely made of cuticle, though it may have a circulation of blood and some sense organs on the surface.

Not all insects have wings. The two most primitive orders (Thysanura and Collembola) have never developed them. In some of the more specialized orders one or both pairs have been lost secondarily; the flies have only the first pair, the second being reduced to a pair of knoblike structures; and the fleas and lice have lost both pairs. Besides, among the orders which typically have two pairs of wings, there are wingless members. In the social insects certain castes lack wings; and in some species the males have wings while the females are wingless, or the opposite.

Typical wings are membranous structures stiffened by veins, and in many groups both wings are of this type and function in flight. In the grasshopper we saw that the first pair of wings were leathery, but function in flight and also as covers for the hind wings. In beetles the front wings are still more specialized, being only stiff horny plates which serve to cover the hind wings and much of the dorsal part of the body. In the less specialized orders the wings have a dense network of veins, as in dragonflies. In the higher orders, as in bees, there are only a few large veins. Since some wings in every insect group can be reduced to a common basic pattern of wing venation, it is thought that they are all homologous, having been derived from a common winged ancestor.

The **abdomen** of insects bears appendages only at the posterior end; these are modified for mating, or as ovipositors, sensory projections, or in other ways. In the female grasshopper they are used for digging a hole in the ground in which to lay eggs. In the ichneumon flies the ovipositor is long and sharp; and when, in some way, the ichneumon fly senses the presence of a beetle larva within a tree, the ovipositor is used to drill a hole in the wood and deposit eggs in the body of the larva. When the eggs hatch, the young ichneumon larvas parasitize the beetle larva. In the honeybee, appendages at the posterior end are modified as a sting connected with poison glands, as some of us know from painful encounters with these insects.

METAMORPHOSIS

The **ant-lion** is a flying insect with biting mouth parts. Its name is derived from the larva, which digs a pit in the sand and lies buried, with only the jaws protruding and held ready to grasp the first ant or other small arthropod that falls into the pit. (See also page 12 of photographs in this chapter)

IN MOST insects the eggs develop after leaving the female, but some forms retain the developing eggs and give birth to fully developed larvas. In the tsetse fly (see figure on p. 45) even the larva is retained within the body of the mother and is fed from special glands; on emerging, it is ready to pupate.

Not all insects undergo a development which involves changes radical enough to be termed a "metamorphosis." The thysanurans and the collembolans ("garden fleas," shown in the drawing which heads the section on variations in structure) hatch from the egg in practically the same condition as the adult. They grow larger and later add joints to the antennas and some other appendages, but such changes are no greater than those undergone by most animals in their development. Some of the more specialized orders, like the lice, have secondarily lost the metamorphosis. The grasshopper is an example of an insect with incomplete metamorphosis. But since the nymph lives in the same habitat as the adult and eats the same food, this type of metamorphosis is often called **gradual metamorphosis,** to distinguish it from the more thoroughgoing changes of the **incomplete metamorphosis** undergone by dragonflies and mayflies.

These have immature stages which resemble the adult in general body form but have adaptive modifications which fit them to live in another habitat and eat food different from that of the adult. The dragonfly adult is terrestrial and catches other flying insects. The young form, or **naiad,** lives in the water and feeds on aquatic animals. The most striking changes of all are those undergone by the insects with **complete metamorphosis,** in which the feeding larval stage is very different from the adult not only in habit but also in details of structure. The most familiar example of this is the caterpillar and the butterfly. But not all larvas are creeping, worm-like forms which eat vegetation. The ant lion larva has a broad, flattened body and digs a pit in the sand. It lies buried just below the center of the pit, with only the pair of large pincer-like jaws protruding (as shown in the drawing which

The **larva of the ant-lion** has long curved jaws and maxillas which form a pair of pincer-like structures for grasping prey and sucking its juices.

heads this section). When an ant or other wingless insect stumbles into the pit, the sand slides under its feet, carrying it down into the waiting jaws of the larva. The adult is a winged form related to the lacewings. In complete metamorphosis the feeding larval stage and the adult are separated by a quiescent pupa stage in which the body of the larva is almost completely broken down and then reorganized. Since there is some tissue reorganization during the molting of the grasshopper and other insects with a gradual or incomplete metamorphosis, the changes in the pupa may be interpreted as a more extreme type of molting. Both metamorphosis and molting are controlled by hormones, chemical substances secreted in one part of the body and then carried in the blood to other regions, where they initiate changes.

The bug *Rhodnius* feeds upon blood and requires only a single meal between each growth period, after which it molts. If a bug is decapitated a few days before it is expected to molt for the last time (before becoming adult), it will still molt; but if it is decapitated early in the last growth period, it fails to molt, though it will live for months. This indicates that the head influences the molting. Now, if the blood of an insect decapitated just before it is expected to molt is transfused into an insect decapitated very early in the growth period, the latter frequently molts. Apparently, then, there is a hormone, produced by glands somewhere in the head, which controls the molting process.

The development, in the larva or other young stages, of structures which do not occur in the adult or in the ancestors of the species is called

cenogenesis or "sidewise evolution." The pit-digging instincts of the ant-lion larva and the modification of its jaws as large pincers for seizing prey are specializations of the young stage which adapt it to its peculiar way of life. It is likely that no adult ant lion ever dug pits in the sand or had such large pincer-like jaws. The specializations of the mosquito larva or the dragonfly nymph which enable them to live in the water, where the adult would only drown, are examples of the same thing. Of course, all structures which adapt the larva to its environment but are absent in the adult cannot be said to be cenogenetic. Many of them have been inherited from the racial history of the group. For example, the chewing jaws of a caterpillar enable it to feed on vegetation, whereas the adult has a long proboscis, representing a modification of the maxillas, for sucking nectar. The jaws can hardly be considered as "sidewise evolution" of the young stage, for they arise as the appendages of the fourth head segment, and we know that primitive insects are all characterized by such simple chewing jaws.

BEHAVIOR IN INSECTS

Social insects are equipped, from the start, with a set of instincts which fit them to their roles in the life of the colony. Here we see termite soldiers (with large jaws) guarding the colony while the workers carry on the other activities.

THE behavior of insects is mostly instinctive and stereotyped, but a limited amount of learning is possible. Grasshoppers have been taught very little, but cockroaches can be trained to go to the right or left by a simple method of punishment for making the wrong turn. As was mentioned before, in connection with color vision and smell, bees can be taught many things. In most species there is, of course, a difference in the instinctive behavior patterns of the two sexes, and the female is equipped with a set of complex habits connected with deposition and protection of the eggs. In the social insects there is even more differentiation, each caste inheriting a set of instincts which fit it to its role in the life of the colony.

THE ORDERS OF INSECTS

The **dung beetle** is a scavenger. It collects excreta of other animals. Here the beetle is shown using its hind legs to push a ball of dung toward its burrow.

THE CLASS **INSECTA** is divided into more than twenty orders, the exact number depending upon whether certain closely similar groups should be placed in separate orders or lumped together in one. Within each order various families are modified to follow almost every insect "walk of life." Among beetles, for example, there are leaf-eaters, grain-feeders, wood-eaters, predators, scavengers (like the dung beetle shown in the drawing which heads this section), parasites, and commensals, to mention a few. In spite of all this diversity, the members of an order have in common the same general structure, including similar wings, mouth parts, and other appendages; and they also show the same type of metamorphosis. In spite of differences in shape, size, and color, they have enough features in common so that typical members of an order can be classified at a glance as flies, dragonflies, termites, and so on. In the photographs in this chapter, the important orders of insects are presented, with illustrations of some of their features, their common representatives, and a few of the ways in which they affect the life of other animals, including man.

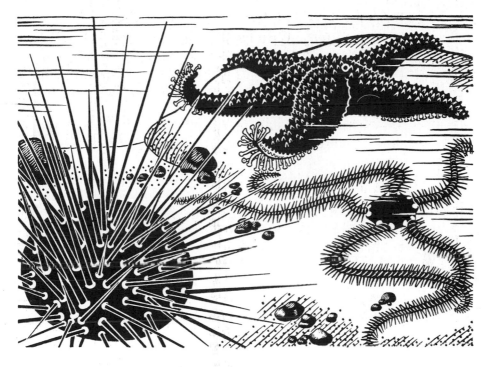

CHAPTER 25

SPINY-SKINNED ANIMALS

WITH the arthropods we reached the peak of invertebrate specialization and should have closed our story. Unfortunately, organic evolution has followed none of the rules of good dramatic style and is full of anticlimaxes. There is one more major phylum to be described, the **ECHINODERMATA** ("spiny-skinned"), a group which very early diverged from the other main lines of evolution. The echinoderm body plan is utterly different from that of any other phylum; and the best clue to its relationship to other groups is the larva, which resembles the larva of acorn worms, allied to the phylum to which man and the other vertebrates belong. Of this, more will be said in the next chapter.

The phylum is exclusively marine and is divided into five classes: starfishes, serpent stars, sea urchins, sea cucumbers, and sea lilies. The echinoderm most familiar to everyone, even to those who have never been near the seashore, is the starfish.

A STARFISH has no resemblance to a fish and should be called a "sea star," from its shape; but the name "starfish" has gained wide usage and is hard to change.

There are a little over one thousand species of starfishes, which differ in certain details but as a class are more similar in structure and habit than the crustaceans or insects, for example.

The body of a starfish consists of a central disk from which radiate a number of **arms.** There are usually *five* of these, but many species have a larger number. Because of their shape, starfishes were once classed with the coelenterates; but now we know that their radial symmetry has nothing to do with that of coelenterates but has been secondarily derived from a basic bilateral symmetry, as will be explained later. There is no head, and the animal can move about with any one of the arms in the lead. The **mouth** is in the center of the disk on the under surface.

The body of the starfish appears to be quite rigid but is capable of a considerable amount of bending and twisting. The rigidity is provided by a meshwork of **calcareous plates** which are imbedded in the soft flesh. From these plates project numerous **calcareous spines,** some of them movable. The flexibility results from the fact that the plates are not united into a single shell but are distinct from each other and are joined by connective tissue and muscles. The skeleton is like our own in that it is an *endoskeleton* imbedded in the flesh, and differs from the exoskeleton of arthropods, which lies outside the body. It does not permit the freedom of movement of vertebrate or arthropod skeletons; but this is not important, as the starfish cannot move rapidly anyway.

Locomotion in the starfish is by means of a kind of hydraulic-pressure mechanism, known as the **water vascular system,** unique to echinoderms. Water enters this system through minute openings in the **sieve plate** (madreporite) on the upper surface and is drawn, by ciliary action, down a tube, the **stone canal** (so-named because its wall is stiffened by calcareous rings), to the **ring canal,** encircling the disk. From the ring canal arise five **radial canals,** one for each arm. Each of these connects, by short side branches, with many pairs of **tube feet,** hollow, thin-walled cylinders which end in suckers. Each tube foot connects with a rounded muscular sac, the **ampulla.** When the ampulla contracts, the water, prevented by a valve from flowing back into the radial canal, is forced under pressure into the tube foot. This extends the elastic tube foot, which attaches to the substratum by its sucker. Next, the longitudinal muscles of the tube foot contract, shortening it, forcing the water back into the ampulla and drawing the animal forward. Of course, one tube foot is a very weak structure; but there are hundreds of them, and their combined effort is capable of moving the starfish slowly over the ocean floor. The tube feet can work as described only when the animal is moving over a rock or some kind of hard substratum. On soft sand or mud the suckers are of little use and the tube feet act as little legs. Moreover, some star-

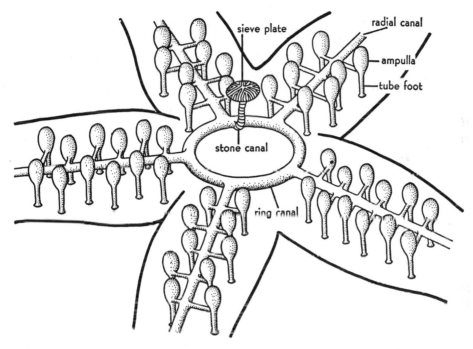

Water-vascular system of starfish, showing essential structures. The ring canal may have one or more sets of vesicles; but as their exact function is not well known, they are omitted here.

fishes have no suckers at the ends of the tube feet. In the diagram of the cross-section of the starfish arm it can be seen that the ampullas lie within the body but that the radial canal and tube feet lie outside the calcareous plates in the center of the under surface of the arm in a V-shaped groove (usually called the "ambulacral groove," from a Latin word meaning a "walk," because the double row of tube feet lining the groove reminded someone of a flower-lined garden walk). On either side of this groove are rows of movable spines, which can be brought together to close the groove and protect the tube feet when the animal is attacked.

Starfishes move rather slowly; but they have no difficulty in running down their prey, for they **feed** mostly on clams, which move still more slowly, or on oysters, which do not move at all. Anyone who has ever tried, barehanded, to open a live clam or oyster will wonder how it can be done by a starfish which is not much bigger than the clam it attacks. The starfish mounts upon the clam in a humped up position, attaches its tube feet to the two shells, and begins to pull. The clam reacts to this by clos-

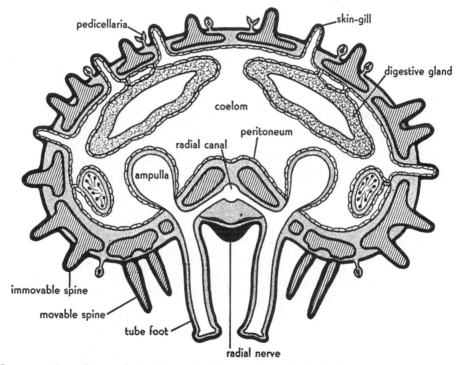

Cross-section of arm of starfish to show large coelom, completely lined by a mesodermal epithelium.

ing its valves tightly. The starfish uses the numerous tube feet in relays, and is therefore able to outlast the clam, maintaining its pull on the shells until the two large muscles which hold the valves together become fatigued and finally relax. When the shells gape, the starfish turns the lower part of its **stomach** inside out and extends it through the mouth, enveloping the soft parts of the clam and digesting them. The partly digested material is taken into the stomach and then into the five pairs of digestive glands (one pair in each arm) which connect with the upper part of the stomach. Very little indigestible material is taken in by this method of feeding, and this accounts for the fact that there is practically no intestine and that the anus, a very small opening near the center of

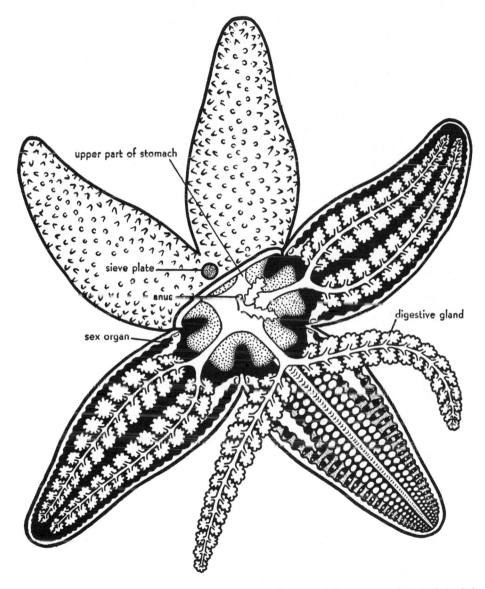

Starfish, with body wall dissected away from the dorsal surface of three arms and most of the disk, to show digestive system. In the arm on the lower right, the two branches of the digestive gland have been spread apart to show the rows of ampullas and the sex organs. The upper part of the stomach is unshaded; the lower part, stippled.

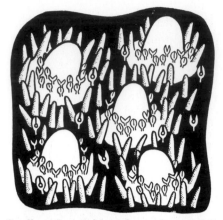

Small portion of the **surface of a starfish** to show the delicate *skin-gills* occupying the spaces between the large protective *spines*. The tiny *pedicellarias* occur in "rosettes" around the bases of the spines and also among the skin-gills.

the upper surface, is used very little, if at all. When the starfish eats a small clam, it may take the whole animal into the stomach; the shell is later ejected from the mouth.

Running beneath the water vascular system are fluid-filled channels of uncertain function, but there is no well-organized circulatory system. The distribution of materials is provided for by the large **coelom** lying between the body wall and the digestive tract and filled with a fluid which bathes all the organs and is kept moving by cilia on the coelomic lining. The thin walls of the tube feet permit gaseous exchanges, but they are not the main organs of respiration. Opening from the coelom are many small finger-like projections, the **skin-gills,** which extend from the surface of the body through spaces between the calcareous plates. They have very thin walls through which oxygen diffuses into the coelomic fluid and carbon dioxide diffuses out.

The delicate skin-gills would be in a dangerously exposed position if it were not that they lie among the heavy spines and are further protected by the **pedicellarias,** small pincers which occur in the spaces between the spines or in clumps around the bases of spines. When a small animal creeps over the surface of a starfish, it is caught and held by the toothed jaws of the pedicellarias, which, together with currents created by cilia of the surface epithelium, serve to protect the skin-gills and keep the surface free of any debris which would interfere with respiration and excretion.

There is no specialized excretory system. **Excretion** is accomplished by ameboid cells in the coelomic fluid which engulf nitrogenous wastes and then escape through the walls of the skin-gills. Probably also excretory are two branched pouches (shown in the diagram of the digestive system) which lie in the coelom and open into the intestine near the anus.

The **nervous system** is simple, as in most slowly moving animals. A ring of nervous tissue encircles the mouth and gives off five radial nerves which extend along the middle of each arm below the radial canals of the water vascular system. There is also a deeper-lying ventral system and one near the upper surface. The **sense organs** are poorly developed. Sensory cells, which occur all over the surface, are sensitive to touch. At the tip of each arm is a pigmented eyespot and a short tentacle, thought to be sensitive to food or other chemical stimuli.

Two **ovaries** or **testes** lie in each arm and open directly to the exterior through small pores in the angles of the arms. The eggs and sperms of starfish, like many other

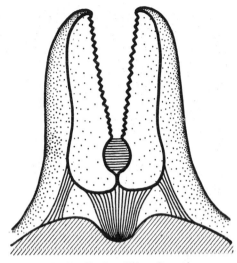

Pedicellaria. (After Cuenot)

marine animals, are shed into the sea water, where fertilization occurs.

Since the eggs and sperms are more easily obtained from mature starfish and sea urchins than from most other animals, they have been much used by embryologists to study the mechanisms of fertilization and early development. Eggs and sperms are single cells which cannot survive long in sea water; unless an egg is fertilized within a short time, it perishes. Sperms surround the egg, and the first one that penetrates the membrane (which it does rapidly, leaving its flagellum outside) starts a physicochemical change such that a large amount of fluid collects beneath the membrane, elevating it and serving as at least one of the factors which prevent other sperms from entering. The sperm nucleus migrates toward the egg nucleus, and both undergo changes preparatory to the first cell division, which is initiated by the entrance of the sperm. Each nucleus brings the hereditary contributions of one parent to the new individual.

By subjecting unfertilized eggs to various stimuli, such as certain acids or concentrated sea water, it is possible to activate the egg and cause it to divide, thus duplicating the effect of the sperm. The larva which develops is normal in every way, though, of course, it has no paternal characteristics. Strong centrifuging of unfertilized eggs causes them to become dumbbell-shaped and finally to break in two. The lighter halves contain the maternal nuclei; the heavier, nonnucleated halves can be collected and fertilized by adding sperms to the dish. The larvas thus obtained have a father but no mother. Such methods make it possible to isolate maternal and paternal characteristics and also to prove that the reproductive cells of both sexes make equivalent contributions to the heredity of the offspring. Finally, fertilized eggs can be centrifuged strongly until they break in two.

The nucleated fragment develops normally. The nonnucleated fragment divides several times but eventually dies (like the *Euplotes* from which the small nucleus was removed; see p. 25). From these experiments we can see that either a male or a female nucleus is sufficient for development, but apparently at least one nucleus is necessary for continued differentiation and life.

After fertilization the zygote divides into two equal cells; and these divide into four, eight, sixteen, and so on until there results a hollow **blastula,** which is ciliated and swims about. By the end of the first day

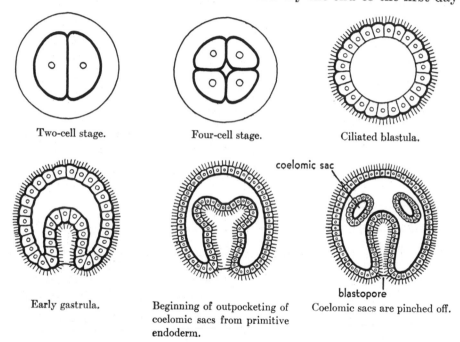

Two-cell stage.   Four-cell stage.   Ciliated blastula.

coelomic sac

blastopore

Early gastrula.   Beginning of outpocketing of   Coelomic sacs are pinched off.
                       coelomic sacs from primitive
                       endoderm.

Early development of a starfish. (After Delage and Herouard)

after hatching from the egg, the free-swimming blastula has been transformed, by an infolding of the cells at one pole, into a two-layered **gastrula** with an outer ectoderm and an inner endoderm. This infolding crowds out most of the old **blastula cavity** (blastocoel) and produces the **primitive gut cavity** (archenteron), which will later become the cavity of the digestive tract. The opening into this new cavity is called the **blastopore.** Through the gastrula stage the embryos of most animals are essentially alike, although they differ in size, in rate of development, in amount of yolk, in the equality of division of the cells, and in the particular way in which the gastrula becomes two-layered. As development proceeds, the number of differences increases. In the starfish the next stage is characterized externally by the elongation of the larva, a decrease of cilia over the general surface, and an increase in the cilia along definite bands. In the meantime important internal changes take place. From the endoderm arise two out-pocketings, which gradually pinch off as two coelomic sacs. These eventually give rise to the **coelom** and its derivative, the water

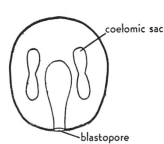

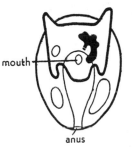

Coelomic sacs divide into anterior and posterior parts.

Left anterior coelom sends outgrowth to dorsal surface; this is forerunner of stone canal of adult.

Left anterior coelom begins to curve around mouth and produce buds. The larval anus is derived from the blastopore.

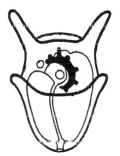
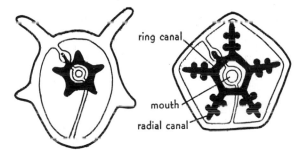

Rudiments of five radial canals are seen as buds from the left anterior coelomic sac.

The right and left posterior sacs form the main body cavity of adult.

A new anus has formed dorsally and does not show in this view of the ventral surface.

Further development of a starfish. (After Delage and Herouard)

vascular system. Up to this point the larva has been bilaterally symmetrical. But after each coelomic sac divides into an anterior and a posterior portion, changes begin which transform the bilateral larva into a radial adult. The left anterior coelomic sac sends a tubular outgrowth to the upper surface, where it opens by a pore in the ectoderm. Cilia lining this tube draw water into the coelom, and this pore canal is the forerunner of the stone canal of the adult. In some starfishes the bilateral symmetry persists somewhat longer, and the right anterior coelom also sends a pore canal to the upper surface, though this canal finally closes up and is lost. This suggests that the ancestral starfish originally produced similar organs on both sides, and that the radial condition of modern starfishes is a result of the degeneration of the right-hand side of the embryo and an overgrowth of the left side. The hind portion of the left anterior coelom soon produces another hollow outgrowth, which encircles the mouth and becomes five-lobed. This is the beginning of the water vascular system (shown in solid black in the diagram). The circular outgrowth becomes the ring canal, and the five lobes represent the rudiments of the future radial canals. During the development of the coelom and its derivatives the endodermal layer differentiates into the esophagus, stomach, and intestine; the blastopore serves as the larval anus. The esophagus bends ventrally and meets a tubular ingrowth from the ectoderm, which breaks through to form the mouth. Meanwhile, the external surface of the larva has been greatly expanded by the extension of lobelike processes which enable the larva to float easily. Certain of the ciliated bands are carried out on these processes, and the beating of their cilia propels the larva about. Food-bearing currents are created by the cilia on bands near the mouth.

The free-swimming starfish larva (called a *bipinnaria*) has bilaterally symmetrical lobes bearing ciliated bands. Other echinoderms have similar but distinctive larvas, some of them with very much longer lobes. At least in the earlier stages, all echinoderm larvas have certain characteristics in common. They are bilaterally symmetrical, swim by means of longitudinal looped ciliated bands, have a similar digestive tract and a coelom which connects with the upper surface by a tubular canal. All this suggests that there must have been some ancestral echinoderm larva which embodied all these features and from which all the different classes of echinoderms have been derived. This hypothetical ancestral larval type (called the *dipleurula*) contrasts strongly with the trochophore-like ancestor from which, it is believed, has come the flatworm-mollusk-annelid stock.

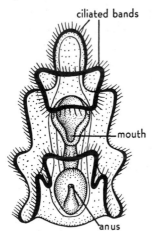

ciliated bands

mouth

anus

The free-swimming **larva of a starfish** is bilaterally symmetrical and has ciliated bands for locomotion and feeding. (After Field)

To return to the starfish development, we see next that the bilateral larva settles down on

some solid object and remains temporarily fixed while it becomes transformed into the adult starfish. The left and right posterior coelomic sacs
form the general body cavity of the adult. The larval mouth and anus
close, and a new mouth breaks through on the original left side, while a
new anus opens on the original right side, thus producing an adult axis
at right angles to the larval axis. The radial canals grow out and develop tube feet. And the first signs of the adult body form appear as five
elevations of the ectoderm.

The symmetry of the adult starfish is called **secondary radial symmetry** because we believe that it is only secondarily derived from an originally bilateral ancestor. This is suggested in the adult by the asymmetrical position of the sieve plate. But the most important evidence comes
from the larval development, in which, by a process of asymmetrical
growth, the bilateral larva is converted into a radial adult.

The fact that the bilateral starfish larva becomes temporarily attached
during the time that it changes over into the radial adult is very interesting in view of what we believe to be the ancestral history of echinoderms.
The earliest-known echinoderm fossils are fixed types. And some of the
modern sea lilies (a class of echinoderms) have a free-swimming bilateral
larva which becomes attached and then metamorphoses into a permanent
ly fixed adult. As we have seen before, bilateral symmetry is the symmetry of fast-moving animals and radial symmetry seems best suited to
sessile animals, which must meet their environment on all sides. It is considered likely, therefore, that the ancestral echinoderm was a bilateral
free-living animal which became radial and took up sessile habits secondarily, and that the modern free-living echinoderms are derived from a
fixed ancestor, whose symmetry they still retain.

SEA URCHINS seem very unlike starfishes; yet they have the same
fundamental structure. A sea urchin looks like a huge animated burr
with long sharp spines that are movable and aid the tube feet in locomotion. Instead of numerous small stony pieces imbedded in a muscular
wall, the sea urchin has closely fitted plates which form a rounded shell
completely inclosing the soft parts. In the center of the lower surface of the
shell is an opening for the mouth, and the anus opens by a small hole on
top. Radiating upward from mouth to anus are five bands of minute
holes through which the tube feet project. These five rows correspond to
the rows of tube feet on the five arms of the starfish. If we imagine the

arms of the starfish bending upward to meet, and if, at the same time, we fill in the angles between them by elevating and reducing the size of the disk so that the sieve plate will lie at the ends of the rows of tube feet, we can see how the round sea urchin is similar to the five-rayed starfish.

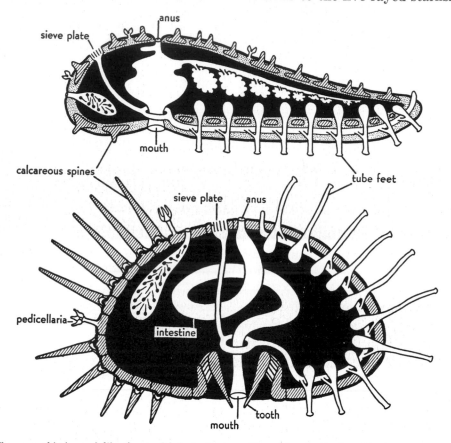

The sea urchin is much like the starfish in basic plan. *Above*, a longitudinal section through the disk and one arm of a starfish. *Below*, a section through a sea urchin.

The sea urchin has the same type of water vascular system that was described for the starfish, but the feet are more slender and much longer, since they must project beyond the spines. The pedicellarias have three jaws and are usually on long stalks. The digestive tract is longer than that of the starfish, and the sea urchin feeds mostly upon vegetation, which requires more prolonged digestion. Around the mouth is an elabo-

Most starfishes live along rocky coasts, where the hard substratum furnishes a place of attachment for the suckers of the tube feet with which these animals pull themselves slowly over the rocks. When the tide goes out, starfishes can be seen stranded in shallow tide pools as shown above. (Mount Desert Island, Maine)

From Maine . . .   **to California . . .**

the seacoast attracts people who like to explore among the rocks for animals that are left behind as the tide goes out. *Left:* Children picking up a starfish at Salsbury Cove, Maine. *Right:* General view of rocky coast at Pacific Grove, California.

Ten-armed starfishes are less common than the typical five-armed kinds, but ones with from 4 to 25 rays are known. Notice the barnacles that cover the rock. (Photo of living animal. Mount Desert Island, Maine)

A starfish rights itself (*top to bottom*) by bending its stiff arms and pulling with the tube feet. Pacific Grove, California)

Six-armed starfish. The larvas of some starfishes, like those of *Leptasterias*, shown here, have no free swimming stage but are sheltered in a brood chamber on the lower side of the female and emerge as little starfishes. (Photo by W. K. Fisher, Pacific Grove, California)

Starfish eating a clam. Humped up over a clam in a tidepool, a starfish applies its tube feet to the two valves and soon pulls the clam open as the exhausted bivalve yields. (Photos of living animal)

Starfish lifted out of the water and held up to show the stomach everted into the opened clam shell. It has secreted digestive juices and is digesting the clam. (Mount Desert Island, Maine)

Mop removes starfishes from oyster bed. The pedicellarias take hold of the threads as the mop is dragged over the sea bottom. (Int. News Photo)

One-armed starfish is regenerating four new arms, of which three appear as tiny buds growing from the disk. (Photo of living animal. Mt. Desert Island, Maine)

Rows of tube feet radiating from the central mouth are characteristic of echinoderms, but starfishes have the most highly developed ones. They are extended by hydraulic pressure but retracted by muscular contraction. At the ends of the tube feet are muscular suckers with which they hold on to surfaces. (Photo of under surface of living animal. Mount Desert Island, Maine)

Pedicellarias and skin gills cover the upper surface of the starfish between the large protective spines. At the end of each arm is a simple pigmented eye. The surface is covered with cilia which create currents that prevent sediment from settling on the animal and interfering with respiration. (Photo of tip of one arm of living animal. Mount Desert Island, Maine)

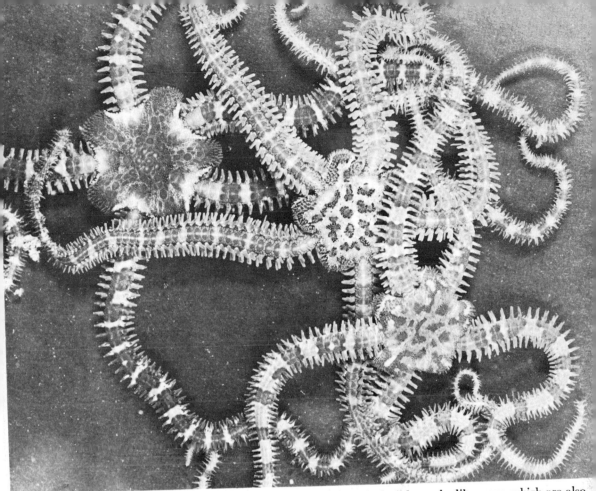

Serpent stars crawl over the ocean floor by agile movements of their flexible, snake-like arms, which are also used to entrap small animals like worms and crustaceans. Most species are five-armed, but ones with six, seven, or eight arms are known. Some have branched arms. Placed together in a restricted space, the animals twine their arms about one another. Their colors are extremely variable, no two being exactly alike. About $2 \times$ natural size. (Photo of living animals. Mount Desert Island, Maine.)

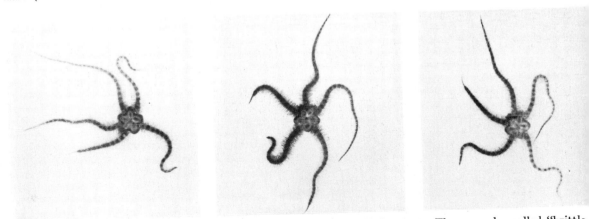

Serpent stars are named from the writhing snakelike movements of the arms. They are also called "brittle stars" because of the ease with which the arms come off when seized by man or other enemies. $\frac{2}{3}$ natural size. (Photos of same animal taken in rapid succession. Pacific Grove, California)

Underside of serpent star to show central mouth from which radiate five arms. The tube feet lack suckers and ampullas and function as tactile sense organs and for respiration. This animal, as the name "brittle star" suggests, broke off its arms when it was handled. Brittle stars can regenerate the missing parts. (Photo of living animal. Mount Desert Island, Maine)

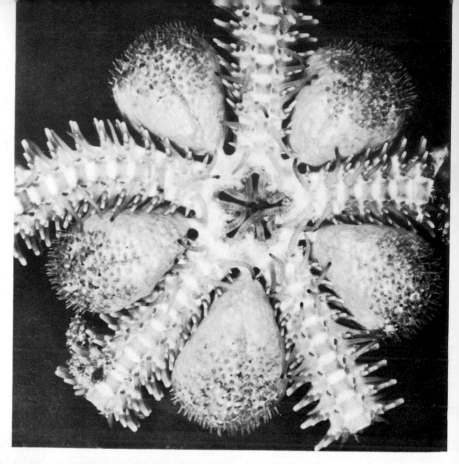

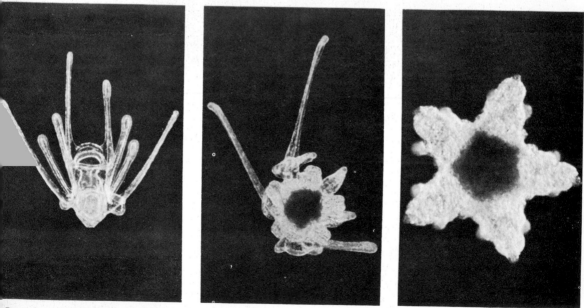

Developmental stages of the brittle star. After gastrulation a bilaterally symmetrical ciliated "ophipluteus" larva with long swimming arms develops (*left*). The arms shorten (*center*) as the animal metamorphoses into a tiny brittle star (*right*) which exhibits a secondarily acquired radial symmetry. (Photomicrographs of living *Ophiocomina nigra* by D. P. Wilson. Plymouth, England)

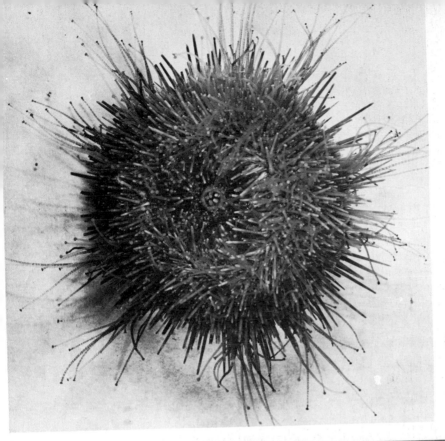

Sea urchins live mostly near rocky shores, feeding on algae and decaying materials and serving as one of the many kinds of scavengers that keep the sea bottom from becoming foul. The tube feet of sea urchins end in suckers and are long and slender, extending beyond the spines. Both are used in locomotion, the tube feet being moved by changes in pressure in the water vascular system and by muscles; the spines are moved by muscles. Five teeth are seen around the mouth. About natural size. (Photo of living animal. Mount Desert Island, Maine)

The test or skeleton of a sea urchin is globular and made up of closely fitted calcareous plates. The spines are removed, but their positions are indicated by the round protuberances on which they were pivoted. The five double rows of holes are openings through which the tube feet protruded. (Pacific Grove, California)

Spines of sea urchins are used in locomotion and offer protection from marauding animals which might otherwise prey on urchins. The spines may be relatively short and stout, as in the short-spined urchin at the *left*, which sits in a hole in the rock made by constant movements of the spines; or they may be longer and very sharp, as in the needle-spined urchin at the *right*. (Photos by Otho Webb. Australia)

Another variation in spines is shown by the slate-pencil sea urchin. (Photo by Otho Webb. Australia)

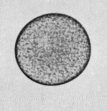

1. Unfertilized egg.

2. Fertilized egg.

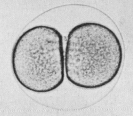

3. Two-celled stage.

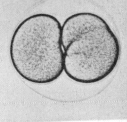

4. Each cell divides.

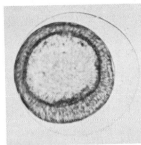

5. Four-celled stage.

6. Eight-celled stage.

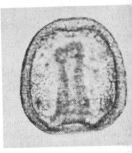

7. Blastula, hollow ball.

8. Gastrula, 2 layers.

The development of a sea urchin is representative of animal development, and equivalent stages occur in all animals from the hydra on. As the sea urchin embryo divides, the cells become progressively smaller, until in the blastula stage, at the time of hatching from the egg membrane, they are barely distinguishable at this magnification (Actual size of unfertilized egg, 1/150 inch). (Photos by D. P. Wilson, Plymouth, England)

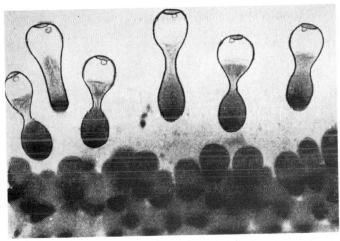

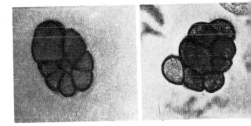

Sea urchin eggs are especially valuable as experimental material for studying embryological and physiological problems. *Above left:* A group of fertilized eggs, photographed as they are being centrifuged at a force equivalent to 10,000 times gravity. The fat is lightest; the nucleus slightly denser. Pigment granules are heaviest. After further centrifuging the cell breaks in two. Nucleated halves develop into normal embryos. Enucleated halves also develop but produce disorganized embryos (*above right*). (Photomicrographs by E. B. Harvey. Woods Hole, Mass.)

Embryos are sensitive to chemical and physical changes in the sea water in which they develop. Fertilized sea urchin eggs reared in sea water diluted with distilled water to 95% of its salt concentration slows down embryonic development (*left*), as compared with the normal development in 100% sea water (*center*). 105% sea water retards embryo (*right*). (Photomicrographs. Woods Hole, Mass.)

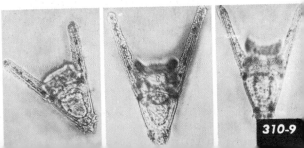

Sand dollars belong to the same class as sea urchins but are flattened and covered with short spines. They move by means of the spines and small tube feet on both surfaces of the body. The large tube feet which protrude from the five double rows of holes on the upper surface are respiratory. The animals swallow sand and digest the organic material contained in it. *Above:* Living sand dollars. *Below:* The dried tests of two sand dollars arranged in positions corresponding to the pair above. The left ones show the upper surface; the right ones the lower surface with the mouth opening in the center and the anus near the margin. (Pacific Grove, California)

Sea cucumbers are fleshy echinoderms in which the usual spines are reduced to minute ossicles imbedded in the skin. When attacked, some throw out their viscera, leaving them for the enemy, meanwhile escaping and regenerating a new set. Others throw out slime threads which entangle the enemy. The animals shown here (in a tide pool off the Australian coast) creep about on the ocean floor by muscular movements of the body wall. The tube feet are little used. The animals swallow sand or mud and digest the organic material. They are collected in great quantities and, when dried, are known as "trepang," or "*bêche-de-mer*," which is then shipped to China to be used in making a soup. (Photo by Otho Webb. Australia)

The sea cucumber in the foreground at left shows three of the five rows of tube feet, by which the animal attaches to rocks. The finely branched tentacles around the mouth are slimy and trap small animals. The sea cucumber at the right has the tube feet contracted. (In the background are two starfishes, one with 5 arms broadly joined to the disk, the other with 8 arms.) (Photo of living animals by F. Schensky. Helgoland)

Sea lily, or crinoid, a stalked echinoderm, brought up from the sea bottom by a dredge. Only a small portion of the stalk is shown. When undisturbed, the branched arms are widely spread, and ciliated grooves on their upper surfaces sweep small organisms into the central mouth. Being deep-water forms, stalked sessile crinoids are seldom seen. More familiar are the feather stars of shallow water. These have a young stalked stage but later break from the stalk and swim about by waving the arms. The tube feet of crinoids are chiefly respiratory, not locomotory. The skeleton is of calcareous plates; there are no spines. (Photo by A. H. Clark. From *Science Service*)

rate set of five teeth, arranged radially and
worked by muscles in such a way as to chew
the food. The stomach cannot be turned inside out, and the intestine is long and coiled.
The anus is functional and is located to one
side of the sieve plate. Other systems are
very much like those described for the starfish. The gonads open from five pores near
the sieve plate.

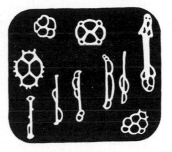

Not all echinoderms have prominent spines. The sea cucumbers
have a leathery skin in which are
imbedded microscopic plates of
calcium carbonate. The plates
shown here were isolated by
mounting a fragment of skin on a
microscopic slide and dissolving
away the living tissue.

THE other echinoderm groups—brittle
stars, sea cucumbers, and sea lilies—all
deviate from the starfish about as much as
the sea urchin but have characteristically a
spiny or leathery skin, a water vascular system based upon a plan of five or multiples of
five, an extensive coelom, and a bilateral larva. These groups are illustrated by photographs.

CHAPTER 26

AMPHIOXIDS, TUNICATES, AND ACORN WORMS

THE only major group we have not yet considered is the phylum **CHORDATA,** composed almost entirely of animals *with* backbones and therefore no proper subject for this book. There are, however, a few "lower chordates" which have no vertebral column. These are mostly inconspicuous animals, seldom seen, or at least not usually noticed. None are of much economic importance, though the amphioxus is so abundant on the seacoast of China that during certain months these small animals are collected by the ton for human consumption.

The amphioxids, and the other invertebrate chordates, the tunicates, are interesting chiefly because they share with the vertebrates certain distinctive characters found nowhere else in the animal kingdom, and so serve to link the vertebrates with the invertebrates.

These chordate characters are all well shown by the **amphioxus** ("sharp at both ends"), a small, laterally compressed, semitransparent animal which lives in shallow marine waters all over the world. It can swim about by fishlike undulations of the body, but spends most of the time buried in the sandy bottom with only the anterior end protruding above the surface. In this position it feeds by drawing into the mouth a steady current of water, from which it strains suspended microscopic organisms.

The most distinctive chordate character, and the one from which the name of the phylum is derived, is the **notochord,** a cartilage-like rod which extends the length of the body and gives support to the soft tissues. It serves as a rigid but flexible axis on which the soft muscles can pull, and so permits powerful side-to-side undulatory movements of the whole body, which carry the animal through the water with a speed unattainable by flabby animals like the planaria or the nereis.

The strong, swift swimming movements of aquatic chordates, made possible by a flexible internal support for the muscles, were probably a major factor in the early success of the group. Besides great muscularity

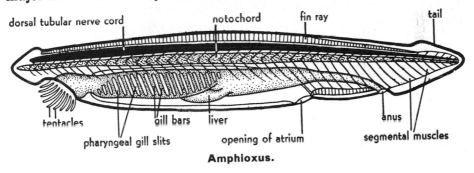

Amphioxus.

(as shown in the cross-section of the amphioxus), chordates are characterized by the prolongation of the body beyond the anus as a *tail,* a region specialized for swimming and containing little else but the skeletal axis, nerves, and muscles. In the amphioxus the tail is very small, and there are no paired fins as in fishes; but running along the entire dorsal surface, and extending around the posterior end onto the ventral surface of the body for a short distance, is a ridge supported by a series of gelatinous *fin-rays.*

The notochord is derived from the roof of the primitive endoderm and is present in the embryos of all chordates including man. In primitive vertebrates (lampreys) and some fishes the notochord persists, and the backbone, which is composed of a series of hard separate vertebras and is stronger though just as flexible, forms around it. In all higher vertebrates the backbone replaces the notochord as the mechanical axis of the body.

A second chordate character is the **dorsal tubular nerve cord. In all** other invertebrates which have been described, the principal nerve cord was ventral or lateral in position; but in the amphioxus and other chordates, it lies between the notochord and the dorsal skin and is hollow. From the cord go a pair of nerves to each of the segmentally arranged bundles of muscles.

The third important chordate character is the structure of the pharynx, which is perforated by pairs of slitlike openings. The **pharyngeal gill slits,**

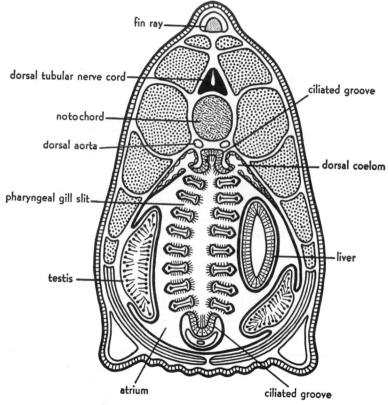

fin ray
dorsal tubular nerve cord
notochord
dorsal aorta
pharyngeal gill slit
testis
atrium
ciliated groove
dorsal coelom
liver
ciliated groove

Cross-section through an amphioxus.

so named because they have mainly a respiratory function in fishes, in the amphioxus serve chiefly as an apparatus for straining food from the water. The pharynx is lined with cilia, which beat inward, producing a steady current of water that enters the mouth and passes out through the pharyngeal slits, leaving behind the suspended particles. The slits do not open directly to the exterior, as in fishes, but into a chamber, the *atrium,* which surrounds the pharynx and opens to the exterior by a pore some distance anterior to the anus. The atrium is lined with ectoderm and is formed, in the embryo, by the outgrowth of two folds of skin, one on each side, which finally fuse in the mid-line. The walls of the pharynx, being perforated from top to bottom by the vertical gill slits, would collapse if it were not for a supporting framework of rods which run in the walls bounding the gill slits. The tissue with its rod, which lies between any two gill slits, is called a *gill bar*.

The amphioxus embryo has about sixty pairs of gill slits (and they increase in number as the animal grows). This large and indefinite number is a primitive condition; and in the shark, for example, there are always six gill slits. Land vertebrates, which breathe by lungs, have no gill slits in the adult stage, but the slits make a fleeting appearance in the embryo. The human embryo develops gill pouches, but slits never break through.

Except for the gill slits, most of the structures associated with the feeding mechanism of the amphioxus are peculiar to the animal and must not be thought of as primitive chordate characters. At the anterior end of the animal is a funnel-like hood fringed with a row of sensory tentacles. At the back of the hood is the *mouth*, a circular aperture bounded by a rim of small tentacles which can be brought together to form a kind of strainer for keeping out large particles. The cilia in the pharynx beat inward and downward, so that water drawn into the mouth is directed toward a *ciliated groove* in the floor of the pharynx, where the suspended organisms are trapped in mucus. By the action of the cilia the food-laden mucus is moved forward to the anterior end of the pharynx, upward on each side, and then backward along a ciliated groove in the roof of the pharynx, to the *intestine*. There the food and the mucus are digested and almost completely absorbed. From the anterior end of the intestine is given off a hollow gland, the *liver*, which extends forward along the right side of the pharynx and can be seen in the cross section. This gland secretes a digestive fluid; and since it arises in the same way as the vertebrate liver, by an outpocketing of the digestive tract, the two organs are thought to be homologous. Further, the blood leaving the capillaries of the intestine of amphioxus is not returned to the general circulation until it has passed through the capillaries of the liver. Such a path for the blood is found nowhere else except in vertebrates, and it furnishes striking evidence that the amphioxus is descended from the same primitive stock which gave rise to the backboned animals.

The *circulatory system* is a closed one; and there is no heart, the blood being pumped by the contractile ventral vessel. In this connection it is interesting to note that in all vertebrate embryos the blood is first pumped by a simple, pulsating tube which only later becomes bent and constricted to form the heart. The blood receives oxygen as it flows through vessels in the gill bars, which are in close contact with the steady current of water passing through the pharynx. After passing through the gill bars, the blood flows into two vessels, the dorsal aortas, which unite behind the pharynx into a single vessel that supplies the intestine. Wastes are extracted from the blood by *excretory organs*, a pair to each pair of gill slits, which lie against the dorsal wall of the pharynx and resemble, strangely enough, the excretory organs of annelids. The *coelom* of the amphioxus has been partly crowded out by the expansion of the atrium. In the pharyngeal region it is represented only by two small cavities, which lie, one on each side of the pharynx, above the atrium. The sexes are separate; and the *reproductive organs* are paired, segmentally repeated pouches which lie along the sides of the body and push the atrial walls inward, so that in the cross-section they appear to lie in the cavity of the atrium.

There are no paired eyes or other well-developed sense organs, though there is a large pigment spot at the anterior tip of the nerve cord and a row of smaller pigment spots along the lower edge of the cord. The nerve cord does not expand at the anterior end into a brain but is tapered to a point. The apparent simplicity of the sense organs and central nervous system is probably not a primitive condition but is, more likely, a degeneration of the head region associated with the sedentary habits of the animal.

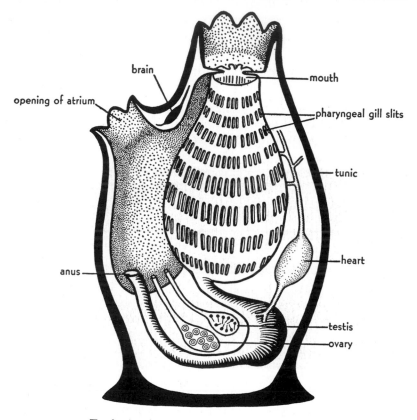

Tunicate. (Modified after Delage and Herouard)

In spite of its poor nervous system and various specializations associated with its particular way of life, the amphioxus is the only living form to which we can look for a concrete idea of what the primitive chordates, which gave rise to the vertebrates, might have been like.

THE **tunicates** are so named because the outer layer of the body wall is a tough, often translucent "tunic" made of a material very much like the cellulose of plants. Some tunicates are free-living and swim near the surface; but most forms are sessile, growing permanently attached to rocks or seaweeds. The simplest kind looks like an upright sac with two openings: one at the top, and one somewhat lower down on one side. When the animal is disturbed, the body wall contracts suddenly, and the water contained within the body is forced out in two jets—hence the common name "sea squirt." The interior of the body is occupied for the most part by a large saclike pharynx perforated by many rows of *pharyngeal slits*. Cilia around the edges of the slits create a current which draws water into the mouth at the top, through the pharyngeal slits, and out into the atrium, a cavity surrounding the pharynx. Food particles are trapped in the pharynx, and water leaves by an opening from the atrium.

The atrium also serves as an exit for feces and sex cells, since the anus and sex organs (the animal is hermaphroditic) open into the atrium. There are few real blood vessels, the blood flowing mostly through spaces among the tissues. The heart is remarkable in

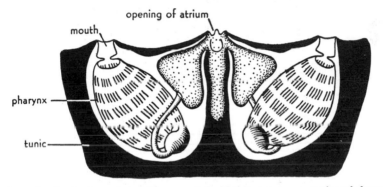

Two members of a **tunicate colony.** They are imbedded in a common tunic and share the same atrial opening, but have separate mouths. (Modified after Delage and Herouard)

that it drives the blood in one direction during several beats and then reverses the direction.

Many tunicates reproduce asexually, as well as sexually. The buds fail to separate and the individuals remain together as a colony, imbedded in a common tunic, which, in the case of sessile tunicates, grows as an incrusting mass over the surface of rocks, shells, or sea weeds. The members may be arranged in small groups; and in the colonial tunicate, shown in the drawing which heads this chapter, there are four such groups represented. Each is star-shaped, and at the tips of the rays are the separate mouths of the several members. At the center of the star is a common opening for the exit of water.

Except for the gill slits there is very little about such a simple, sessile animal to suggest any reason for including it in the same phylum with fishes or mammals. But the development of a tunicate tells another story. The larva is a free-swimming animal that reminds one of a tadpole. It has a large tail which contains, besides muscles, a well-developed *notochord*

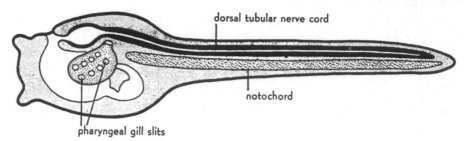

dorsal tubular nerve cord

notochord

pharyngeal gill slits

The free-swimming **larva of a tunicate** has all three chordate characters. (Based on several sources)

and a *dorsal tubular nerve cord.* The trunk contains, in addition to other organs, a *pharynx with slits.* The larva finally settles down on a rock, loses the tail containing the notochord and nerve cord, and revamps its whole structure in adaptation to its sessile life. The adult has no trace of a notochord, and the nervous system is represented only by a ganglion in the dorsal region of the body between the two openings. Here we have another striking example of how animal relationships can be established through a study of the young stages, even though the adult is a degenerate form whose similarities to other animals are quite obscure.

M ORE primitive, but related to the invertebrate chordates, are the **acorn worms,** soft, elongate animals which burrow in the sand or mud of seashores. At the anterior end of the body is a muscular *proboscis* joined by a narrow stalk to a short, wide *collar.* The worm shown in the accompanying drawing has a long cylindrical proboscis,

proboscis

mouth

collar

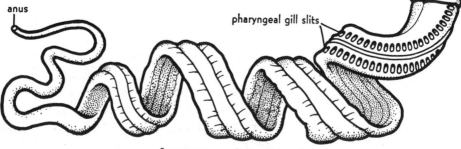

anus

pharyngeal gill slits

A corn worm. (After Bateson)

but in many forms (as in the ones in the drawing at the head of the chapter) the proboscis is ovoid. Its shape, and the way it fits into the collar when not extended, reminded someone of an acorn—hence the name of the worm. The proboscis and collar are muscular and are used in burrowing. The proboscis is forced through the sand, with the collar following. The distention of the collar firmly anchors the anterior end of the worm, so that contraction of the muscles in the trunk region draws the trunk forward.

The mouth is in the middle of the ventral surface, at the base of the proboscis and concealed by the edge of the collar. As the animal burrows,

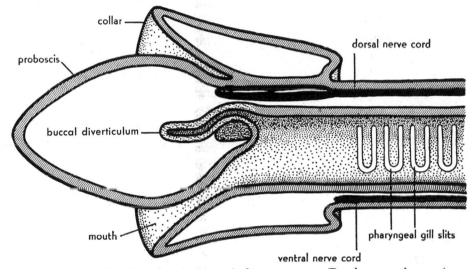

Longitudinal section through **anterior end of acorn worm.** (Based on several sources)

sand or mud passes into the mouth, through the digestive tract, and out the anus at the posterior tip. Organic materials present in the sand or mud are digested. Water taken in is supposed to pass out through the *pharyngeal gill slits* which open through the dorsal wall of the anterior part of the trunk.

The nervous system consists of a network of nerve cells extending under the whole of the surface ectoderm and in the trunk region is concentrated along the mid-dorsal and mid-ventral lines of the body as two nerve cords. Only the *dorsal cord* extends into the collar, where it is especially thick, and in some species of acorn worms it is hollow, suggesting a resemblance to the tubular, dorsal nerve cord of a typical chordate.

In almost all respects the nervous system of acorn worms is the most primitive of any group of animals having an organ-system level of construction. It does not show any of the centralization found in the ganglionated cords of even the simplest flatworms. It

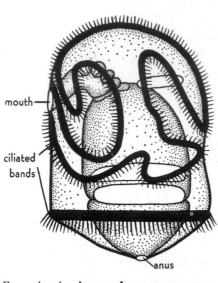

mouth

ciliated
bands

anus

Free-swimming **larva of an acorn worm.**
(Modified after Morgan)

is most closely comparable to the diffusely organized nerve nets of coelenterates and ctenophores but fails to show even the variety of response of many members of those two groups. The two nerve cords are almost entirely simple conduction pathways for nervous impulses; they cannot be considered a central nervous system in the same sense as this name is applied to the integrated ganglionated nervous systems of most invertebrate groups.

A short rodlike outgrowth (the buccal diverticulum) of the anterior end of the digestive tract into the base of the proboscis strengthens the proboscis and is composed of vacuolated cells like those in the notochord of the amphioxus, but that it really corresponds to the notochord of typical chordates seems unlikely.

Aside from their interest as animals which appear to have branched from the early chordate stock, the acorn worms, through their larvas, furnish one of the few real clues that link the chordates with any other phylum. The larva of the acorn worm looks so surprisingly like the larvas of certain echinoderms that they can be mistaken for them. Moreover, similarities extend beyond the structure of the two organisms and include several details in the formation of the coelom and other parts. The later development of the larvas is very different, for the echinoderm larva metamorphoses into an adult with a kind of radial symmetry, while the adult acorn worm is bilateral.

MEMBERS of the phylum Chordata all have, at some time in their life-history, a notochord, dorsal tubular nervous system, and pharyngeal slits. But in most other respects they fall into three groups— vertebrates, amphioxids, and tunicates—so different that each is designated a subphylum. Of the two groups of invertebrate chordates, the amphioxus is most, and the tunicates least, like the vertebrates. Since acorn worms have no unquestionable notochord or dorsal tubular nerve cord, zoölogists place these wormlike animals with pharyngeal slits in a phylum (the **HEMICHORDATA**) by themselves.

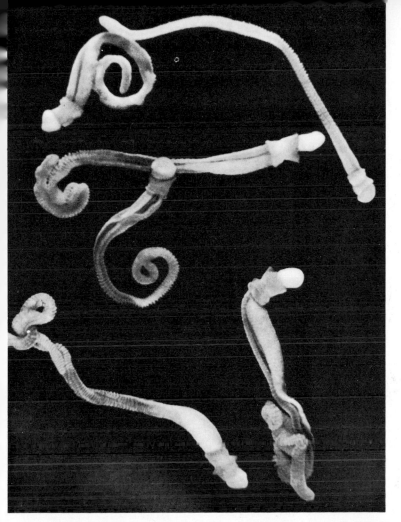

Acorn worms are sluggish burrowers on sandy or muddy bottoms in shallow marine waters. Once settled in their mucus-lined burrows, they seldom exert themselves except to extend or retreat from the mouth of the burrow. They feed by passing through the digestive tract a steady stream of mud or muddy sand from which they extract the contained organic matter. While acorn worms have certain structures like those of chordates, they have neither the musculature nor the rigidity of vertebrates and are so flabby that they burrow slowly. They are very delicate and practically always fragment if brought up in a dredge. The worms shown here (*Ptychodera bahamensis*) were brought up intact by going down in a diving helmet in about 20 feet of water and gently scooping them up by hand from the sandy bottom. (Photo of living animals. Bermuda)

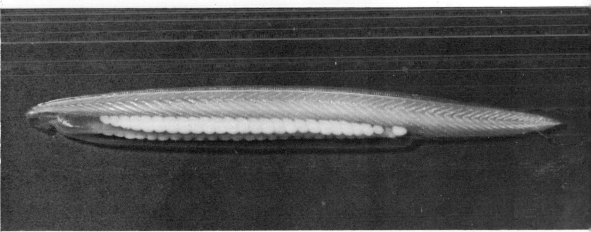

An amphioxus, (*Branchiostoma*), living, magnified, shows practically all the significant structures of its anatomy (compare with diagram in text). Clearly seen are the pharyngeal slits, notochord, and dorsal tubular nerve cord. This archaic form is world-wide in its distribution, being found on sandy shores in temperate and tropical regions. The living adult animal is about 5 cm long.

A typical tunicate is the "sea peach" which lives on rocks and wharf pilings along the North Atlantic Coast. It is a baglike, sessile animal completely inclosed in a tough covering or "tunic." The tunic is the thickened outer layer of the body wall and is made of a material like the cellulose that gives rigidity and strength to the cell walls of plants. Like all animals that give up the free-living habit and revert to a sessile life, the adult tunicate is a degenerate form. It has no sense organs, and the basic bilateral symmetry is overlain by a superficial radial symmetry. The tunicate is a filter-feeder, and the interior of the body is occupied mostly by a large saclike pharynx pierced by rows of gill slits that furnish a good clue to the affinities of the animal. (Photo of living animal. Mount Desert Island, Maine)

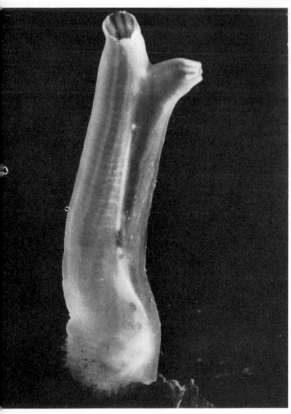

The tube tunicate, *Ciona intestinalis* (¾ nat. size) has a semitransparent tunic through which can be seen the muscles of the body wall. Water enters the upper opening and leaves by way of the lower opening. (Photo by D. P. Wilson. Plymouth, Eng.)

Clustering tunicates are all separate individuals that grow close together. *Dendrodoa grossularia*, shown here about natural size, is a small red tunicate that grows in berry-like bunches on rocks near low water mark. (Photo by D. P. Wilson. Plymouth, England)

CHAPTER 27

RECORDS OF THE INVERTEBRATE PAST

ALTHOUGH animals have left no birth certificates, marriage contracts, tombstone inscriptions, or written documents, upon which students of human history depend so much, there are abundant records of the invertebrate past—not just of the past few thousand years but of some 550,000,000 years or more.

Any evidence in the materials or rocks of the earth's crust that gives some idea of the size, shape, or structure of the whole or any part of a plant or animal that once lived is called a **fossil**. The name is derived from the Latin "to dig" because it was, at one time, applied to almost anything of interest that was dug up. Coal is an indication of past life, but it cannot be considered a fossil because by itself it gives no idea of the organisms which were responsible for its formation. Nor would we classify as a fossil any empty snail shell that turned up while we were digging about in the garden, for not only must a fossil have something of the character of an organism but it must have age. The exact age at which animal remains become fossils is not fixed; most fossils are evidences of organisms that lived at least 25,000 years ago; and most, though not all of them, are of species now extinct.

Fossils can be formed in a variety of ways. The rarest, but also the most interesting, kinds are **animals preserved with little or no change**

from their condition at the time of death. Such are found in considerable numbers in the amber of the Baltic region of Europe, which in early Oligocene times (about 38,000,000 years ago) was covered with a dense coniferous forest. Drops of gummy resin, dripping from the trees, trapped spiders and mites scurrying across the forest floor or insects on the wing. The sticky resin, on fossilization, became hard amber; and the arthropods imbedded in it are preserved intact, sometimes even with their colors unchanged. So perfectly embalmed are some of the specimens that they have been dissected and their intestinal parasites examined. Less complete, but much more common, unaltered fossils are the shells of mollusks and brachiopods, which, even after millions of years of preservation in the rocks, may still retain their colors or luster.

Most fossils, however, have undergone more or less change since the death of the organism. Horny coverings, such as the chitinous exoskeletons of arthropods, leave only thin **films of carbon** in the rocks. Calcareous shells and other skeletal structures are slowly dissolved by water percolating through the ground and are gradually replaced by minerals, such as calcite, silicon dioxide, or iron sulphide, which are deposited in the cavities left by the slow dissolution of the original materials. Shells may be completely dissolved away, leaving only a cavity, on the wall of which is a **mold** of the external surface of the shell. Or the cavity later may be filled with some mineral, forming a **cast** of the original fossil.

As we would expect, the vast majority of animal fossils are of forms with hard parts: calcareous shells or spicules; horny coverings, exoskeletons, or jaws; and silicious coverings or spicules. But even very soft animals like jellyfishes may leave **impressions** on soft mud which, if soon covered by a layer of fine sediment, may eventually be preserved when the mud hardens into rock. In a similar way there have been formed many of the **indirect evidences** of the former activities of animals. *Tracks or trails* impressed on muddy or sandy bottoms sometimes indicate the kind of appendage that left the record. *Burrows* in mud or sand and *borings* in rocks can be identified if, as in the case of sea urchins and pelecypods, the shells or parts of the shells are left in the cavities. Certain fossil markings, either straplike in form or resembling the pellets eliminated by living animals, are interpreted as fossilized animal excreta. In some cases these *coprolites*, as they are called, have revealed what the extinct animals fed on.

In order to be fossilized, an animal must be buried at the time of death or very shortly thereafter; otherwise the body is likely to be eaten by scavengers or disintegrated by bacteria. **Fossils formed on land** are relatively rare, for, even when land animals are covered by wind-blown earth or sand rapidly enough to escape destruction by scavengers or bacterial decay, the potential fossils are soon removed by erosion. Only under exceptional conditions, such as prevail when an eruption traps animals under heavy layers of volcanic ash, or dripping resin accumulates over long periods, or land animals are caught in floods and are carried out to sea, do we get fossils good enough to contribute to our sketchy picture of ancient terrestrial life. **Fossils of marine invertebrates,** however, are very abundant, for rapid burial is easily accomplished in shallow sea water, where there is constant shifting of the mud or sand at the bottom. Later, when by geological processes these marine sediments have been uplifted from the sea bottom, they present a legible record, with the earliest animals preserved in the lowermost layers and the most recent types in the uppermost strata. In most places we can examine only the layers fairly close to the surface, and we have no access to the hundreds of millions of fossils in the underlying rocks. Over certain large areas, however, the older rocks have been exposed by uplift and erosion. They can also be examined in places like the walls of the Grand Canyon of the Colorado, where the river has cut a cross-section a mile deep through the earth's crust.

A systematic study of the fossils found in the successive layers of rock reveals, not only that animals have been present on the earth during at least 550,000,000 years for which we have good fossil records, but that *the deeper we go into the rocks the less and less familiar are the fossils which we find.* Those a mere million years old are of animals much like living forms. Those several million years old are more different, and those still older belong to orders, and even to whole classes, of animals which no longer exist. In other words, the fossil record furnishes direct evidence that animals were not always as they are today, and that modern forms, which do not occur as fossils, must be descended, by a process of gradual modification which we call *organic evolution,* from the earlier and simpler animals whose remains we find in the rocks. In some cases the record is so complete that we are able to trace the gradual evolution, from one layer of rock to the next, of a definite structure or set of structures, all intermediate stages from the most primitive type to the modern living form being present.

Strangely enough, the science of ancient life, or paleontology, has been developed not by biologists, who have been busy enough studying living animals, but by geologists, who have demonstrated that the rocks of the various periods in geologic time are characterized by a distinctive assemblage of animal and plant fossils. Certain of these fossils, which are world-wide in distribution, distinctive, and restricted to limited periods in geologic time, serve as markers or *index fossils* by which rocks of any period can be recognized no matter where they occur.

GEOLOGICAL time has been divided into six **eras,** the end of each marking the time of some significant geological change, such as continental uplift and mountain-building. Shorter and less profound changes in the face of the earth form a basis for dividing each of the eras into a number of **periods.**

The first, or **Azoic** ("no life"), era marks the origin of the earth from the sun and the formation of rocks; no life was present. This period lasted over 1,000,000,000 years (a time estimate based on several lines of evidence, mostly on the rate of disintegration of radioactive substances). In the second, or **Archeozoic** ("primitive life") era, there was extensive volcanic activity and mountain-building. If life had evolved, the rocks show little evidence of it, though they do contain carbonaceous material, probably an indication of primitive algae. Perhaps one-celled animals appeared at this time. Rocks of the **Proterozoic** ("first life") era have only rarely yielded a recognizable fossil; yet this era must certainly have been a time of great evolutionary development, for by the *Cambrian* period, the first period of the **Paleozoic** ("ancient life") era, the animal kingdom is already highly diversified. Nearly all phyla which leave any kind of a fossil record are well represented in Cambrian rocks—many of them by several groups, which already show the distinctive characters of modern classes. Why pre-Cambrian fossils are so rare is not yet understood, though many possible explanations have been suggested. Two of the most plausible ones are that animals did not evolve hard, preservable parts until the Cambrian and that the pre-Cambrian rocks, being very ancient, have been subjected to so many stresses that any fossils which they once contained have been destroyed. In any case, the earliest well-preserved, abundant fossils are those laid down in the Cambrian period; and from that time on they increase steadily with the expansion of the various groups. By the *Ordovician*, the second period of the Paleozoic, many invertebrate groups are at their peak of abundance, and vertebrates are already on the scene. At the close of the Paleozoic, many of its most important invertebrate groups become extinct. The **Mesozoic** ("middle life") is an era of waning influence for many marine invertebrate groups, though certain large, shelled cephalopods reach their climax during this time. At the beginning of the **Cenozoic** ("recent life") era there appear many of the more modern types of invertebrates.

TABLE OF GEOLOGIC TIME

Eras (Millions of Years Ago*)	Periods (Duration in Millions of Years*)	Principal Evolutionary Events among the Invertebrates
CENOZOIC (60)	Quaternary (2)	Arthropods and mollusks most abundant; all other phyla well represented
	Tertiary (58)	Modern invertebrate types appear
MESOZOIC (200)	Cretaceous (70)	Extinction of ammonoids
	Jurassic (38)	Ammonoids abundant; modern types of crustaceans appear
	Triassic (32)	Marine invertebrates decline in numbers and importance. *Limulus* present
PALEOZOIC (550)	Permian (35)	Last of the trilobites and eurypterids. Extinction of most paleozoic invertebrate types
	Pennsylvanian (45)	First fossil insects (although insects probably evolved as early as the Devonian)
	Mississippian (35)	Climax of crinoids and blastoids
	Devonian (35)	Brachiopods still flourishing. Marked decline of graptolites and trilobites
	Silurian (25)	Many graptolites. First extensive coral reefs. Brachiopods at peak. Eurypterids abundant. Trilobites begin to decline. First land invertebrates
	Ordovician (70)	Peak of invertebrate dominance. Climax of trilobites and nautiloid cephalopods. Brachiopods abundant. (First vertebrate fossils)
	Cambrian (105)	First abundant fossils. Nearly all invertebrate phyla represented. Trilobites and brachiopods most numerous
PROTEROZOIC (1,200)	(650)	Most of the invertebrate phyla probably evolved, but fossils are rare and poorly preserved
ARCHEOZOIC (2,000)	(800)	No fossil record, but simplest living organisms (one-celled plants and animals) probably arose
AZOIC (3,000)	(1,000)	No evidence of living organisms. (Formation of the earth)

* Based on Croneis and Krumbein, *Down to Earth.*

THE major time units, their estimated age and length of duration, together with some of the invertebrate groups characteristic of the various periods, are summarized in the accompanying table. The rest of this chapter is devoted to a brief description of some of the better-known fossil groups.

AMONG the **protozoa** one would naturally not expect to find fossils of flagellates, amebas, or ciliates, but only of the ameboid protozoans with hard parts. Although the silicious skeletons of *radiolarians* are known from the Paleozoic to the present, the calcareous shells of the *foraminifers* furnish by far the best record of ancient protozoan life. Good fossil foraminifers occur in the Cambrian rocks, increase steadily both in numbers of species and of individuals, until in the Cretaceous they finally become very abundant; and there appear many families which are still represented today. From the Tertiary to the present there has been a gradual extinction of the earlier types and a replacement by those more like the modern foraminifers. This gradual change in type, plus the fact that the calcareous shells are abundant, widespread, easily preserved, and have distinctive shapes and markings, makes fossil foraminifers extremely valuable as index fossils. There are, of course, many different kinds of index fossils; but most of them, like trilobites and brachiopods, are large and can only be obtained from rocks which are exposed. Because of their minute size, foraminiferan shells can be obtained undamaged, from rocks very far below the surface, in the borings brought up by a drill. For this reason oil companies have found it profitable to employ paleontologists who do nothing but study the borings from oil-well drills. By comparing the shells found at different levels with those present in layers known to be oil-bearing, they are able to direct the well-drilling operations.

THE fossil record of **sponges** is not an abundant one, but it stretches over a very long time, silicious spicules being known even from the pre-Cambrian, and whole specimens from the Cambrian. The fibers of horny sponges have practically no chance of being preserved, and the delicate spicules of calcareous sponges are usually dissolved under the local acid conditions created by the decay of a dead sponge. Silicious spicules are more readily preserved but may later be dissolved away by ground water percolating through the rocks, and in such cases the cavities left may be filled with calcareous or other materials. Consequently, the composition of a sponge fossil does not necessarily indicate the original composi-

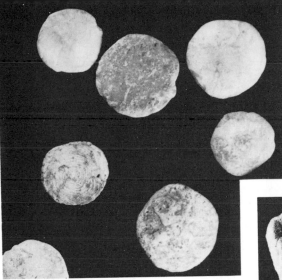

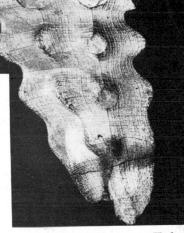

Foraminifers of the past dwarf the largest modern forms, which, though about the size of a dime, are giants among living protozoa. *Nummulites*, shells of which are shown about natural size, flourished on the sea bottom in the early Tertiary and formed great beds of limestone, now exposed in the Alps and northern Africa. The pyramids of Gizeh, near Cairo, are built of nummulitic limestone.

Small cup sponge, *Palaeomanon*, from the rocks laid down in the seas of middle Silurian times. (⅔ actual size)

Large silicious sponge, *Hydnoceras*, lived on the bottom of a Devonian sea in what is now New York State. (About ½ actual size)

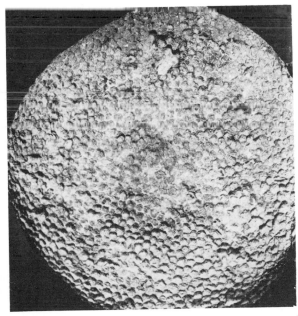

Tetracoral, *Columnaria*, a colonial coral with internal partitions in multiples of 4. Before becoming extinct at the end of the Paleozoic, the tetracorals probably gave rise to the hexacorals, with partitions in multiples of 6, to which belong modern sea anemones and reef corals. (½ actual size)

Graptolites are carbonized fossils left by extinct colonial forms. *Diplograptus*, from the Ordovician, has stems, each with a double row of horny cups occupied by the members of the colony. It is not clear whether they are coelenterates or hemichordates. (3/4 actual size)

The earliest brachiopods had thin horny shells which were unhinged like those of *Lingulella* (*left*). After the Cambrian these unhinged types were outnumbered by the hinged brachiopods with calcareous shells like *Terebratula* (*right*) and *Spirifer* (*below*). (Photo of *Lingulella* by Mrs. C. L. Fenton)

Brachiopod shells, when very numerous, form solid beds of limestone called "lamp-shell coquina." This piece of Ordovician coquina contains *Dalmanella* shells. Notice the small hole in one of the shells, probably the result of an attack by a carnivorous gastropod. (Photo by Mrs. C. L. Fenton)

Cystoid, *Pleurocystis*, from the Ordovician. Note the broken bases of the two arms. The most primitive of echinoderms, cystoids became extinct at the end of the Paleozoic. (Slightly less than actual size)

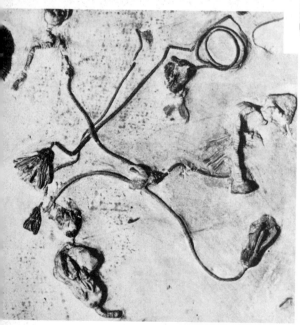

Stalked sessile crinoids (*above*) were the dominant members of their class during the Paleozoic; about ⅓ actual size. Sessile crinoids still live in deep waters, but the free-swimming types are much more common now. *Right.* Crown and part of stem of a crinoid, *Dizygocrinus*, from the Mississippian; natural size.

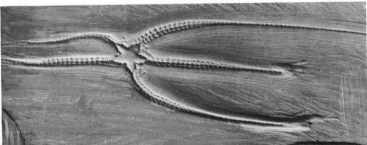

Upper left: **sea urchin** test, *Cidaris,* slightly enlarged, from the Jurassic. *Above:* **starfish,** *Petraster,* from the Ordovician; actual size. *Left below:* **serpent star,** *Furcaster,* from the Devonian; ⅔ actual size.

326-3

Shelled cephalopods started out with straight chambered shells shaped like a cone, as in *Orthoceras* (about ⅔ actual size) at *left*. These finally reached lengths approaching 20 feet. Some of them gradually evolved curved shells, as in *Jolietoceras* (slightly less than ½ actual size) (*above right*). Both nautiloids and ammonoids finally developed coiled shells, as in the ammonite, *Coeloceras* (about actual size) (*below right*). Certain coiled ammonites of the Cretaceous had shells almost 7 feet in diameter—an all-time record for shelled invertebrates.

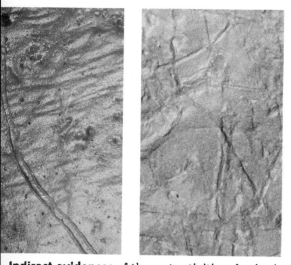

Indirect evidences of the past activities of animals are the fossil snail tracks on a rock (*right*) from the Cambrian. Compare these with tracks of a living snail (*left*). (Photos by Mrs. C. L. Fenton)

An ancestor of Limulus is *Prestwitchia* from the Pennsylvanian, a fossil member of the primitive arachnid-like group of which *Limulus* is the sole survivor. (Slightly enlarged)

tion of the living sponge, and identification must be based on the general structure and on the shape of the spicules.

FOSSILS of **coelenterates** are known from the pre-Cambrian to the present. Although pre-Cambrian fossils of any kind are extremely rare, and after many years of search few other well-authenticated fossils have ever been found in the pre-Cambrian rocks exposed in the Grand Canyon, those rocks have yielded a well-preserved imprint of one of the most fragile of animals, a jellyfish. Finds like this one serve to emphasize the fortuitous nature of any fossil record.

The **Hydrozoa** are represented in the lower Cambrian layers; and in the upper Cambrian rocks we find the beginnings of the **Graptolithina,** an interesting group of colonial animals, all extinct, that have been claimed for the coelenterates and the hemichordates. Graptolites are most often found as thin carbonaceous films which look like saw blades with one or both edges toothed. A blade is the remains of a horny colonial skeleton, and each notch in the toothed edge represents a small cup which formerly housed a single member of the colony. From the blade there comes off a long thread by which, it is thought, the colony was attached to floating seaweed. Some graptolites are found in large groups radiating from some central body, which may have been a gas-filled float. If so, they were independent floating colonies comparable with our modern siphonophores. The types just described flourished in the Ordovician and Silurian and died out before the beginning of the Devonian—possibly because of something related to the rise of the fishes. The more primitive branching graptolites, most of which are thought to have lived attached to the bottom, appeared in the Cambrian and had a few representatives which managed to survive throughout the Devonian. This situation is what we find in most groups; the primitive forms not only precede the more specialized and "progressive" ones, but, being less dependent upon special conditions, usually also outlast them.

Fossil **Scyphozoa** are known from the lower Cambrian, but they are rare and the authenticity of some is questioned. They consist of what appear to be molds of the upper and lower surfaces and mud fillings of the pouches which comprise the large central part of the gastrovascular cavity.

The **Anthozoa,** as one might predict from their highly developed structure, are the latest class to appear in the fossil record. The sea anemones, lacking hard parts, have left no recognizable traces. But the forms which secrete skeletons, particularly those made of limestone, provide an abundant and informative record of the past activities of their delicate and perishable polyps. Two classes of extinct corals are known. One of these, the *Tetracoralla* (named for the fact that the stony partitions are in multiples of four), occurs as solitary cup corals or as dome-shaped or branching colonies. The skeletons appear in the Ordovician, attain their maximum in the Silurian, when certain ones contribute abundantly to the fossil reefs, and disappear by the end of the Palaeozoic. The *Hexacoralla*, to which modern anemones and stony corals belong, are not represented until after the beginning of the Mesozoic. Their sudden, though late, appearance is accounted for by assuming that they are descended directly from the exclusively Paleozoic Tetracoralla. The stony corals were important contributors to the Mesozoic and Cenozoic reefs and are the dominant reef-builders of modern seas. The *Octocoralla*, or alcyonarians, anthozoans with eight tentacles and eight internal partitions, are represented from the Ordovician on.

THE wormlike invertebrates have left a very poor fossil record. They may be discussed together because part of the record consists of trails, tracks, tubes, and burrows; and it is impossible to say which of the various groups is responsible for certain of them. The **flatworms** have left no recognizable remains. Fossil **nemerteans** are unknown; but, since the living forms do burrow, it is possible that some of the fossil trails and burrows were made by members of this group. Fossil parasitic **roundworms** are reported to have been found in fossil insects.

OF THE **annelids,** only the polychetes have left a record of much importance. From the pre-Cambrian we have tracks and burrows, many of which must have been left by this group. From the mid-Cambrian we have well-preserved fossils with segmental bundles of bristles and other characteristic features of the class, showing that the animals must have had a long pre-Cambrian development. The free-swimming polychetes, modern representatives of which have a protrusible pharynx armed with hard jaws, are thought to be the source of the numerous small toothed jaws found in practically all rocks from the Paleozoic on. Beginning with the Ordovician, we find evidence of tube-dwelling polychetes in the form of tubes which are calcareous or are made of sand grains or other particles cemented together by a secretion. Some of them are found attached to brachiopod shells, often in clusters.

THE **Bryozoa** appear first in the Ordovician, and apparently have been quite abundant from that time to the present, for over 1,500 species are known from the Paleozoic, 1,000 from the Mesozoic, and there are additional thousands of Cenozoic and modern species. While whole colonies are sizable (a number of fragments of the skeletons of branching colonies appear in the drawing which heads this chapter, and in the drawing look like pieces of broken twigs), the individual cases are microscopic and therefore have many of the same advantages for the paleontologist as foraminiferan shells. Two whole classes of Paleozoic bryozoans became extinct at the end of that era. One class, already represented in the Ordovician, and which has never been very abundant, still has living representatives. The class to which most modern bryozoans belong did not appear until the Jurassic.

WE ARE accustomed to think of the **brachiopods** as a small and unimportant phylum, which in this book has been described as one of the "lesser lights." But if we were to view the animal kingdom in terms of the past as well as the present, we would have to reserve a more prominent place for the group that has left one of the most abundant, most complete, and most beautifully preserved of all fossil records. Not only do brachiopods have hard shells which lend themselves readily to preservation, but the animals live in shallow seas where the chance of fossilization is great. Consequently, their remains are often so abundant as to form most of the rock in which they occur. The group left its earliest record in the Cambrian; and before the end of that period, four of the five orders were already established, with the dominant position occupied by the relatively primitive order still represented by *Lingula*. Brachiopods with hinged valves were present in the Cambrian but did not become of major importance until the beginning of the Ordovician. Something of the early abundance and later decline of the brachiopods can be judged from the numbers of fossil genera recorded for the several eras. We know about 450 genera from the Paleozoic, about 180 from the Mesozoic, and only about 75 from the Cenozoic to the present. Although 3,000 species are recorded from the Ordovician and Silurian periods, which represent the peak of brachiopod abundance, only a little over 200 species are living today.

BEING shelled animals and second only to the arthropods in numbers of species, the **mollusks** have left an abundant, unbroken, and extremely legible record from the Cambrian to the present. The *amphi-*

neurans are certainly the most primitive of modern mollusks, but they do not appear as fossils until the Ordovician. And from that time to the present they have left only about 100 fossil species, so that there is nothing in their record to make us think that they were ever of much importance. The *scaphopods*, too, do not get well started until the Ordovician and have never amounted to much. The three major classes—gastropods, pelecypods, and cephalopods—all appear in the Cambrian. The first two groups increase steadily from that time to the present; and the **gastropods** are probably very close to their peak of development now, comprising 90,000 of the 110,000 species of living mollusks. The **cephalopods,** represented by fewer than 1,000 living species, were once far more important than the gastropods; and they have left 10,000 fossil species to attest to their past glory. The dominant cephalopods of modern seas are squids and octopuses, which are probably more numerous at the present than they ever were. But it is the cephalopods with external shells, now represented by only 3 species of *Nautilus*, that made cephalopod history. The shells of the early *nautiloids* were straight cones internally divided into a series of chambers, as in the modern nautilus. In the Ordovician these shells were over 15 feet long, a length never again reached by any shelled invertebrate. Later the shells became slightly curved, then finally coiled. The nautiloids rose to a peak in the late Ordovician and Silurian and then declined; but they gave rise, it is believed, to another great group of coiled cephalopods, the *ammonoids*, which also went through an evolution from straight to coiled shells. The ammonoids became the dominant animals of Mesozoic seas and then died out in the Cretaceous. Some of the cretaceous ammonoids reverted to an uncoiled condition, so that the evolution of this group parallels that of gastropods (see p. 187). The advantage of coiling is probably the same in the two groups; it converts a long, unwieldly, straight cone into a compact, manageable coil.

E VEN the most advanced of invertebrates, the **arthropods,** must have had a long pre-Cambrian history, for at the beginning of the Cambrian we find three classes of arthropods already well started.

The most numerous of these are the **trilobites,** which constitute over half of all Cambrian fossils. Aside from the undoubted success of these early arthropods, the abundance of their fossil remains is probably due in part to the fact that they molted frequently and discarded numerous exoskeletons which were capable of fossilization. The name trilobite means "three-lobed" and refers to the fact that the dorsal surface of the body is divided, by two longitudinal furrows, into three lobes. The body is also divided transversely into three regions: a head; a middle flexible portion, the thorax; and a posterior region, the abdomen, which consists of a number of fused segments and in some trilobites is prolonged into a spine. The head bears a pair of compound eyes, a pair of antennas, and four pairs of similar, jointed, two-branched appendages. The outer branch, which is flattened and has a row of bristles along its posterior edge, is thought to have served for respiration and swimming. The inner branch was probably used for walking. Similar appendages occur on all segments of the body, and they all have inwardly directed projections from the basal part of the limb. On the head appendages these projections are modified for chewing food, as in modern horseshoe crabs and arachnids. Most trilobites are from 1 to 3 inches long, but some forms reached a length of over 2 feet. Their habits can only be inferred; but it is thought that they lived in the sea, since their remains are always found with corals, crinoids, brachiopods, and other exclusively marine animals. Most trilobites probably inhabited shallow waters and were bottom-crawling types, either feeding on the various seaweeds, sponges, coelenterates, brachiopods, and mollusks which are known to have lived in the same places, or perhaps scavenging organic debris by ploughing through the mud. The trilobites were the dominant invertebrates of the Cambrian, and they continued to flourish during the Ordovician, but then declined. They were rare after the Devonian, and the last few survivors finally died out in the Permian. It is perhaps no mere coincidence that their decline followed the rise of the giant Ordovician cephalopods and the hordes of Devonian fishes, both of which could have fed on trilobites.

The trilobites probably gave rise to no other group of arthropods, but they seem most closely related to **crustaceans.** The branchiopods, ancestors of the more primitive modern crustaceans like the fairy shrimps and cladocerans, are well represented in lower Cambrian rocks. But the large crustaceans, such as lobsters and crabs, do not appear until the middle of the Mesozoic.

The **aquatic arachnid-like arthropods** appeared in the Cambrian; but their only living representative is *Limulus*, the horseshoe crab, which has changed little since its appearance in the Triassic, about 200,000,000 years ago. The most interesting of these extinct arachnid-like animals are the **eurypterids,** some of which attained a length of 10 feet, the largest size known for any arthropod. It is not certain whether they were marine or inhabited fresh-water streams, from where they were washed out to sea and buried in marine sediments. They probably lived mostly on the bottom, but their two large paddle-like appendages suggest that they could swim. Eurypterids are known from the Cambrian, flourished in the Silurian and Devonian, but, like the trilobites, became extinct at the end of the Paleozoic.

The **first land animals** may well have been certain primitive air-breathing arachnids (Palaeophonus and Proscorpius from the Silurian), which resemble eurypterids in certain respects. The transition from book gills, like those of a limulus, to book lungs, like those of a scorpion, requires very few changes.

The fossil record of land arthropods, like that of other land animals, is a scanty one and does not necessarily give a true picture of either the numbers of species or the time of the earliest appearance of the different groups. Although about half a million species of living insects have been described, only a few thousand fossil insect species have been found. Most of these are from a few special regions where fossilization took place under very unusual conditions. The insect-containing amber of the Baltic region of Europe has already been mentioned; one of the best American localities for collecting insect fossils is at Florissant, Colorado, where falling volcanic ash from an ancient eruption carried land insects down into a lake and buried them in the mud at the bottom. In spite of their slim chances of fossilization, **millipedes** are known from the Devonian and **centipedes** and **insects** from the Pennsylvanian. All of the early insects are now extinct, and only in a few cases can modern insects be assigned to an order which was already represented in the Paleozoic.

SINCE most **echinoderms** have skeletons made of calcareous plates and live in shallow marine waters, the group has left one of the most informative of fossil records. The earliest echinoderms were, typically, sessile types which lived attached to the bottom, either directly or by a stem. The oldest and least specialized ones were the **cystoids,** which had an ovoid or globular shell composed of tightly fitting plates. On the surface of the shell were ciliated grooves (usually five) for food-collecting, and these extended onto the arms which were attached at the upper end of the shell. From the cystoids, which appear in the Cambrian and become abundant in the Silurian, are thought to have come the other two classes of stalked echinoderms, the **blastoids** and **crinoids.** Not all stalked

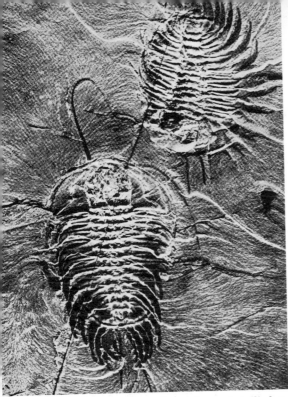

Trilobites are extinct aquatic arthropods allied to crustaceans. These two beautifully preserved fossils of *Neolenus* from mid-Cambrian rocks clearly show the trilobed body. Notice the antenna on the one at the left. (Photo by C. E. Resser)

Eurypterids are extinct aquatic arthropods related to arachnids. *Eurypterus*, from the Silurian, shows some small appendages and two large ones. The long segmented abdomen has no appendages and ends in a spine. (About ⅔ actual size)

Restoration of a Silurian sea bottom shows a large eurypterid, *Pterygotus*, which reached a length of 9 feet, the greatest size known for any arthropod. These heavy animals probably spent most of their time on the bottom (whether in the sea or in fresh water is an open question), but the two large paddle-like appendages suggest that they could swim. They were probably carnivorous. Like the trilobites and many other invertebrate groups, they became extinct at the end of the Paleozoic. (Photo courtesy Buffalo Museum of Science)

The largest insect that ever lived had a wing spread of 2½ feet. It belonged to the order Protodonata, extinct since the end of the Triassic period, which probably gave rise to modern dragonflies. This restoration is of a specimen found in Permian rocks near Elmo, Kansas. (Photo courtesy Chicago Nat. Hist. Museum)

Fossil ant from the Tertiary rocks of Colorado. Note the large compound eyes. (Photo from F. M. Carpenter)

Above: **An ant** imbedded in Tertiary amber (fossil resin) from the Baltic region. (Photo by E. Magdeburg)

Two termites, imbedded in amber from the Middle Tertiary (about 38,000,000 years ago) look as if they had died but yesterday. (Specimen lent by A. E. Emerson. Photo by P. S. Tice)

Above: **Termite wing** on a piece of Middle Tertiary rock from Spokane, Washington. Even the most delicate veins are easily seen. (Photo courtesy A.E. Emerson)

forms lived attached to the bottom; but the stalked, sessile types were the dominant echinoderms of Paleozoic times. The cystoids and blastoids became extinct by the end of the Paleozoic, and only the crinoids (sea lilies) have survived to the present. The earliest crinoids were all attached, with stems usually from 1 to 3 feet long, or over 70 feet long in at least one case. Fossil crinoids were known long before any living forms had been seen; and the class was believed to be extinct until, only a little over 60 years ago, a dredge brought up some living specimens. Although 90 per cent of modern forms are stemless and free-swimming, in all cases where the development has been studied there is an early stalked stage.

During the Paleozoic the free-living echinoderms—**starfishes, serpent stars, sea urchins,** and **sea cucumbers**—were inferior to the stalked types both in numbers of individuals and in species. But in the early Mesozoic they expanded rapidly and have maintained their superiority ever since.

TO THE geologist the fossil record serves not only as a means of determining the time of deposition of rocks in widely separated parts of the world but also as a key to the study of ancient geography and climate. Fossil corals, echinoderms, brachiopods, and cephalopods always indicate the former presence of salt water. The occurrence of fossil coral reefs in Chicago is clear evidence that this region was once covered by a sea and that the climate at one time must have been much warmer than it is now. To biologists the fossil record furnishes abundant and direct evidence of the evolution of modern animals from simpler types which have preceded them. By examining one layer of rock after another, we can follow the early appearance of a group as a few simple, adaptable forms, which gradually increase in number, specialize, and radiate out into a variety of habitats, then finally degenerate into bizarre, overspecialized forms which, with the first radical change in the environment, die out altogether. Such is the evolutionary history, clearly recorded in the rocks, of the trilobites and ammonoids—animals which dominated the seas for millions of years, yet have left not a single descendant. There is no reason to doubt that many of the invertebrate groups flourishing at the present time are heading for the same fate.

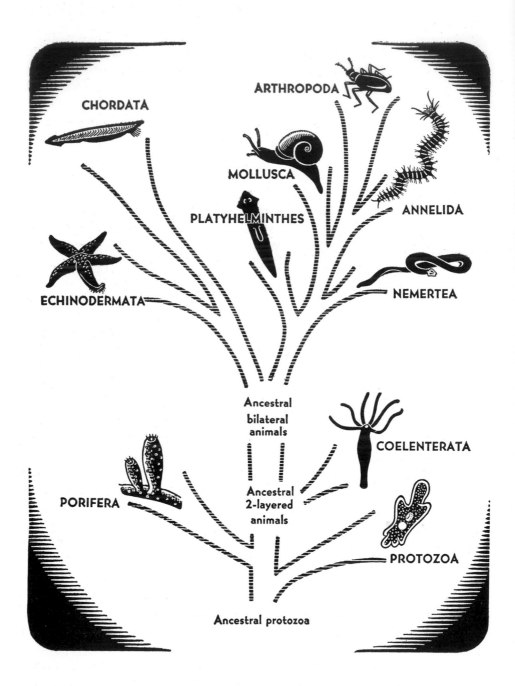

CHAPTER 28

INVERTEBRATE RELATIONSHIPS

EVERYONE enjoys the unraveling of a good mystery, but no one would like to read on from clue to clue, until the earliest and most important events seemed about to be disclosed, only to find that the rest of the pages in the book were missing. Just this kind of exasperating situation confronts us when we try to relate animals to one another in an orderly scheme. Anyone can see that honeybees are much like bumble bees, that bees resemble flies more than they do spiders, and that spiders are more like lobsters than like clams. But when we attempt to relate the phyla, which, by definition, are groups of animals with fundamentally different body plans, there is little we can say with certainty. Arthropods are clearly allied to annelids, but how they are related to such utterly different animals as starfishes or vertebrates is quite obscure. The fossil record, which in many cases provides us with a whole series of gradually changing specimens from which we can work out the evolution of one small group from another, is of practically no use in relating the phyla to each other. For, as we dig deeper and deeper into the rocks, expecting to find a level at which the most recently evolved phyla no longer appear, we find instead that the fossil record is obliterated. The oldest rocks have stood the longest time and have been subjected to the greatest number of stresses, including those caused by the weight of the rocks above them; any fossils which they once contained have been changed beyond recognition. In the rocks of the earliest period for which we have good fossils (the Cambrian period), all the important invertebrate phyla are already represented. Thus, while the fossil record tells us a great deal about what the early representatives of the phyla were like, it has nothing to say about the order in which the phyla arose. Despite this, the situation is by no means hopeless. Good detectives have been known to reconstruct, in detail, the events leading up to a crime to which there were no witnesses. And biologists have been able to find some definite clues to events that happened considerably more than 500,000,000 years ago!

The most important kind of evidence is that based on a comparative study of the structure and development of the various groups. The use of such evidence is based on the assumption that the more closely the body plans of two phyla resemble each other, the closer their relationship and the more recent their common ancestor. This is the *principle of homology*, discussed in chapter 22. Sometimes the adult structure of two groups is so highly modified in adaptation to their different ways of life that the groups

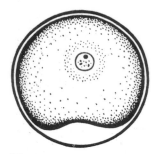  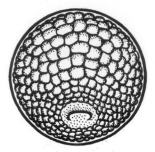

The fertilized egg of the lob-   ... divides répeatedly be-   ... a proliferation of cells in-
ster, a single cell ...   coming a blastula ...   to the interior results in a
                                                                   gastrula ...

show very little similarity, and yet the early embryonic stages are almost
identical. Here, also, we must assume a relationship, though a somewhat
more remote one, for the early stages of development tend to be more
conservative than the later ones, and a study of the embryology of animals often reveals basic similarities which otherwise would not have been
suspected.

Related evidence is that based upon the **principle of recapitulation,**
which states that "ontogeny tends to recapitulate phylogeny." Trans-
lating this into everyday English it reads: "Every animal, in its individual development from egg to adult, passes through a series of stages
which correspond to stages in the long evolutionary history of its group."
This does not mean that the embryo of a man at any time resembles an
adult fish or an adult reptile, but that it goes through stages which correspond to those undergone by the *embryos* of fishes and reptiles. The
reason for this, apparently, is that the early developmental stages of an
animal, in which the basic body structure is laid down, are less subject to
modification than are the later stages in which the more superficial structures appear. Thus, the animals of a single phylum tend to look alike in
their early embryonic stages and become gradually differentiated only
later, as the structures peculiar to the different classes, and finally of the
orders and still smaller categories, are produced. Moreover, even animals
of different phyla appear so much alike in their very early development
that they cannot be distinguished.

For example, a lobster starts out as a fertilized egg, a single cell which
looks like the egg of any other invertebrate and is spherically symmetrical,
showing no more differentiation than the simplest protozoan. The egg
divides repeatedly, resulting in a blastula, a hollow ball, composed of a
single layer of cells. We can find its counterpart among adult animals in

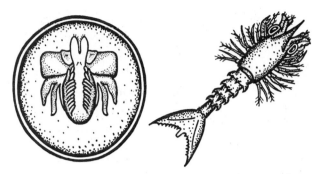

... with mesoderm, the bilateral embryo attains the flatworm level of construction ...

... segments remind one of the annelid grade of construction ...

... two-branched appendages, all similar, are like those of a primitive crustacean.

colonial organisms like the volvox, a colony of flagellated cells arranged in a single layer on the surface of a hollow ball of jellylike material (p. 40). The lobster blastula, by a proliferation of cells into the interior, soon becomes converted into a two-layer gastrula, much like that of any other animal. Since the gastrula has a depression at one end, which makes it radially symmetrical, and since it has two layers of cells, we may compare it with the coelenterate stage in evolution, though, of course, it has none of the specialized features of adult coelenterates. With the development of the mesoderm, and a differentiation of anterior and posterior ends, besides dorsal and ventral surfaces, the bilateral embryo has achieved the structural level of a flatworm. Next, segments appear, and pairs of similar appendages grow out, as in annelids. The appendages become two-branched; but, since they are mostly alike, the embryo reminds us of a primitive crustacean. With the differentiation and specialization of the appendages we finally recognize the developing animal to be a lobster. Thus, the development of a single individual is a condensed and modified recapitulation of what we believe was the evolutionary history of its phylum. In the same way we can hope to learn, from a study of the development of a planaria or a nereis, something of the evolutionary history of flatworms and annelids.

However, it should be pointed out at once that the **principle of recapitulation has very definite limitations.** In the first place, no development occupying a few days or weeks could possibly go through every stage in an evolutionary history stretching over at least a billion years. In the second place, not only adults but also embryos undergo evolution in adaptation to their environment. The mosquito larva, which lives in the water and feeds on debris, is a young stage modified for a specialized way of life that was never followed by any ancestor of the mosquito. Mosquitoes are descended from land insects, and the adults have had no connection with aquatic life since their primitive arthropod ancestors left the water. Moreover, many embryos develop special membranes and other structures which serve to protect or to nourish the young stages and have no counterpart in any adult animal. It has also been pointed out that, in the purely mechanical matter of changing from a single-celled zygote into a multicellular, many-layered animal of complex structure, all embryos must go through stages which are similar but do not necessarily indicate common descent for all the forms which exhibit them. Thus, in order to get a three-layered animal from a single cell, the cell has to form a one-layered organism, then a two-layered one, and, finally, a three-layered one; there is no other easy way to achieve such a result. And in the development from a spherical egg with all axes alike to a bilateral adult with three differentiated ones, there would have to be some intermediate radial stage with only one differentiated axis, even if the group had never had an ancestor which was radially symmetrical in the adult stage. Such arguments warn us that we must be cautious in assigning evolutionary significance to every step in embryonic development. Still, the principle of recapitulation, when correctly interpreted, has explained the appearance and subsequent disappearance in the embryo of many seemingly useless structures, such as the tail or gill pouches of man. And it has contributed greatly to our understanding of the relationships of animals to one another.

A CRITICISM often directed at the method of determining animal relationships from structural similarity is that the method lacks objectivity. Any animal group differs from or resembles another group not in any one structure but in a multiplicity of structures. And relationships assigned by various authors have differed according to which structural differences or similarities received the most emphasis. There is one method for determining animal relationships which restricts itself to a single criterion of similarity and is more objective. It consists of making a quantitative measure of the degree of biochemical similarity between the distinctive proteins that characterize each animal species. This method has, at the present time, definite limitations in its applicability. It is most useful in determining very close relationships, but it becomes less useful as we go up the scale toward those more distant rela tionships which are also the most difficult to ascertain by other methods. Wherever the biochemical methods for testing protein specificities have been successfully applied, they have, on the whole, supported the classifications that have already been worked out from structural and developmental lines of evidence and have given assurance that these older and less objective methods are generally very reliable.

An example of the use of a biochemical method for determining the relationships between animals will serve to illustrate the method. An investigator desiring to determine which two of three species of tapeworms are more closely related prepared a fat-free, sterile extract (mostly proteins) of some of the tissues of *Moniezia expansa*. This extract was injected into a vein of a healthy rabbit in several gradually increasing doses. About 10 days after the last injection, blood was withdrawn from the rabbit, and the serum was separated from the rest of the blood. This serum now contained **antibodies,** protective substances formed by the rabbit in response to the introduction into its blood of a foreign material like *Moniezia* extract; and these antibodies were specific for *Moniezia expansa* proteins. When the serum was placed in a test tube and to the same tube was added, without mixing, a layer of tissue extract from *Dipylidium caninum*, a white precipitate formed at the junction of the two layers of fluid. Extract from tissues of *Taenia* produced a precipitate also, but only with a more concentrated extract, indicating a less close relationship of *Moniezia* with *Taenia* than with *Dipylidium*. More distantly related genera would require still more concentrated extracts to produce a visible precipitate.

By a similar procedure the investigator prepared anti–*Dipylidium caninum* serum and also anti–*Taenia taeniaformis* serum. When the experiment was repeated utilizing anti–*Dipylidium* serum and anti–*Taenia* serum with *Moniezia expansa* extract, the relationship manifested in the first test was confirmed.

BASING our ideas on the principles of homology and recapitulation, we are able to construct **animal trees** or other schemes which attempt to show the order of evolution and the relationships between the phyla. Considering the remoteness of the events with which we are dealing, and the inconclusive nature of much of the evidence, it is clear that any "invertebrate tree" must be considered highly speculative. The "tree" presented at the beginning of this chapter is only one version; it will have to be changed in the future if new evidence turns up. At the same time, it ties together a great body of facts which would otherwise have less meaning and serves as a framework on which to hang what appears to be a fairly plausible account of the evolution of the main phyla of invertebrates.

It is highly probable that the capacity for photosynthesis was a characteristic of the ancestors of primitive organisms. From a hypothetical ancestral type of "plant-animal," the exact nature of which is unknown, came at least two main lines of descent, the animal kingdom and the plant kingdom (except the simplest plants, such as the bacteria, as was discussed in chap. 1). The reason for this belief, as explained at the beginning of this book, is the similarity between primitive plants and primitive animals. By a **loss of chlorophyll** (perhaps at several different times for different protozoan groups) and the development of a variety of locomotory and food-catching mechanisms, the animal kingdom arose. The most primitive animals are **single cells,** but we must remember that the modern *protozoa* have had a long evolutionary history and have undergone many changes before arriving at the condition in which we find them today.

The exact manner in which **multicellularity** arose cannot now be determined. But it is easy to understand how it could have evolved through the failure of individual cells to separate completely after division. Such colonies of attached but relatively independent cells are known to occur among the protozoa (see colonial collar flagellates on p. 43). In the volvox colony, already mentioned, there is a certain amount of co-operation in locomotion and in reproduction, but not enough to elevate the colony to the ranks of multicellular organisms. The volvox colony is highly specialized and must not be thought of as ancestral in any sense, but it gives us some idea of what one type of primitive multicellular organism might have been like.

In the *sponges*, the least integrated of the truly many-celled animals, the cells show considerable division of labor, but tissues are poorly developed, and the animals cannot be said to have gone much beyond a **cellular level of organization.** The porous construction, peculiar method of feeding, and the lack of a definite mouth and digestive cavity are among the reasons for thinking that sponges have no direct relationship to other animals. Perhaps they evolved from primitive collared flagellates, whose modern representatives are the only animals besides sponges which have collar cells.

Since sponges are not on the main line of evolutionary advance, the stage beyond the first multicellular organisms which led to the higher phyla can only be imagined. By the passage of some of the cells from the surface into the interior, a **two-layered animal** was formed. This hypothetical two-layered ancestor probably evolved from a different group of protozoa than that which gave rise to sponges. Constructed on the **tissue level of organization,** with an outer ciliated ectoderm specialized for locomotion, protection, and sensation, and an inner endoderm specialized for digestion, it was master of the ancient seas as it swam about, feeding on protozoans and unicellular plants. Just what it looked like we do not know, but it probably resembled the radially symmetrical, ciliated, two-layered, free-swimming larva of the obelia and most other marine *coelenterates*. The wide occurrence of such a larva in the embryology of coelenterates indicates that this phylum probably arose from a simple two-layered ancestor by the outgrowth of tentacles around the mouth.

Bilateral symmetry seems to have originated very early, because it is present in all higher phyla; indeed, as explained in chapter 11, there were beginnings of bilaterality in the coelenterates. It is not known just how

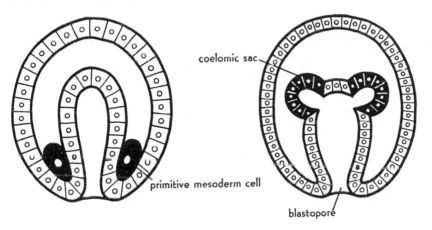

Origin of the mesoderm in arthropod and chordate lines. *Left*, the embryo of a mollusk showing the primitive mesoderm cells budded off from the primitive endoderm. *Right*, an echinoderm larva showing the mesoderm being budded off as pouches from the sides of the primitive endoderm.

the radial gastrula-like ancestor became bilateral, but specialization of one end and a bottom-feeding crawling habit were probably stages in the process.

The next stage in evolutionary advance seems to have been the formation of the **mesoderm**, a tissue between the ectoderm and endoderm, from which more definite organs and **organ-systems** could be made, resulting in greater size and complexity of construction. Mesoderm develops in animals by one of two main ways. In the flatworms, nemerteans, mollusks, annelids, and arthropods it usually originates from two special cells, known as "primitive mesoderm cells," set aside in the early gastrula. In the Echinodermata, the phylum to which the starfish belongs, and in the Chordata, the phylum to which man belongs, the mesoderm comes from outpocketings of the primitive endoderm. Because of this difference in mesoderm formation, among other reasons, we recognize two great lines of evolution known as the *arthropod* and the *chordate lines*, and indicated on the animal tree by a main branching.

Following the **arthropod line** of evolution, we see the *flatworms* exhibiting the beginning of the importance of the mesoderm. The flatworms are the first animals to make substantial use of the mesoderm, incorporating it into the development of organ systems. We come next to the *nemerteans*. From many details of adult structure and early development, nemerteans can be regarded as closely related to flatworms, but they show two advances: the development of a digestive tract with two openings, and the beginning of a circulatory system. Much more advanced are the *mollusks* (snails, slugs, clams, oysters, squids, octopuses, etc.), which have further perfected the systems pioneered by nemerteans. Mollusks probably diverged from the main line of evolution about the time **segmentation** arose, but the *annelids* more fully exploited it. Thus, on the animal tree they are shown to branch off at a point below that from which *annelids* diverge. In spite of the great lack of structural similarity, mollusks are closely related to annelids—one of the few relationships between any two invertebrate phyla for which we have indisputable evidence. The early embryos of annelids and mollusks are almost identical, cell for cell. The mesoderm arises from a corresponding cell in both groups; and their free-swimming larvas, the **trochophores,** are very much alike (chap. 19). A trochophore-like larva, with a ciliated band around the equator, is characteristic not only of annelids and mollusks but also of a number of minor phyla (discussed in chap. 16) not shown on the "tree." And the larvas of flatworms and nemerteans are certainly much like the trochophore. Thus, the trochophore type of larva serves to link together a whole series of phyla.

The *arthropods* have no trochophore, and they show few similarities to annelids in the early stages of development; but their adult structure is similar in so many respects that there can be no doubt that the two groups had a common segmented ancestor with a pair of appendages to each segment and a nervous system which encircled the anterior end of the digestive tract and passed backward along the ventral surface as a double cord with segmental ganglia.

The **chordate line** includes only three phyla: echinoderms, hemichordates and chordates. This means that the invertebrate phyla most closely related to the vertebrates include such animals as starfishes and acorn worms. The reasoning by which we arrive at this conclusion is based mainly on a comparison of the developing embryos of certain members of the three groups. In the first place, as already mentioned, the mesoderm of echinoderms, hemichordates, and chordates arises in the same way. Also, the **coelom** is formed in these groups from the hollow mesodermal pouches, whereas in the arthropod line it arises by splits in the bands of mesoderm budded off from the primitive mesoderm cells. In either case the end result is the same, and, by inspecting the adult animals, it cannot be told that their coeloms arise in different ways. In addition, the free-swimming larvas of echinoderms and hemichordates resemble each other but are quite different from the trochophore larvas characteristic of annelids, mollusks, and some of the other phyla of the arthropod line. The bilateral echinoderm and hemichordate larvas are more flattened than the trochophore and have longitudinal, looped ciliated bands for locomotion. Since similar types of larvas occur in all classes of echinoderms, they are believed to resemble a hypothetical ancestral type (dipleurula) from which all modern echinoderms have been derived. Finally, by the use of biochemical tests (such as were described earlier in this chapter for determining the relative degree of relationship between different groups) it has been shown that the proteins of certain acorn worms and tunicates resemble more closely the proteins of certain starfishes and sea cucumbers than they do those of certain annelids and arthropods.

The phylum *Chordata* consists mainly of vertebrates but includes two groups of invertebrates: tunicates and the amphioxus. These groups have no backbone but are classed with the vertebrates because, as described in chapter 26, they possess, at some time in their life history, a stiffening rod (the notochord), pharyngeal gill pouches or slits, and a dorsal tubular nerve cord. Now, it must be clearly understood that echinoderms have none of these structures; their affinity with chordates is based on similarities in the development of the mesoderm and coelom mentioned before and on the slender clue afforded by a striking resemblance between the (dipleurula) type of larva found in echinoderms and the larva of the acorn worm. These are so much alike that the larva of a New England acorn worm was described, by a specialist on echinoderms, as an echinoderm larva. As was pointed out in the discussion of recapitulation, larval resemblances may be misleading, because larvas themselves undergo evolution in adaptation to their environment. Since the larvas of echinoderms and acorn worms both live in the surface waters of the ocean, feeding on microscopic organisms, their similarities may be inde-

pendent responses to the same conditions of life. On the other hand, the
larvas of flatworms, nemerteans, mollusks, and annelids have for over
half a billion years lived in the same places and fed in the same way; yet
all conform to the trochophore type. This leads us to believe that the
differences between these two larval types have real evolutionary significance and are the result of a divergence of two main stocks from some
ancestral bilateral animal in the remote past.

Since the echinoderms are not clearly segmented, the segmentation of
chordates must have arisen after the two groups which finally gave rise to

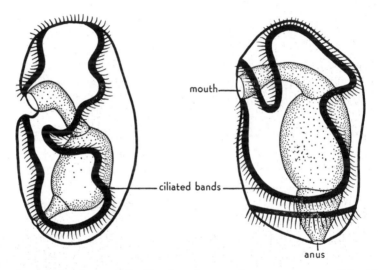

mouth

ciliated bands

anus

Larva of an echinoderm. (After Bury)   **Larva of an acorn worm.** (After Morgan)

the modern echinoderms and chordates had already separated. The segmentation of chordates results from the budding of a series of mesodermal
pouches from the primitive endoderm. In annelids and arthropods it
forms by a crosswise breaking-up of continuous bands of mesoderm. For
this reason, and the fact that the chordate and arthropod lines are thought
to have separated from each other long before segmentation arose, the
segmentation of the chordates and of the annelids and arthropods is not
considered homologous but must have evolved independently in the two
groups.

In the earliest schematizations of animal evolution, the animals were
usually pictured as ascending, on a vertical ladder, directly "from ameba
to man," with the other animals placed on intermediate rungs according

Marine laboratories are located on both our coasts, usually at points where there is a variety of habitats to furnish marine invertebrates of different types. The *Marine Biological Laboratory* situated on Cape Cod at Woods Hole, Massachusetts, is the largest institution of its kind and has played a very important role in the development of American biology and in the training of biological investigators. The laboratory is occupied the year around by a small group of scientists, but in the summer their number swells to hundreds, who come primarily to do research but also to attend lectures and exchange ideas with fellow-scientists. Classes are offered to undergraduate and graduate students. The main building (*left*) houses laboratories, offices, and a very large and excellent library. (Photo courtesy Marine Biological Laboratory)

Marine laboratories offer opportunities for study all over the world. The *Plymouth Laboratory of the Biological Association of the United Kingdom* is the largest of the British laboratories and has contributed very importantly to our knowledge of marine invertebrates. It maintains research facilities for many scientists, classes for students, and an aquarium for the public. There is a very fine library. Many of the photographs of marine invertebrates which appear in this book were made at Plymouth by Douglas P. Wilson, of the resident staff. (Photo by Miss E. J. Batham)

Rocky seacoasts provide the best collecting grounds for marine invertebrates, but many can be found only in marine mud flats or on sandy beaches. Some animals must be collected by special methods requiring boats, dredges, trawls, or other equipment. But anyone can see a great variety of marine invertebrates merely by visiting exposed rocks and rocky tidepools between high tide, *left*, and low tide, *right*, marks. (Photos made at the *Mt. Desert Island Biological Laboratory*, Salsbury Cove, Maine)

Open marine bays offer a combination of exposure to the surf and protection from wave shock. The *Hopkins Marine Station* on Monterey Bay, at Pacific Grove, California, has excellent research facilities for investigators and summer classes for students. The extraordinarily rich fauna is especially accessible. Typical of the many large tidepools is the one above, filled with large sea urchins among which are huge sea anemones. Other excellent marine laboratories are located on the West Coast from Friday Harbor, Washington, to La Jolla, California. (Photo of tidepool by W. K. Fisher)

Subtropical shores offer opportunities to see the larger colonial coelenterates. These can be studied best where special facilities are provided by marine laboratories, of which there are several on our Florida coast. Almost as accessible is the *Bermuda Biological Station*, which offers facilities for research and for seeing the fauna of subtropical coral reefs. On brief visits to subtropical shores, the best place to view representative forms is in a public aquarium. The photograph above, made in the *Bermuda Aquarium*, shows in the center an expanded gorgonian and below it to the left an anemone with long tentacles. To the right are nugget and brain corals. Above the gorgonian is a bouquet-shaped "stinging coral" or millepore (a hydrozoan). It is the only kind of "coral" with stinging capsules painful to man.

Fresh-water lakes are studied as intensively as the ocean. Here a specialist on small crustaceans is using a net to strain small animals from the surface water at a marshy spot on the shore of Lake Erie.

Lake bottoms can be studied by several methods. Here a specialist in lake biology is demonstrating to students the use of a dredge which can be lowered to take samples of mud from the bottom of a small lake.

Inland field stations for the study of land invertebrates occur all over the world. One of the best for the study of tropical biology is the *Barro Colorado Island Biological Laboratory* in the Canal Zone. Situated on Barro Colorado Island in Gatun Lake, it affords an isolated sanctuary for many animals that are no longer easy to find on the mainland. There are research facilities for investigators. *Right:* An interior view of a corner of the main building. *Below left:* Night collecting of insects that come to a light at the side of the laboratory. *Below right:* Insects collected in about 2 hours on a single evening.

Field trips to any woods or meadow offer a chance to see representatives of practically every important group of terrestrial invertebrates. A net is an indispensable tool for catching insects on the wing. But, to see a variety of terrestrial invertebrates one would search the branches of shrubs and trees, in the grass, under logs, in rotting wood, and in the ground.

to their order of increasing complexity. Now it is realized that animals do not form a continuous series and that their relationships to one another are more correctly represented by a branching "tree." The ancestral groups, long extinct, are placed on the main stems. Extinct forms, now represented only by fossils, are placed on dead side branches (not shown in the "tree" given here). On the ends of the living side branches are placed the modern forms. A "tree" not only fits the facts more closely than a ladder, but its many separate branches, representing independent lines of specialization, help to suggest why we find such endless variety among the animals without backbones.

CHAPTER 29

FURTHER KNOWLEDGE

AS WE approach the end of this introduction to the invertebrates, it becomes necessary to say something to the inquiring student who wants to know more than has already been told about the sources of the materials presented in earlier chapters as well as how to gain more direct access to the large and ever growing volume of knowledge about the invertebrates.

How new facts are discovered, who performs the observations and experiments, where such work is done, and how it is made available to everyone interested in learning about it are questions that cannot be completely answered within the scope of an introductory book. Here it is possible only to allude briefly to these questions and to erect a few signposts that indicate the general direction in which students may set out to find what they wish to know.

MOST of the facts, principles, and problems which have been presented in this book are based on the work of **zoölogists,** almost all of them members of the staff of a college or a university, a research institute or a museum, a field laboratory or an aquarium, a government agency or an industrial laboratory.

Modern zoölogists are usually specialists who work within a limited field and can be classified roughly according to the nature of the problems they attack. **Taxonomists** describe, name, and classify animals, but the number of animal species is so great that a taxonomist can become expert on no more than one or a few small groups. **Morphologists** describe the structure of animals, and here again this field is so large that each worker becomes restricted to some particular group of animals or to a comparative study of some particular organ or organ system as it occurs throughout the animal kingdom. Taxonomists must necessarily know a great deal about the morphology of the group they work on, and morphologists may make original contributions to taxonomy, so that these two fields often overlap. Specialists on the morphology and taxonomy of certain groups are called *protozoölogists, helminthologists, malacologists, entomologists*, etc. **Developmental biologists** study the embryology, growth, and regeneration of animals. **Cell biologists** study the microscopic and ultramicroscopic anatomy and function of tissues and cells. **Physiologists** are concerned with how animals carry on their life-activities such as digestion, respiration, excretion, etc. **Molecular biologists** elucidate the biochemical processes of cells. **Geneticists** study the mechanism of heredity. They commonly use invertebrates, especially the fruit fly because of its small size and short life-cycle. **Paleontologists** study fossils. **Ecologists** are concerned with the relationships between animals and their environment and necessarily must constantly consider all the different kinds of animals as well as the plants. Some ecologists stress a physicochemical analysis of environmental factors; others are primarily interested in the interactions of organisms with one another. Many of the branches of zoölogy have accumulated a good descriptive background and are now largely experimental, so that specialization in these fields is on particular problems rather than on particular animals.

Specialists in all types of invertebrate zoölogy are represented on the staffs of academic institutions, and, in general, they devote themselves largely to the analysis of problems, most of which require special laboratory facilities. Zoölogists in government institutions are mostly concerned, directly or indirectly, with field or laboratory problems of economic importance. The essential work of describing and classifying animals goes on most actively in museums, some of which have facilities for sending out field expeditions to collect animals from all over the world.

Zoölogists who wish to study animals that cannot be obtained in their own locality often have them shipped in by special arrangements with their colleagues in other localities or may buy them from local collectors or from the various biological supply houses, whose facilities are available to anyone who wishes to buy living or preserved animals of a great variety. In addition, there are special **field stations** or laboratories maintained all over the world to provide laboratory facilities to scientific investigators at places where there is ready access to a rich and varied fauna. Some of these laboratories offer classes for students and opportunities to see an abundance of invertebrates of many kinds. A list and a brief description of these field stations are given in *Biological Field Stations of North America* by Hunsaker and Dalgleish and in *World Directory of Hydrobiological and Fisheries Institutions* by R. W. Hiatt.

K NOWLEDGE that is gained by scientific investigation is made available in several ways. Every year various scientific groups hold meetings at which members present their original researches in short talks, often illustrated with slides or motion pictures. Or they may prepare exhibits of their work. At such meetings they discuss the progress of their researches and exchange ideas with other scientists investigating similar problems. But by far the most important medium of communication between scientists is the publication of articles in special **scientific journals.** There are many hundreds of such journals in biology alone, and in most of them there are papers on some phase of the biology of the invertebrates. A few of the journals most accessible to and most useful for the student interested in the invertebrates are listed in the Bibliography. They can be found in the libraries of large academic or research institutions.

Zoölogists read regularly in order to learn as much as possible about what is known in the fields of their investigations, as well as to ascertain that they are not trying to find out something that has already been discovered. A major part of the reading consists of articles in scientific journals because the most complete and reliable sources of information are the published accounts written by the investigators who did the original research. Most scientific papers are primarily reports of original investigations, but almost all of them begin by summarizing what has already been accomplished on the immediate problem under consideration. Scientific articles are the units of which the body of scientific literature is built. The majority of modern scientific articles are difficult to understand without considerable background.

The scientific literature, the whole of published writings of original scientific research, is so voluminous and is increasing at such a rate that scientists find it difficult to scan all the journals to find pertinent articles or to read in their entirety all the articles covering even a limited area of biology. To make easier their task, biologists collectively support, write, and edit a special journal, *Biological Abstracts,* which consists of short summaries or abstracts of currently published articles, all classified according to subject matter and indexed by authors and subject. There are also present in biology libraries a number of special indexes for finding references to scientific work by subject matter, by name of author, or by the group of animals concerned.

An important aid to scientific workers in keeping up with the broad front of advancing zoölogical knowledge outside the limited field in which they read most intensively are the **reviews**—long summarizing articles which synthesize and critically evaluate all the evidence bearing on some problem. Such reviews, written by leading scholars in various fields, coordinate and compact the large and often unwieldy literature (sometimes more than a thousand papers) dealing with a single problem. Some journals are devoted entirely to reviews. In some fields reviews are published annually in a special volume, such as *Annual Reviews of Physiology*. Since most reviews have extensive bibliographies, they can serve as good starting-points from which students can find their way about in the literature of some field that attracts their interest.

Other useful publications are books or specialized **monographs,** which are devoted to a single subject, such as *The Medusae of the British Isles* by F. S. Russell. Broader but often just as intensive treatments are the advanced systematic **treatises** written by one or by a number of co-operating specialists and often consisting of many volumes covering some or all of the invertebrate groups. Good accounts of the various animal groups can be found in some of the better **encyclopedias.**

Finally there are **textbooks,** advanced and elementary, in which the authors have selected what they regard as the essential and better established aspects of the subject suitable for presentation to students. *Advanced textbooks* are usually less detailed than treatises, but they have the same systematic treatment in which every group is considered in its various aspects. Such books hold very accessible answers to many of the more difficult questions about animals that are asked by more advanced students. Some of these books are well documented with footnotes or bibliographies that give the sources in the scientific literature on which the authors have based their statements and to which the reader is referred for a more detailed treatment of the topic.

Elementary textbooks are not intended to be complete, systematic treatments of the various groups. Certain animals may be emphasized or omitted according to whether the author chooses them to illustrate certain facts or principles. For good reasons there is a minimum of documentation of the text, and this anonymity may give some students the erroneous impression that the materials presented are based entirely on the personal experiences and investigations of the author. On the contrary, these materials are based in larger part on the scientific literature. There are, in the literature, hundreds of published articles that bear on the few facts and the theory presented, for example, on pages 14 and 15, and the author consulted a very large number of them.

STUDENTS interested in pursuing a particular problem should look first in an advanced textbook, a treatise, or an encyclopedia. If the textual treatment in these is not satisfactory, students should scan the bibliographies of such publications. Each pertinent reference will lead to other sources in the literature. These may reveal that there is a thoroughly worked out answer to the question, that there are conflicting views on the subject, or that it has not yet been investigated.

SELECTED BIBLIOGRAPHY

JOURNALS CONTAINING TECHNICAL ARTICLES ON THE INVERTEBRATES

American Journal of Tropical Medicine and Hygiene. Williams & Wilkins Co., Baltimore.
American Microscopical Society, Transactions. Allen Press, Lawrence, Kansas.
The American Naturalist. Amer. Soc. of Naturalists. University of Chicago Press, Chicago.
American Zoologist. American Soc. of Zoologists. Utica, N.Y.
Animal Behaviour. Assn. for the Study of Animal Behaviour. Bailliere Tindall, London.
Archiv fur Protistenkunde. Gustav Fischer, Jena.
Biological Bulletin. Marine Biological Laboratory, Woods Hole, Mass. Science Press, Lancaster, Pa.
Biological Reviews. Cambridge Physiological Society. Cambridge University Press, London.
Bioscience. American Institute of Biological Sciences, Arlington, Va.
Bulletin of Marine Science. Allen Press, Lawrence, Kansas.
Crustaceana. E. J. Brill, Leiden, Netherlands.
Deep-Sea Research. Pergamon Press, New York.
Ecology. Duke University Press, Durham, N.C.
Entomological Society of America, Annals. Columbus, Ohio.
Journal of Animal Ecology. Blackwell Scientific Publications, Oxford.
Journal of Ecology. Blackwell Scientific Publications, Oxford.
Journal of Embryology and Experimental Morphology. Cambridge University Press, New York.
Journal of Experimental Biology. Cambridge University Press, London.
Journal of Experimental Marine Biology and Ecology. North Holland Publishing Co., Netherlands.
Journal of Experimental Zoology. Wistar Institute of Anatomy and Biology, Philadelphia.
Journal of Marine Biological Association, Cambridge University Press, Cambridge, England.
Journal of Morphology. Wistar Institute of Anatomy and Biology, Philadelphia.
Journal of Natural History. Taylor & Francis Ltd., London.
Journal of Parasitology. American Society of Parasitologists. Allen Press, Lawrence, Kansas.
Journal of Zoology. Academic Press, New York.
Marine Biology. Springer-Verlag, New York.
Nature. Macmillan Journals, London.
Physiological Zoology. University of Chicago Press, Chicago.
Proceedings of the Malacological Society. Blackwell Scientific Publications, Oxford.
Proceedings of the Royal Society of London. Royal Society, London.
Quarterly Journal of Microscopic Science. Clarendon Press, London.
Quarterly Review of Biology. Williams & Wilkins Co., Baltimore.
Science. American Association for the Advancement of Science, Washington, D.C.
Systematic Zoology. Society of Systematic Zoology. Smithsonian Institution, Washington, D.C.

JOURNALS CONTAINING POPULAR ARTICLES ON THE INVERTEBRATES

Audubon. National Audubon Society, New York.
Endeavour. North Block, Thames House, Millbank, London.
Natural History. Museum of Natural History, New York.
Oceans. Menlo Park, California.
Scientific American. New York.
Wildlife. London.

ENCYCLOPEDIAS WITH ARTICLES ON THE INVERTEBRATES
Articles on various phyla and in many cases on separate classes or orders.

Encyclopaedia Britannica. Encyclopaedia Britannica, Inc., Chicago.
McGraw-Hill Encyclopedia of Science and Technology. McGraw-Hill Book Co. Inc., New York, London.

TREATISES PRIMARILY ON MORPHOLOGY AND TAXONOMY OF INVERTEBRATES

Bronn, H. G. (ed.). *Klassen und Ordnungen des Tierreichs.* Leipzig. 1866—.
Cambridge Natural History. The Macmillan Co., New York. 1895-1909.
Delage, Y. and Hérouard, E. *Traité de zoologie concrète.* Schleicher Fréres, Paris. 1896-1903.
Florkin, M. and Scheer, B. T. (eds.). *Chemical Zoology.* Vol. 1. *Protozoa* (George W. Kidder, ed.), 1967. Vol. 2. *Porifera, Coelenterata, Platyhelminthes,* 1968. Vol. 3. *Echinodermata, Nematoda, Acanthocephala,* 1969. Vol. 4. *Annelida, Echiura, Sipuncula,* 1969. Vol. 5. *Arthropoda,* Part A, 1970. Vol. 6. *Arthropoda,* Part B, 1971. Vol. 7. *Mollusca,* 1972. Vol. 8. *Deuterostomians (Hemichordata, Pogonophora, Chaetognatha, Cephalochordata, Tunicata),* 1974. Academic Press, New York.
Giese, A. C. and Pearse, J. S. *Reproduction of Marine Invertebrates.* Vol. 1. *Acoelomate and Pseudocoelomate Metazoans,* 1974. Vol. 2. *Entoprocts and Lesser Coelomates,* 1975. Other volumes to follow. Academic Press, New York.
Grassé, Pierre-P., ed. *Traité de Zoologie.* Vol. I—XI. Masson et Cie. 1952—.
Hyman, L. H. *The Invertebrates.* Vol. 1. *Protozoa through Ctenophora,* 1940. Vol. 2. *Platyhelminthes, Rhyncocoela,* 1951. Vol. 3. *Acanthocephala, Aschelminthes, Entoprocta,* 1951. Vol. 4. *Echinodermata,* 1955. Vol. 5. *Smaller Coelomate Groups,* 1959. Vol. 6. *Mollusca I,* 1967. Aplacophora, Polyplacophora, Monoplacophora, Gastropoda. McGraw-Hill Book Co. Inc., New York.
Kukenthal, W. and Krumbach, T. (eds.). *Handbuch der Zoologie.* Walter de Gruyter & Co., Berlin, 1923—.
Lankester, E. R. *Treatise on Zoology.* Adam & Charles Black, London. 1900-1909.

TEXTBOOKS ON THE INVERTEBRATES

Barnes, R. D. *Invertebrate Zoology.* W. B. Saunders Co., Philadelphia. 1972.
Barrington, E. J. W. *Invertebrate Structure and Function.* Houghton Mifflin Co., Boston. 1967.
Bayer, F. M. and Owre, H. B. *The Free-Living Lower Invertebrates: Porifera, Coelenterata, Ctenophora, Platyhelminthes, Nemertea.*
Beklemishev, W. N. *Principles of Comparative Anatomy of Invertebrates.* Vols. 1 and 2. University of Chicago Press. 1969.
Grassé, P., Poisson, R. A., and Tuzet, O. *Zoologie I: Invertebres.* Masson et Cie, Paris. 1961.
Hegner, R. W. and Engemann, J. G. *Invertebrate Zoology.* The Macmillan Co., New York. 1968.
Hickman, C. P. *Biology of the Invertebrates.* The C. V. Mosby Co., Saint Louis, 1973.
Kaestner, A. *Invertebrate Zoology.* Vol. I, 1967. Vol. II, 1968. Vol. III, 1970. Tr. by Herbert W. Levi and Lorna R. Levi. Interscience (John Wiley & Sons), New York.
Marshall, A. J. and Williams, W. D. *Textbook of Zoology: Invertebrates.* (7th ed. of Textbook of Zoology; Vol. 1 by J. Parker and W. A. Haswell.) American Elsevier Publishing Co. 1921.
Meglitsch, P. A. *Invertebrate Zoology.* Oxford University Press, New York. 1972.
Russell-Hunter, W. D. 1968. *A Biology of Lower Invertebrates.* 1969. *A Biology of Higher Invertebrates.* Annelida, Arthropoda, Echinodermata, Chordata, etc. The Macmillan Co., New York.
Wells, M. *Lower Animals.* McGraw-Hill Book Co., New York. 1968.

BOOKS DEALING PRIMARILY WITH NATURAL HISTORY OF INVERTEBRATES

Buchsbaum, R. and Milne, L. *The Lower Animals—Living Invertebrates of the World.* Doubleday & Co., Garden City, N.Y. 1960.

Grzimek, B. *Grzimek's Animal Life Encyclopedia.* Vol. 1. *Lower Animals,* 1974. Vol. 2. *Insects,* 1975. Vol. 3. *Mollusks and Echinoderms,* 1974. Van Nostrand & Reinhold Co., New York.

NATURAL HISTORY AND IDENTIFICATION OF INVERTEBRATES OF VARIOUS REGIONS

Brusca, R. C. *A Handbook to the Common Intertidal Invertebrates of the Gulf of California.* Univ. of Arizona Press, Tucson. 1973.

Burch, J. B. *How To Know the Eastern Land Snails.* Pictured-Keys for determining the Land Snails of the United States occurring east of the Rocky Mountain Divide. Wm. C. Brown, Dubuque, Iowa. 1962.

Dakin, W. J., Bennett, I., and Pope, E. *Australian Seashores.* Angus and Robertson, London. 1953.

Gillett, K. *The Australian Great Barrier Reef.* A. H. & A. W. Reed, Sydney. 1968.

Gosner, K. L. *Guide to the Identification of Marine and Estuarine Invertebrates Cape Hatteras to the Bay of Fundy.* Keys. Wiley-Interscience, New York. 1971.

Haunau, H. W. and Mock, B. H. *Beneath the Seas of the West Indies.* Argos, Inc., Miami.

Johnson, M. E. and Snook, H. J. *Seashore Animals of the Pacific Coast.* The Macmillan Co., New York.

Keen, M. *Sea Shells of Tropical West America.* Stanford University Press. 1971.

Kozloff, Eugene N. *Seashore life of Puget Sound, the Strait of Georgia, and the San Juan Archipelago.* Natural history. University of Washington Press, Seattle. 1973.

Morton, J. and Millar, M. *The New Zealand Sea Shore.* Collins, London. 1968.

Ricketts, E. F. and Calvin, J. *Between Pacific Tides.* Rev. by J. W. Hedgpeth. 1968.

Smith, R. I. and Carlton, J. T. *Light's Manual Intertidal Invertebrates of the Central California Coast.* 3rd ed. University of California Press, Berkeley. 1975.

BOOKS ON MARINE BIOLOGY

Davis, R. A., Jr. *Principles of Oceanography.* Addison-Wesley Pub. Co., Menlo Park, Calif. 1972.

Hardy, Alister. *The Open Sea. The World of Plankton.* Collins, London. 1956.

Hermann, Friedrich. *Marine Biology.* University of Washington Press, Seattle. 1969.

Ingmanson, D. G. and Wallace, W. J. *Oceanology: An Introduction.* Wadsworth Publishing Co., Belmont, Calif. 1973.

MacGinitie, G. E. and N. *Natural History of Marine Animals.* 2nd ed. McGraw-Hill Book Co., New York. 1968.

McConnaughey, B. H. *Introduction to Marine Biology.* The C. V. Mosby Co., Saint Louis. 1970.

Marshall, N. B. *Aspects of Deep Sea Biology.* Hutchinson, London. 1958.

Marshall, N. and Marshall, O. *Ocean Life.* The Macmillan Co., New York. 1971.

Stephenson, T. A. and Stephenson, A. *Life Between Tidemarks on Rocky Shores.* W. H. Freeman & Co., San Francisco. 1972.

Thorson, G. *Life in the Sea.* McGraw-Hill Book Co. 1971.

BOOKS ON FRESHWATER INVERTEBRATE BIOLOGY

Edmondson, W. T. (ed.). *Ward & Whipple's Fresh-water Biology.* John Wiley & Sons, New York. 1959.

Klots, E. B. *Freshwater Life.* G. P. Putnam Sons, New York. 1966.

Needham, J. G. and Needham, P. R. *A Guide to the Study of Freshwater Biology.* Holden-Day, Inc., San Francisco. 1962.

Pennak, R. W. *Fresh-water Invertebrates of the United States.* Ronald Press Co., New York. 1953.

BOOKS DEALING WITH SPECIAL ASPECTS OF INVERTEBRATE BIOLOGY

Bullock, T. H. and Horridge, G. A. *Structure and Function in the Nervous System of Invertebrates.* W. H. Freeman & Co., San Francisco. 1965.

Carthy, J. D. *An Introduction to the Behaviour of Invertebrates.* George Allen & Unwin Ltd., London. 1958.

Chandler, A. C. and Read, C. P. *Introduction to Parasitology.* John Wiley & Sons, New York. 1961.

Cheng, T. C. *Biology of Animal Parasites.* W. B. Saunders Co., Philadelphia. 1964.

Halstead, B. W. *Dangerous Marine Animals.* Cornell Maritime Press. 1959.

Horsfall, W. R. *Medical Entomology. Arthropods and Human Disease.* Ronald Press, New York. 1962.

James, M. T. and Harwood, R. F. *Medical Entomology.* The Macmillan Co., New York. 1969.

Moore, R. C., Lalicker, C. G., and Fischer, A. G. *Invertebrate Fossils.* McGraw-Hill Book Co., New York. 1952.

Needham, J. C. (ed.). *Culture Methods for Invertebrate Animals.* Comstock Publishing Co., Ithaca, N.Y. 1937.

Noble, E. R. and Noble, G. A. *Parasitology.* Lea & Febiger, Philadelphia. 1964.

Olsen, O. W. *Animal Parasites. Their Life Cycles and Ecology.* University Park Press, Baltimore. 1974.

Russell, F. E. *Marine Toxins and Venomous and Poisonous Marine Animals.* Crown Publishers. 1971.

Scheer, B. T. *Comparative Physiology.* John Wiley & Sons, New York. 1948.

BOOKS AND MONOGRAPHS ON VARIOUS INVERTEBRATE GROUPS

Protozoa:

Corliss, J. O. *The Ciliated Protozoa.* Pergamon Press, New York. 1961.

Cushman, J. A. *Foraminifera, Their Classification and Economic Use.* 4th ed. Harvard University Press. 1948.

Grell, K. G. *Protozoology.* Springer-Verlag, New York. 1973.

Hall, R. P. *Protozoa.* Holt, Rinehart & Winston, Inc., New York. 1964.

Jahn, T. L. and Jahn, F. F. *How to Know the Protozoa.* Wm. C. Brown Co., Dubuque, Iowa. 1949.

Jeon, K. W. (ed.). *The Biology of Amoeba.* Academic Press, New York and London.

Kudo, R. R. *Protozoology.* 5th ed. Springfield: C. C. Thomas. 1966.

Lwoff, A. *The Biochemistry and Physiology of Protozoa I, II.* Academic Press, New York. 1951-5.

Pitelka, D. *The Electron Microscopic Structure of Protozoa.* Pergamon Press, New York. 1962.

Tartar, V. *The Biology of Stentor.* Pergamon Press, New York.

Wichterman, R. *The Biology of Paramecium.* Blakiston Co., New York. 1953.

Porifera:

Fry, W. G. (ed.). *The Biology of the Porifera.* Academic Press, New York. 1970.

Mesozoa:

McConnaughey, B. H. The Mesozoa. *In* E. C. Dougherty, ed., *The Lower Metazoa.* University of California Press. 1963.

Coelenterata (Cnidaria):

Mayer, A. G. *Medusae of the World.* Vol. III. *The Scyphomedusae.* Carnegie Inst. Wash. Publ. *109:* 499-735. 1910.

Rees, W. J. (ed.). *The Cnidaria and Their Evolution.* Academic Press, New York. 1966.

Russell, F. S. *The Medusae of the British Isles.* Anthomedusae, Leptomedusae, Limnomedusae, Trachymedusae, Narcomedusae. Cambridge University Press, Cambridge, England. 1953.

Russell, F. S. *The Medusae of the British Isles.* Vol. II. *Pelagic Scyphozoa....* Cambridge Univ. Press, Cambridge, England. 1970.

Stephenson, T. A. *The British Sea Anemones.* London: Royal Society. Vol. I, 1928. Vol. II, 1935.

Ctenophora:

Mayer, A. G. *Ctenophores of the Atlantic Coast of North America.* Carnegie Institution of Washington Pub. 162.

Platyhelminthes:

Brondsted, H. V. *Planarian Regeneration.* Pergamon Press, New York. 1969.

Dawes, B. *The Trematoda.* Cambridge University Press, Cambridge, England. 1946.

Wardle, R. A. and McLeod, J. A. *The Zoology of Tapeworms.* University of Minnesota Press, Minneapolis. 1952.

Nemertea:

Gibson, R. *Nemerteans.* Hutchinson University Library, London. 1973.

Nematoda:

Chitwood, B. G. et al. *An Introduction to Nematology.* Monumental Pub. Co., Baltimore. 1937.

Crofton, H. D. *Nematodes.* Hutchinson University Library, London. 1966.

Lee, D. L. *The Physiology of Nematodes.* W. H. Freeman Co., San Francisco. 1965.

Mollusca:

Abbott, R. T. *American Seashells: The Marine Mollusca of the Atlantic and Pacific Coasts of North America.* Van Nostrand Reinhold Co., New York. 1974.

Lane, F. W. *Kingdom of the Octopus.* Jarrolds, London. 1957.

Mead, A. R. *The Giant African Snail.* University of Chicago Press. 1961.

Morton, J. E. *Molluscs.* Hutchinson University Library, London. 1963.

Purchon, R. D. *The Biology of the Mollusca.* Pergamon Press, 1968.

Solem, G. A. *The Shell Makers. Introducing Mollusks.* John Wiley & Sons, New York.

Wilbur, K. M. and Yonge, C. M. (eds.). *Physiology of Mollusca.* Academic Press. Vol. I, 1964. Vol. II, 1966.

Yonge, C. M. *Oysters.* Collins, London. 1960.

Sipuncula and Echiura:

Stephen, A. C. and Edmonds, S. J. *The Phyla Sipuncula and Echiura.* London: British Museum (Natural History). 1972.

Annelida:

Brinkhurst, R. O. and Jamieson, B. G. M. *Aquatic Oligochaeta of the World.* Edinburgh: Oliver Boyd. 1971.

Dales, R. P. *Annelids.* Hutchinson University Library, London. 1967.

Edwards, C. A. and Lofty, J. R. *Biology of Earthworms.* Halsted Press (John Wiley & Sons), New York. 1972.

Mann, K. H. *Leeches (Hirudinea) Their Structure, Physiology, Ecology and Embryology.* Pergamon, Elmsford, N.Y. 1962.

Stephenson, J. *The Oligochaeta.* Clarendon Press, Oxford. 1930.

Arthropoda:

Bodenheimer, F. S. *Insects as Human Food.* Dr. W. Junk, Publishers, The Hague. 1951.

Borror, D. J. and De Long, D. M. *An Introduction to the Study of Insects.* 3rd ed. Holt, Rinehart, and Winston. 1971.

Bristowe, W. S. *The World of Spiders.* Collins, London. 1958.

Butler, C. G. *The World of the Honeybee.* Collins, London. 1962.

Carthy, J. D. *The Behavior of Arthropods.* W. H. Freeman & Co., San Francisco. 1965.

Cloudsley-Thompson, J. L. *Spiders, Scorpions, Centipedes and Mites.* Pergamon Press, New York. 1958.

Comstock, J. H. *A Textbook of Entomology.* Comstock Publishing Assoc., Ithaca, N.Y. 1940.

Comstock, J. H. and Gertsch, W. J. *The Spider Book.* Doubleday, Doran & Co., New York. 1940.

Corbet, P. S., Longfield, C., and Moore, N. W. *Dragonflies.* Collins, London. 1960.

Darwin, C. *A Monograph of the Sub-class Cirripedia.* Part I, *Lepadidae*, 1851; Part II, *Balanidae*, 1854. London: Royal Society (reprinted Cramer, 1964).

Davey, K. G. *Reproduction in the Insects.* W. H. Freeman & Co., San Francisco. 1965.

Essig, E. O. *Insects and Mites of Western North America.* The Macmillan Co., New York. 1958.

Ford, E. G. *Moths.* Collins, London. 1955.

Free, J. B. and Butler, C. G. *Bumblebees.* Collins, London. 1959.

Frisch, K. von. *Bees: Their Vision, Chemical Senses, and Language.* Cornell Univ. Press, Ithaca, N.Y. 1950.

Green, J. *A Biology of Crustacea.* Quadrangle Books, New York. 1961.

Hogue, C. L. *The Armies of the Ant.* World Publishing, New York. 1972.

Horsfall, W. R. *Mosquitoes. Their Bionomics and Relation to Disease.* Ronald Press, New York. 1955

Hughes, T. E. *Mites, or the Acari.* University of London. The Athlone Press, London. 1959.

Imms, A. D. *Insect Natural History.* Collins, London. 1947.

Jacques, H. E. *How To Know the Insects.* Brown, Dubuque, Iowa. 1947.

Klots, A. B. *The World of Butterflies and Moths.* George G. Harrap & Co. Ltd., London. 1958.

Klots, A. B. *A Field Guide to the Butterflies of North America, East of the Great Plains.* Houghton Mifflin Co., Boston. 1951.

Lanham, U. *The Insects.* Columbia University Press, New York. 1964.

Linsenmaier, W. *Insects of the World.* McGraw-Hill Book Co., New York. 1972.

Snow, K. R. *The Arachnids: An Introduction.* Columbia University Press, New York. 1970.

Tinker, S. W. *Pacific Crustacea.* Charles E. Tuttle Co., Rutland, Vt. 1965.

Urquhart, F. A. *The Monarch Butterfly.* University of Toronto Press, Toronto. 1960.

Wenner, A. M. *The Bee Language Controversy.* Educational Programs Improvement Corp. 1971.

Wigglesworth, V. B. *The Principles of Insect Physiology.* Methuen, London. 1953.

Wigglesworth, V. B. *The Life of Insects.* London: Weidenfeld and Nicolson. 1964.

Williams, C. B. *Insect Migration.* Collins, London. 1958.

Bryozoa:

Larwood, G. P. *Living and Fossil Bryozoa.* Academic Press, New York. 1973.

Ryland, J. S. *Bryozoans.* London: Hutchinson University Library. 1970.

Brachiopoda:

Rudwich, M. J. S. *Living and Fossil Brachiopods.* London: Hutchinson Univ. Library. 1970.

Echinodermata:

Boolootian, R. A. (ed.). *Physiology of Echinodermata.* John Wiley, New York. 1966.
Nichols, D. *Echinoderms.* London: Hutchinson University Library. 1962.

Hemichordata and Chordata:

Barrington, E. J. W. *The Biology of Hemichordata and Prochordata.* W. H. Freeman & Co., San Francisco. 1965.
Berrill, N. J. *The Tunicata. With an Account of the British Species.* London: Royal Society. 1950.

APPENDIX: CLASSIFICATION

SYSTEMS of classification, whether arrived at by consensus of many biologists or by a single highly opinionated author, are somewhat arbitrary, based on incomplete information, and in need of constant rethinking and changes. Consequently, they vary from one source to another. Even the spelling of group names (above the level of family) varies from one scheme to another. The scheme of classification used throughout this text was selected for its familiarity and wide acceptance. It omits all extinct groups, except for a few mentioned in the chapter on fossils, and even among living groups, it does not include certain small or rare ones.

The classification presented below represents much recent thinking, but is not yet widely accepted. It includes some groups not described in the text, but some rare and most extinct groups have been omitted.

CLASSIFICATION OF ORGANISMS

(Numbers are page references in the text)

Kingdom **ARCHETISTA**—Acellular; saprobic. Viruses, phages, rickettsias.

Kingdom **MONERA**—No definite nuclei; autotrophic and/or saprobic. Blue-green algae, bacteria, spirochetes.

Kingdom **PROTISTA**—Nucleated cells or colonies; photosynthetic, saprobic, and/or ingestive. (Only those groups are listed here which include members that may be considered to be "protozoans.")

Phylum **Euglenophyta:** Euglenoids (38)
Phylum **Pyrrophyta:** Cryptomonads, dinoflagellates (42)
Phylum **Zoomastigina:** Animal-like flagellates (38-47)
Phylum **Opalinata:** Opalinids (55)
Phylum **Sarcodina:** Ameboids, radiolarians (12-21; 47-51)
Phylum **Sporozoa:** Sporozoans (51-54)
Phylum **Cillophora:** Ciliates, suctorians (21-30; 54-58)

Kingdom **FUNGI**—Nucleated cells, usually organized into tissues and organs sometime during life cycle; mainly saprobic. Slime molds, yeasts, molds, rusts, and mushrooms.

Kingdom **METAPHYTA (PLANTAE)**—Nucleated cells usually organized into tissues and organs; mainly photosynthetic: Red, brown, and green algae; liverworts and mosses; ferns; gymnosperms and angiosperms.

Kingdom **METAZOA (ANIMALIA)**— Nucleated cells organized into tissues and organs; mainly ingestive.

GROUP 1. ACOELOMATE PHYLA

Phylum **Archaeocyatha:** Cup sponges (extinct) (326-1)

Phylum **Porifera:** Sponges (59-68)

Phylum **Placozoa:** Only one species, *Trichoplax adherens.* Dorosoventrally flattened and differentiated; ciliated dorsally and ventrally. No permanent anterior-posterior polarity. Two layers with middle layer of mesenchyme. Asexual and sexual reproduction. 1-5 mm in diameter. So far found only in marine aquariums containing rocks and animals from coral reefs.

Phylum **Cnidaria (Coelenterata)**—(69-102)
    Class **Hydrozoa:** hydromedusas, hydroids
    Class **Scyphozoa:** jellyfishes
    Class **Anthozoa:** corals, sea anemones, sea fans, etc.

Phylum **Ctenophora:** comb jellies (103-108)

Phylum **Mesozoa:** Microscopic, flattened, ciliated parasites consisting only of a small group of outer cells enclosing one or more reproductive cells. Complicated life cycle of alternating asexual and sexual generations. Dicyemid metazoans common in excretory organs of squids and octopuses. Orthonectid mesozoans have been found in internal spaces and tissues of flatworms, nemerteans, annelids, a clam, and serpent stars.

Phylum **Platyhelminthes:** (109-150)
    Class **Turbellaria:** free-living flatworms
    Class **Trematoda:** flukes
    Class **Cestoda:** tapeworms

Phylum **Gnathostomulida:** gnathostomulids (150)

Phylum **Nemertea:** ribbon worms (151-155)

GROUP 2. PSEUDOCOELOMATE PHYLA

The first five or six phyla are sometimes grouped as a single phylum, the "Aschelminthes," but they are usually considered as separate phyla.

Phylum **Nematoda:** thread worms (156-164)

Phylum **Nematomorpha:** horsehair worms (165-166)

Phylum **Gastrotricha:** gastrotrichs (172)

Phylum **Rotifera:** rotifers (168-171)

Phylum **Acanthocephala:** spiny-headed worms (166-167)

Phylum **Kinorhyncha:** kinorhynchs (172)

Phylum **Entoprocta (Endoprocta):** endoprocts (176)

GROUP 3. PHYLA WITH COELOM FROM SPLITTING OF MESODERM BANDS; MOUTH
DEVELOPS FROM BLASTOPORE (PROTOSTOME)

Phylum **Mollusca:** mollusks (molluscs) (181-206)
  Class **Aplacophora:** solenogasters (184)
  Class **Monoplacophora:** *Neopilina* (184)
  Class **Polyplacophora:** chitons
  Class **Scaphopoda:** tooth shells
  Class **Gastropoda:** limpets, snails, and slugs
  Class **Pelecypoda (Bivalvia, Lamellibranchia):** clams, oysters, mussels
  Class **Cephalopoda:** squids, octopuses, nautilus

Phylum **Annelida:** annelid or segmented worms (207-234)
  Class **Polychaeta:** mainly segmented marine worms
  Class **Oligochaeta:** earthworms and others
  Class **Hirudinea:** leeches
Phylum **Sipuncula:** peanut worms (235-5)

Phylum **Echiura:** spoon worms (234-5)

Phylum **Pogonophora:** beard worms (180)
Phylum **Tardigrada:** water bears (180)
Phylum **Pentastomida:** pentastomids
Phylum **Onychophora:** velvet worms (235-238)
Phylum **Arthropoda:** arthropods (239-266)
  Class **Trilobita:** trilobites (extinct) (352-1)
  Class **Crustacea:** crustaceans
    Order **Branchiopoda:** brine shrimps, fairy shrimps, water fleas or cladocerans
    Order **Ostracoda:** ostracods
    Order **Copepoda:** copepods
    Order **Branchiura:** parasitic copepods
    Order **Cirripedia:** barnacles
    Order **Malacostraca:** shrimps, lobsters, crabs, isopods, amphipods,
      euphausids, mysids, mantis shrimps, etc.
  Class **Chelicerata:** chelicerates
    Order **Pycnogonida:** sea spiders
    Order **Merostomata:** horse-shoe crabs, eurypterids (extinct)
  Class **Arachnida:** scorpions, mites, spiders, etc.
  Class **Myriapoda:** millipeds, centipedes, etc.
  Class **Hexapoda (Insecta):** insects

GROUP 4. LOPHOPHORATE PHYLA

Phylum **Phoronida:** phoronids (179)
Phylum **Bryozoa:** bryozoans or moss animals (173-176)
Phylum **Brachiopoda:** lamp shells (176-178)

GROUP 5. PHYLA WITH COELOM FROM OUTPOCKETING OF PRIMITIVE MESODERM
  POUCHES; MOUTH DOES NOT DEVELOP FROM BLASTOPORE (ᴅEUTEROSTOME)

Phylum **Chaetognatha:** arrow worms (180)

Phylum **Echinodermata:** echinoderms (299-311)
  Class **Crinoidea:** feather stars and sea lilies
  Class **Asteroidea:** sea stars
  Class **Ophiuroidea:** serpent or brittle stars
  Class **Echinoidea:** sea urchins, sand dollars, etc.
  Class **Holothuroidea:** sea cucumbers

Phylum **Hemichordata**
  Class **Pterobranchiata:** pterobranchs
  Class **Enteropneusta:** acorn worms (312; 318-320)

Phylum **Graptolithina:** graptolites (extinct) 326-1) Assigned to the coelenterates
  by some.

Phylum **Chordata:** chordates
  Subphylum **Urochordata (Tunicata):** tunicates (316-318)
  Subphylum **Cephalochordata:** amphioxus or lancelet (312-315)
  Subphylum **Vertebrata:** fishes, amphibians, reptiles, birds, and mammals

Numbers in *italics* refer to the photographic inserts; the first number is that of the location of the insert, the second is the number of the page in the group. Generic and specific names of animals are printed in italics.

H

Limestone; *see also* Calcareous
  cups of corals, 100
  Indiana building-stone, 50
Limnoria lignorum, 268–5
Limpet, 185, *204–1*
Limulus; see King crab
Lingula, 178, 325
Lingulella, 326–2
Liriope, 92, 93
  luminescence of, 205
Lithobius, 273
Littorina, 204–2
Liver of amphioxus, 313, 314, 315
Liver fluke, 140, 141, *136–2*
Lobster, 256, *268–7, 8*
  covered with barnacles, *268–4*
  development of, 335, 336
  fossil, 331
  Homarus, 268–8
  Panulirus, 268–7
Locomotion, 6
  of ameba, 14
  of arthropod, 243
  of chiton, 182
  of clam, 190
  of earthworm, 221
  of hydra, 81
  of nereis, 208
  of paramecium, 21
  of roundworms, 157
  of squid, 200
  of starfish, 300, *310–4*
Locust, *292–17*
Loligo, 204–9
Lophodella, 172–3
Lophohelia, 100
Lophophore
  of brachiopod, 177, *172–4*
  of bryozoan, 174
  of Phoronidea, 179
Louse, body, *292–2*
Luciferase, 205
Luciferin, 205
Lumbricus terrestris, **228**
Luminescence, 204
  of bacteria, 205
  of ctenophores, *108–2*
  of firefly, *292–19*
  of fireworms, 231
  of fungi, 205
  of *Noctiluca,* 42
  of polychetes, 231
  of squid, 204
Lung book of spider, 270
Lung of gastropod, 188

M

Macronucleus, **25,** 54, *56–4*
Macroperipatus geayi, 238–1
Macrophage, *10–1*
Macrotermes natalensis, 292–10
Madrepora, 100–16
Madreporite; *see* Sieve plate
Maggot of flies, as food of man, *292–31*
Malaria
  of birds, 53
  mosquito, 53, *292–29*
  symptoms of, 54
  treatment and control of, 54
Malarial parasite, 52, *56–1*
Malpighian tubules of grasshopper, 281
Mandible; *see* Jaw
Mantis, praying, *292–9*
Mantle
  of chiton, 182
  of clam, 191
  of squid, 200
Manubrium, 90
Marine Biological Laboratory, *344–1*
Marine shrimp, *268–6*
Mastax, 170
Mating of earthworms, 229, *234–3*
Maxillas
  of arthropod, 246
  of butterfly, 290
  of cicada, 291
  of grasshopper, 278
  of lobster, 258, 261
  of moths and butterflies, *292–4*
Maxilliped of lobster, 258
Mayfly, 294, *292–13*
Mealy bug, *292–16*
Measly beef, *144–2*
Mechanical models of ameba, 20
Mediolateral gradient of bilateral animals, 131
Mediterranean fruit fly, *292–30*
Medusa, 90; *see also* Jellyfish
  of obelia, 84, 86, 87, *100–2*
Megathura, 204–1
Membrane, cell, 9
Membranelles, 56
Mesenchyme
  blood cavities in, 250
  cells, 71
  of planaria, 114
  of sponges, 62, 63, 64
  of three-layered animals, 110
Mesentery of nereis, 211
Mesoderm, 110
  of developing grasshopper, 286
  evolution of, 341